Assay of
Calcium-regulating
Hormones

Assay of Calcium-regulating Hormones

Edited by Daniel D. Bikle

With 21 Contributors

With 72 Figures

Springer-Verlag
New York Berlin Heidelberg Tokyo

Daniel D. Bikle, M.D., Ph.D.
Department of Medicine
University of California, San Francisco
Veterans Administration Medical Center
4150 Clement Street (111N)
San Francisco, California 94121
U.S.A.

Sponsoring Editor: Jerry L. Stone, Ph.D.
Production: Anthony Buatti

Library of Congress Cataloging in Publication Data
Main entry under title:
Assay of calcium-regulating hormones.
    Includes index.
    1. Vitamin D—Analysis.   2. Parathyroid hormone—Analysis.   3. Calcitonin—
Analysis.   4. Radioimmunoassay.   I. Bikle, Daniel D. [DNLM: 1. Parathyroid
hormones—Analysis.   2. Calcitonin—Analysis.   3. Vitamin D—Analysis.
4. Biological assay—Methods. QY 330 A844]
QP772.V53A87   1983   599'.01927   83-10477

© 1983 by Springer-Verlag New York Inc.
Softcover reprint of the hardcover 1st edition 1983

Typeset by University Graphics, Inc., Atlantic Highlands, New Jersey.

9 8 7 6 5 4 3 2 1

ISBN-13:978-1-4612-5555-0   e-ISBN-13:978-1-4612-5553-6
DOI: 10.1007/978-1-4612-5553-6

Preface

The ability to measure accurately the hormones regulating calcium homeostasis is the fundamental first step toward understanding the roles these hormones play in health and disease. Techniques for such measurements have only been available for the past 10 years or so and remain in a state of rapid development. Sensitive parathyroid hormone (PTH) radioimmunoassays appeared in the early 1970s, and with them came a whole new appreciation for the prevalence and implications of hyperparathyroidism, primary or secondary, in the population. The calcitonin (CT) radioimmunoassay came later and achieved rapid success in the diagnosis of a previously poorly understood cancer, medullary carcinoma of the thyroid, frequently associated with the familial multiple endocrine neoplasia type 2 syndromes (a and b). As the sensitivity of the calcitonin radioimmunoassay has improved, our understanding of the role of calcitonin in normal physiological processes has increased. The knowledge that vitamin D must be metabolized to produce its biologic effects is only 15 years old. This has had profound implications in our understanding of a variety of metabolic bone, kidney, and gastrointestinal diseases. Assays to measure the major circulating form of vitamin D, 25-hydroxyvitamin D, were described 10 years ago. Assays for the other metabolites, in particular, 1,25-dihydroxyvitamin D, were described even more recently. As of today, we know of many vitamin D metabolites and have developed the techniques to measure most of them; however, many questions remain concerning their physiological role. As these assays are applied to clinical situations and research investigations, answers should be forthcoming.

In this syllabus we have attempted to provide detailed protocols of established methods used to measure PTH, CT, and the vitamin D metabolites. The individuals describing these methods are at the forefront of this rapidly developing technology. The intent is to provide the reader with sufficient detail to permit the establishment of these techniques in the laboratory.

As will become apparent on reading the different chapters, a variety of approaches can be used to achieve the same accurate results. The advantages and disadvantages of the different approaches are discussed, but the reader is encouraged to pick the ones that work best in his or her own laboratory.

Contributors

Claude D. Arnaud, M.D., Professor of Medicine, University of California, San Francisco, and Chief, Endocrine Section, Veterans Administration Medical Center, 4150 Clement Street (111N), San Francisco, California 94121, U.S.A.

Sara B. Arnaud, M.D., Assistant Professor of Medicine and Pediatrics, University of California, San Francisco, and Director, Endocrine Unit Clinical Laboratory, Veterans Administration Medical Center, 4150 Clement Street (1B37), San Francisco, California 94121, U.S.A.

Bayard D. Catherwood, M.D., Assistant Professor of Medicine, University of California, San Diego, and Research Associate, Veterans Administration Medical Center, 3350 La Jolla Village Drive, San Diego, California 92161, U.S.A.

Leonard J. Deftos, M.D., Professor of Medicine, University of California, San Diego, and Chief, Endocrine Section, Veterans Administration Medical Center, 3350 La Jolla Village Drive, San Diego, California 92161, U.S.A.

Irene Y. Fu, M.S., Senior Research Biochemist, Case Western Reserve School of Medicine, Veterans Administration Medical Center, 10701 East Boulevard, Cleveland, Ohio 44106, U.S.A.

David Goltzman, M.D., Associate Professor of Medicine, McGill University, Allan Research Building, Room 117, 1033 Pine Avenue West, Montreal PQ H3A 1A1, Canada

Richard W. Gray, Ph.D., Associate Professor of Medicine, Medical College of Wisconsin, and Director, Clinical Research Center, Core Laboratory, Froedtert Memorial Lutheran Hospital, Milwaukee, Wisconsin 53226, U.S.A.

John G. Haddad, M.D., Professor of Medicine and Chief, Endocrine Section, The University of Pennsylvania School of Medicine, 531 Johnson Pavillon G/2, 36th and Hamilton Walk, Philadelphia, Pennsylvania 19104, U.S.A.

Bernard P. Halloran, Ph.D., Assistant Professor of Medicine, University of California, San Francisco, and Director, Vitamin D Laboratory, Veterans Administration Medical Center, 4150 Clement Street (111N), San Francisco, California 94121, U.S.A.

Charles D. Hawker, Ph.D., National Manager, Research and Development, Smith-Kline Clinical Laboratories, Inc., 11636 Administration Drive, St. Louis, Missouri 63146, U.S.A.

Hunter Heath III, M.D., Associate Professor of Medicine, Mayo Clinic, Endocrine Research Unit, 5-164 W. Joseph Building, Rochester, Minnesota 55905, U.S.A.

Bruce W. Hollis, Ph.D., Assistant Professor of Nutrition, Department of Nutrition, Case Western Reserve School of Medicine, 2121 Abington Road, Cleveland, Ohio 44106, U.S.A.

Ronald L. Horst, Ph.D., Research Physiologist, National Animal Disease Center, P.O. Box 70, Dayton Avenue, Ames, Iowa 50010, U.S.A.

David M. Kaetzel, Ph.D., NIH Postdoctoral Research Fellow, Case Western Reserve School of Medicine, Veterans Administration Medical Center, 10701 East Boulevard, Cleveland, Ohio 44106, U.S.A.

Phillip W. Lambert, M.D., Associate Professor of Medicine, Case Western Reserve School of Medicine, Vitamin D Research Laboratory, Veterans Adminstration Medical Center, 10701 East Boulevard, Cleveland, Ohio 44106, U.S.A.

L. E. Mallette, M.D., Ph.D., Assistant Professor of Medicine, Baylor Medical College, and Director, Calcium Metabolism Research Laboratory, Veterans Administration Medical Center, 2002 Holcombe Boulevard, Houston, Texas 77030, U.S.A.

Stavros C. Manolagas, M.D., Ph.D., Assistant Professor of Medicine, University of California, San Diego, and Department of Medicine–Endocrinology, Veterans Administration Medical Center, 3350 La Jolla Village Drive, San Diego, California 92161, U.S.A.

Juliann Meger, B.S., Research Technologist, 9365 Pratt Circle, Omaha, Nebraska 68134, U.S.A.

Robert A. Nissenson, Ph.D., Assistant Professor of Medicine, University of California, San Francisco, and Assistant Chief, Divison of Endocrinology, Veterans Administration Medical Center, 4150 Clement Street (111N), San Francisco, California 94121, U.S.A.

Glen W. Sizemore, Professor of Medicine, Mayo Clinic, 200 First Street S.W., Rochester, Minnesota 55905, U.S.A.

Anne P. Teitelbaum, Ph.D., Research Endocrinologist, Veterans Administration Medical Center, 4150 Clement Street (111N), San Francisco, California 94121, U.S.A.

Contents

Assay of
Calcium-regulating
Hormones

Chapter 1

Hormonal Regulation of Calcium Homeostasis

Claude D. Arnaud

A highly integrated and complex endocrine system acts to maintain calcium, phosphate, and magnesium homeostasis in all vertebrates (Figure 1). It involves an interplay between the actions of two polypeptide hormones, parathyroid hormone (PTH) and calcitonin (CT), and a sterol hormone, 1α, 25-dihydroxyvitamin D [1α,25-(OH)$_2$D$_3$]. The regulation of the biosynthesis and the secretion of the polypeptide hormones is accomplished by a negative feedback mechanism involving the calcium ion activity of the extracellular fluids (Figure 2). The biosynthesis of 1α,25-(OH)$_2$D$_3$ from the major circulating metabolite of vitamin D, 25 hydroxyvitamin D$_3$ (25-OHD$_3$) takes place in the kidney and is regulated by PTH and CT as well as the extracellular fluid concentrations of calcium and phosphate. Other hormones, such as cortisol, aldosterone, growth hormone, thyroxin, epinephrine, estrogen, and testosterone, in addition to as yet unknown compounds and certain physical phenomena, undoubtedly play roles separately and in concert in conditioning organ responses to PTH, CT, and 1α,25-(OH)$_2$D$_3$.

Parathyroid hormone, CT, and 1α-(OH)$_2$D$_3$ act on the intestine, kidney, and bone to regulate the flow of minerals into and out of the extracellular fluid compartment (Figure 1). The target cells of these organs constitute a cellular barrier which functions to separate the extracellular fluid compartment from the intestinal and renal tubular lumina and from the bone fluid compartment (adjacent to mobilizable bone mineral) (Figure 3). These target cells are highly specialized to carry out solute transport against a concentration gradient and, because of this, are often described as being polarized.

Parathyroid hormone functions to prevent oscillations of the serum calcium below and serum phosphate above the physiologic concentrations. It acts to increase the translocation of calcium from intestinal and renal tubular lumina and from the bone fluid compartment into the blood. Whereas the effects of PTH on bone and kidney are direct, it is generally agreed that its effects on the intestine are indirect and mediated by 1α,25-(OH)$_2$D$_3$. The hormone stimulates (directly and through its hypophosphatemic effects) the conversion of 25-OHD$_3$ to 1α,25-(OH)$_2$D$_3$ in the kidney by a mitochondrial 25-OHD 1α-hydroxylase in the renal tubule. The 1α,25-(OH)$_2$D$_3$ thus formed stimulates intestinal calcium absorption. The effect of parathyroid hormone on the renal tubule to increase phosphate excretion and decrease the concentration of phos-

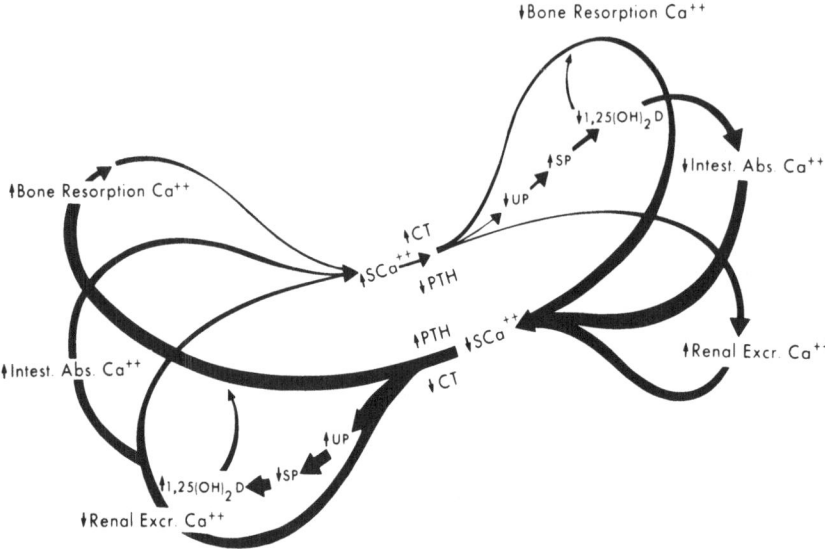

Figure 1. Schema of the regulation of calcium homeostatis, consisting of three over-lapping control loops that interlock and relate to one another through the level of blood concentrations of ionic calcium, parathyroid hormone, and calcitonin. Each loop involves a calciotropic hormone target organ (bone, intestine, or kidney). The limbs of the three loops that describe physiological events that increase blood concentrations of calcium *(SCa)* are on the left, and the limbs that describe events that decrease blood concentrations of calcium *(SCa)* are on the right. See text for detailed descriptions. (From Arnaud CD, Fed Proc 37:2558, 1978.)

phate in the blood has an important physiologic function beyond the regulation of renal $1\alpha,25\text{-}(OH)_2D_3$ production. It prevents the development of hyperphosphatemia which would be caused by the obligate removal of phosphate from bone with calcium during PTH-induced bone resorption. The importance of this latter function is most apparent in patients with end-stage renal failure associated with severe hyperparathyroidism. In these patients, large quantities of phosphate are released from bone resulting in hyperphosphatemia because the kidney is not available to excrete phosphate.

Calcitonin functions to prevent oscillations of the serum calcium above physiologic concentrations and, like PTH, is hypophosphatemic. It acts to decrease the translocation of calcium from the renal tubule and bone fluid compartments into the blood and thus can be considered to be a counterregulator of PTH. The effects of CT on vitamin D metabolism and on the intestinal absorption of calcium are uncertain at the present writing, although effects on both have been described.

At the level of bone and, possibly, kidney, it is likely that PTH and CT regulate the entry of calcium into polarized cells which comprise the surface membrane envelope of bone. $1\alpha,25\text{-}(OH)_2D_3$ may act primarily to support the cellular calcium transport system that subserves the active extrusion of calcium, against a concentration gradient, from the interior of the cell, across the

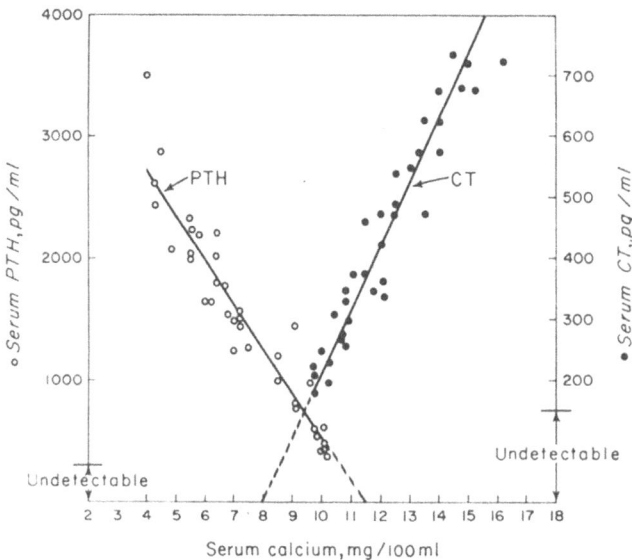

Figure 2. Plasma immunoreactive PTH (iPTH) and CT (iCT) as a function of plasma total calcium in pigs given either EDTA or calcium infusions to either decrease or increase plasma calcium; r = -0.942 for iPTH and 0.964 for iCT; P for both < 0.001. (From Arnaud CD et al., in: *Calcitonin: Proceedings of the Second International Symposium on Calcitonin,* S. Taylor (Ed.). Heinemann, London, 1969, p. 99.)

antiluminal membrane into the blood (Figure 4). Thus, there is an important dependency relationship between the calciotropic hormones, especially parathyroid hormone and $1\alpha,25\text{-}(OH)_2D_3$. Whereas, on the one hand, the renal production of $1\alpha,25\text{-}(OH)_2D_3$ is regulated by the prevailing concentration of parathyroid hormone in the blood, the effect of parathyroid hormone to

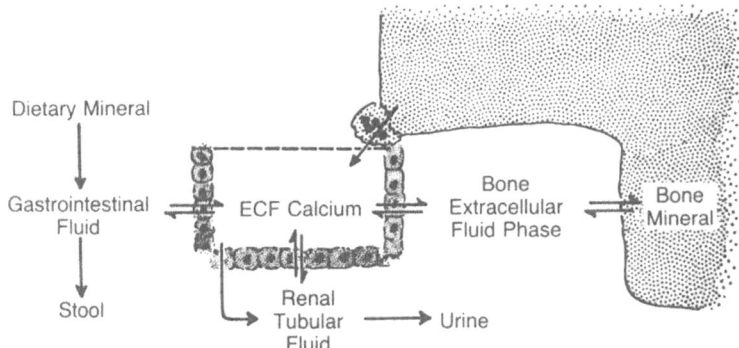

Figure 3. Schema of the cellular barrier separating the extracellular fluid compartment from the intestinal and renal tubular lumina and from the bone fluid compartment. PTH, CT, and $1,25(OH)_2D_3$ act on these cells (directly or indirectly) to regulate the flow of calcium into and out of the extracellular fluid compartments (see Figure 1). (From Rasmussen et al., Fed Proc 29:1190, 1970.)

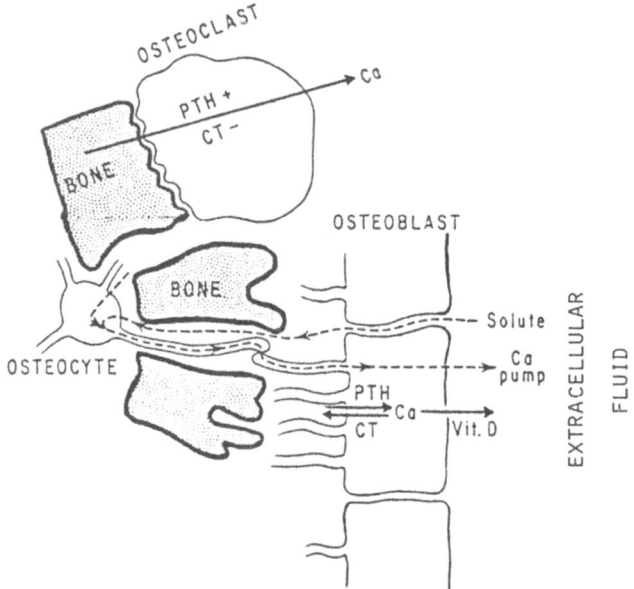

Figure 4. Schematic diagram of bone and bone cells. Bone crystal is represented by the shaded areas. The osteoclast, with active "ruffled border," is shown resorbing bone. The cells labeled "osteoblasts" (perhaps better called "surface osteocytes") are not forming bone but are actively extruding Ca from the bone fluid (between cells and crystals) under the influence of hormones. Processes connecting deep osteocytes are shown participating in the transport of Ca *(dashed arrows)*.

increase the plasma calcium, by virtue of its ability to stimulate entry of calcium into cells, depends upon $1\alpha,25\text{-}(OH)_2D_3$ to maintain the integrity of the calcium transport system responsible for the pumping of calcium from the cell into the extracellular fluids.

The interrelationships among the numerous elements involved in maintaining mineral homeostasis are illustrated in Figure 1[1] which depicts three overlapping feedback loops, each involving one of the target organs of the calciotropic hormones. Separately, the loops all involve the controlling elements, plasma calcium, PTH, and CT. Therefore, the loops have been made to interlock at the level of these elements. The limbs of the loops describing physiologic events that increase plasma calcium are on the left and those describing events that decrease plasma calcium are on the right. Under physiologic conditions, there are small oscillations in the plasma calcium. Decreases in plasma calcium increase parathyroid hormone secretion and decrease calcitonin secretion. The effects of these changes in hormone secretion lead to increased bone resorption, decreased renal excretion of calcium, and increased intestinal calcium absorption by virtue of the ability of PTH to stimulate the production of $1\alpha,25\text{-}(OH)_2D_3$. A small increase in the concentration of plasma calcium above its physiologic concentration probably occurs and results in the inhibition of PTH secretion and the stimulation of calcitonin secretion. These changes in blood hormone concentration result in (right limbs) decreased bone resorption, increased renal excretion of calcium, and decreased intestinal absorption of calcium resulting in a small decrease in the plasma calcium below the physiologic level. This sequence of events is probably repeated on a millisecond time scale so that oscillations of the plasma calcium are minimized and maintained as close to physiologic levels as possible. This "butterfly" scheme of mineral homeostasis not only demonstrates the interrelationships between the control elements in mineral homeostasis under physiological circumstances, but also allows for the rapid visualization of potential pathogenetic schemes and a detailed but overall appreciation of the multiple possible adaptive responses which might be caused by perturbations of the system by disease states and potential treatment modalities.

I. Parathyroid Hormone

A. Structure and Biosynthesis

Parathyroid hormone is an 84-amino-acid linear polypeptide with a molecular weight of 9500. Its biosynthesis and intracellular processing are complex (Figure 5). The original gene product of the parathyroid cell is a 113-amino-acid precursor termed preproparathyroid hormone (Figure 6). The hydrophobic 23-amino-acid "pre" sequence acts to bind the polyribosome precursor complex to the endoplasmic reticulum providing access to the cisternal space and, presumably, to the enzyme ("clipase") which removes the "pre" sequence leaving the 90-amino-acid proparathyroid hormone structure. Proparathyroid hormone is converted to parathyroid hormone in the Golgi apparatus by proteolytic removal of the remaining six amino terminal acid sequence (tryptic clipase). Here, the 84-amino-acid polypeptide is readied for secretion either in a secretory granule or in its free form. There is no evidence that either of the parathyroid hormone precursor molecules or the "pre" or "pro" peptide sequences normally find their way into the circulation. In contrast to the rapid regulation of secretion, PTH biosynthesis is only slowly influenced by changes in the concentration of extracellular ionic calcium.

Intracellular stores of parathyroid hormone may be regulated by a degradative pathway stimulated by high and inhibited by low extracellular calcium. This degradative pathway is of considerable current interest. Not only may it provide an important mechanism for regulating parathyroid hormone economy, but also there is good evidence that the fragments of the hormone produced during intracellular degradation are secreted and may provide a major source of the multiple immunoreactive forms of parathyroid hormone known to circulate in the blood (see below).

The amino acid sequences of bovine, porcine, and human parathyroid hormone have been determined (Figure 6). They differ sufficiently that immuno-

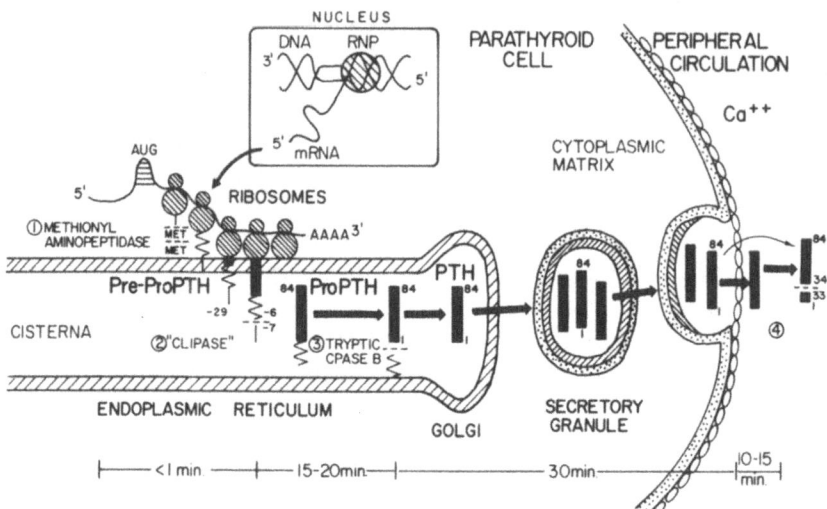

Figure 5. Schema depicting the proposed intracellular pathway of the biosynthesis of parathyroid hormone. Preproparathyroid hormone *(Pre-ProPTH)*, the initial product of synthesis on the ribosomes, is converted into proparathyroid hormone *(ProPTH)* by removal of (*1*) the NH_2-terminal methionyl residues *(methionyl aminopeptidase)* and (*2*) the NH_2-terminal sequence (-29 through -7) of 23 amino acids *("clipase")* during or within seconds after synthesis, respectively. The conversion of PrePropTH to PropTH probably occurs during transport of the polypeptide into the cisterna of the rough endoplasmic reticulum. By 20 min after synthesis, PropTH reaches the Golgi region and is converted into PTH by (*3*) removal of the NH_2-terminal hexapeptide [*tryptic cpase* (clipase) *B*]. PTH is stored in the secretory granule until released into the circulation in response to a fall in the blood concentration of calcium. The time needed for these events is given below the schema. (From Habener JF, Recent Prog Horm Res 33:287, 1977.)

logic cross-reactivity between them is incomplete. This probably accounts for the difficulties encountered in developing radioimmunoassays for the measurement of human parathyroid hormone using antisera directed against the more readily available bovine and procine peptides. All of the structural information required for full biological activity of the native 84-amino-acid parathyroid hormone lies in the amino terminal 34-amino-acid sequence. This active fragment of both bovine and human hormones has been synthesized and is commercially available for investigational use. Studies of the carboxyl-terminal fragments, including amino acids 53–84, have shown it to be biologically inert.

B. Control of Secretion

Parathyroid hormone is rapidly released from the parathyroid gland in response to a decrease in the plasma ionic calcium and acts on kidney, intestine (indirectly, see below), and bone to restore the concentration of this cation to

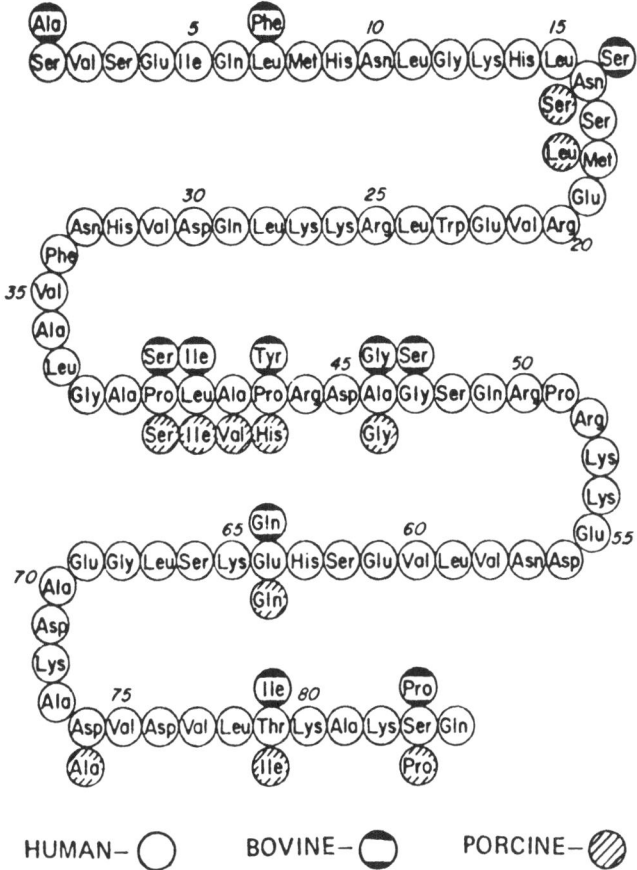

Figure 6. Parathyroid hormone. The structure of bovine and porcine PTH is shown with the positions indicated above and below the human structure. (From Keutman HT et al., Biochemistry 17:5723, 1978.)

just above the normal set point (Figure 2). This, in turn, inhibits the secretion of the hormone. This negative feedback cycle is visually depicted in the "butterfly" diagram shown in Figure 1. The concentration of extracellular ionic calcium is the major regulator of parathyroid hormone secretion. Other factors such as the plasma phosphate or blood pH influence secretion only indirectly by altering the degree to which calcium is complexed (phosphate) or bound to albumin (pH) (normally about 50% of the total). The effects of extracellular magnesium concentrations on secretion are qualitatively similar to ionic calcium but are physiologically less important. Paradoxically, severe, prolonged hypomagnesemia markedly inhibits secretion of parathyroid hormone and may be associated with hypocalcemia. Known direct parathyroid hormone secretagogues of questionable physiological importance include β-adrenergic agonists, prostaglandins, and histamine. These agents, as well as decreased ionic calcium, stimulate the production of cyclic 3'5'-adenosine monophosphate (cyclic

AMP) in parathyroid cells in vitro, and it is presumed that this compound mediates their actions on hormone secretion.

C. Metabolism and Circulating Forms

Circulating parathyroid hormone is heterogenous (Figure 7). It consists of the intact 84-amino-acid polypeptide and multiple fragments of the hormone. These fragments are derived mostly from the carboxyl region of the hormone molecule and, therefore, are likely to be biologically inert. The relative quantities of intact parathyroid hormone and its fragments in serum are not known precisely at present, but there are grossly more circulating fragments than intact hormone. This difference is due primarily to longer survival time of the fragments in the circulation. The fragments are probably derived both from the parathyroid gland release and from degradative metabolism of intact 84-amino-acid parathyroid hormone (liver and kidney); the quantitative importance of these sources is uncertain.

Biologically active forms of parathyroid hormone circulate in the blood normally at extremely low concentrations (less than 100 pg/ml). There are prob-

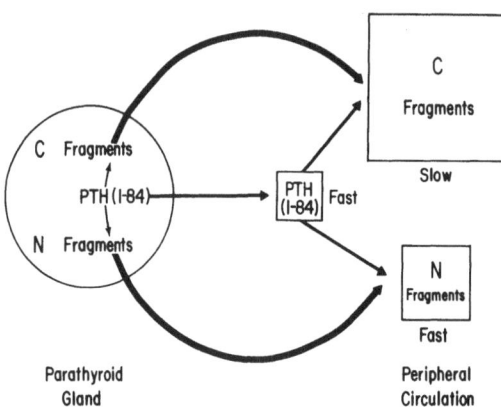

Figure 7. Simplified model of the metabolic alterations in PTH(1–84) that result in its immunoheterogeneity in gland extracts and biologic fluids. This scheme is based on data obtained from the study of abnormal parathyroid tissue and hyperparathyroid serum. Data on normal glands and serum are not available at this writing. An uncertain proportion of intraglandular PTH(1–84) is proteolytically cleaved into amino (N) and carboxyl (C) region fragments. All of these PTH molecular species are released into the circulation (approximately 80% PTH(1–84) and 20% fragments). Once in the circulation, PTH(1–84) undergoes proteolytic cleavage in liver and kidney, generating N and C fragments which are added to the circulating pool of hormone fragments. The rate of disappearance of C fragments from the circulation is slow (T½ ≃ 20–40 min) and that of PTH (1–84) and of N fragments is fast (T½ < 10 min). These differences in metabolic turnover of circulating PTH molecular species probably account for the large pool of C fragments and small pools of PTH(1–84) and N fragments in the blood under steady state conditions (see text). (From DiBella FP et al., Excerpta Med Int Congr Ser No 421:338, 1977.)

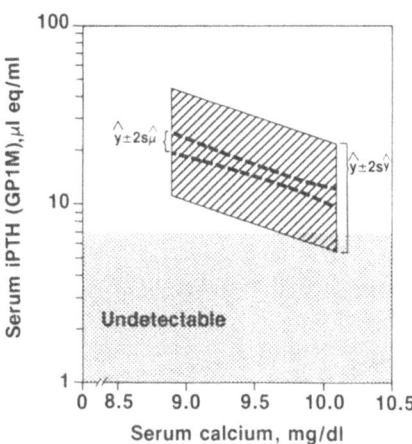

Figure 8. Serum immunoreactive PTH (iPTH; log scale) as a function of total serum calcium in 150 normal human subjects (r = −0.424; P < −0.001). y ± 2sμ = mean ± 2 SEM; y ± 2sy = mean ± 2 SD. GP1M is the antiserum (carboxyl region) used in the radioimmunoassay. (Reproduced, with permission, from Purnell D et al., Am J Med 56:801, 1974.)

ably individual constitutionally determined "set-point" values for the plasma ionic calcium above which glandular secretion rates are decreased and below which they are increased. However, the steady state levels of parathyroid hormone are probably determined primarily by the degree to which the parathyroid glands must adapt (by changes in the amount of functioning tissue) to individual, chronic, environmentally induced changes in the level of plasma ionic calcium (i.e., dietary calcium and phosphate). In normal humans (Figure 8) as in lower animals (Figure 2), there is an inverse relationship between the fasting levels of serum calcium and serum immunoreactive parathyroid hormone.

D. Actions

The major function of PTH is to defend against hypocalcemia. As illustrated in Figure 1, it carries out this function by virtually all of the actions which could be teleologically postulated: (1) release of calcium from bone, (2) conservation of calcium by the kidney, (3) enhanced absorption of calcium from the gut (indirectly via 1,25(OH)$_2$D), and (4) reduction in the plasma phosphate.

Parathyroid hormone acts on the kidney to (1) increase renal tubular reabsorption of calcium and magnesium and (2) to increase phosphate and bicarbonate excretion by inhibiting their proximal tubular reabsorption. These latter effects have several important, although indirect, effects on the homeostasis of extracellular calcium. Hormone-induced bicarbonaturia tends to produce acidosis which decreases the ability of circulating albumin to bind calcium, thus increasing ionic calcium by physiochemical means. Hormone-induced phosphaturia assures that the increased release of phosphate from bone which occurs during hormone-induced calcium mobilization from bone does not produce hyperphosphatemia. The latter would complex calcium tending to counterbalance the physiologic effect of parathyroid hormone to increase plasma ionic calcium.

The most important indirect effect of the phosphaturic action of the hormone is illustrated in the intestinal feedback loop in Figure 1. Parathyroid hormone, either directly or indirectly by its hypophosphatemic action, stimulates the activity of renal tubular 25-OH 1α-hydroxylase to convert the major circulating metabolite of vitamin D_3, 25-OHD$_3$ to its major biologically active metabolite, $1\alpha,25$-(OH)$_2$D$_3$. This latter compound acts directly on the intestinal mucosal cells to increase calcium absorption and on bone to increase resorption (short feed-forward element in the diagram placed between the intestinal and bone loops in Figure 1.

In the process of stimulating adenylate cyclase in renal tubular cells (see below), parathyroid hormone increases urinary excretion of cyclic AMP. Presumably cyclic AMP is released into the tubular fluid following its intracellular synthesis. Urinary cyclic AMP is increased by the administration of other hormones, including epinephrine and glucagon, but its renal production and excretion is almost entirely due to parathyroid hormone. It can, therefore, be used as a measure of PTH action on the kidney as noted below.

The major action of parathyroid hormone on bone is to enhance the net release of calcium and phosphate into extracellular fluid (Figure 4). This occurs as a result of a direct effect of the hormone on the differentiation and/ or activities of all bone cells (osteogenic precursors, osteoblasts, osteoclasts, and osteocytes). These events appear to depend upon a permissive effect of $1\alpha,25$-(OH)$_2$D$_3$, but the precise cellular mechanism of this important hormonal interrelationship is poorly understood. The physiology and pathophysiology of mineral metabolism demonstrate a wide range of interactions between parathyroid hormone (tropic) and vitamin D (permissive).

E. Mechanism of Action

Parathyroid hormone binds to specific plasma membrane receptors of target cells. These occupied receptors interact with a guanyl nucleotide-regulated membrane-bound protein which in turn activates membrane-bound adenylate cyclase to convert ATP to cyclic AMP. Cyclic AMP, by virtue of its ability to initiate a cascade of enzyme-activated intracellular phosphorylations, is considered to be at least one intracellular "second messenger" responsible for mediating the final expression of the action of the hormone. The details of these enzyme activations and the manner in which they relate to discrete effects of the hormone are unknown. Other potential "second messengers" of parathyroid hormone (i.e., calcium itself) which might act in concert or modulate the actions of cyclic AMP are currently under active investigation.

There is important new information concerning the structural requirements for the action of PTH at the side of its receptor in plasma membrane. It appears that the two amino acids at the amino terminus of the molecule are essential for the formation of a high-affinity state of the receptor, which in turn is required for activation of adenylate cyclase. Interestingly, a synthetic analogue of PTH, comprising the sequence 3–34, acts as a competitive antagonist of the 1–34 fragment in activating adenylate cyclase in vitro. Unfortunately, this synthetic PTH antagonist, for unknown reasons, has weak agonist activity in vivo. Future work in the search for a pure parathyroid hormone antagonist active in vivo is of obvious importance in the treatment of acute hypercalcemic disorders due to excess parathyroid hormone.

F. Assay in Biological Fluids

The major tool for the assay of parathyroid hormone in biological fluids has been radioimmunologic. Most radioimmunoassays of human parathyroid hormone have used cross-reacting antisera directed against porcine or bovine parathyroid hormone and ^{125}I-labeled bovine parathyroid hormone as a radioligand. Although standard preparations of PTH vary, the heterogeneity of circulating parathyroid hormone has generally necessitated the use of different hyperparathyroid sera as assay standards. Thus, depending upon the hyperparathyroid serum used in individual laboratories, the ranges of normal values differ.

Most antisera are multivalent and contain antibodies directed at multiple regions of the hormone molecule. Thus, the antibody present in a given antiserum in highest concentration and with highest affinity for the hormone molecule will determine the specificity of an assay. Another consideration in this regard is the degree of difference between the amino acid sequence of the region of [^{125}I] bovine PTH(1–84) and the same region in human PTH against which the antibody is directed.

It has become clear that specificity of parathyroid hormone radioimmunoassay is of paramount importance in the interpretation of the results of such assays. This is because of the heterogeneity of circulating PTH and the presence in serum of large quantities of biologically inert fragments of the hormone (see above). Assays using antisera directed against the carboxyl region of the molecule recognize both inactive carboxyl-region fragments and intact biologically active parathyroid hormone, whereas assays using antisera directed against the amino region of the molecule recognize only intact PTH and amino-region fragments. Because much larger quantities of the carboxyl-region fragments are present in the circulation than intact PTH (Figure 7), carboxyl-region-specific assays will give higher values for the serum immunoreactive PTH (iPTH) than amino-region-specific assays. Most important, but paradoxically, carboxyl-region-specific assays appear to be superior to amino-region-specific assays in separating normal subjects from patients with hyperparathyroidism. This observation is poorly understood but probably relates to the secretion of carboxyl-region fragments from the parathyroid gland and the relatively slower disappearance of carboxyl-region fragments from the circulation than intact PTH (Figure 7).

We have developed certain specific criteria which can be used in evaluating carboxyl-region-specific radioimmunoassays for their suitability for clinical

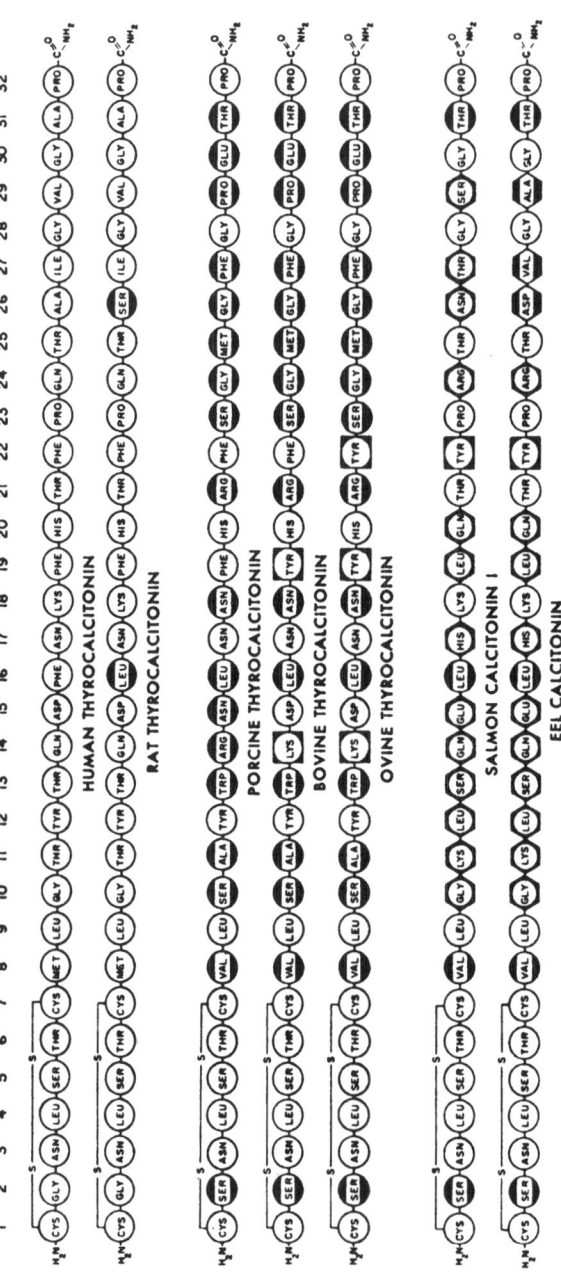

Figure 9. Amino acid sequence of calcitonins. Note the similarity between human and rat calcitonin and the substantial differences in amino acid sequence from the porcine, bovine, and ovine calcitonins. Salmon and eel calcitonin are actually more similar to human calcitonin than are hormones from ungulates. In the 7 amino acid N-terminal ring, the only substitution is serine for glycine. (From Hirsali PF, J Exp Zool 178:139, 1971)

application. These include (1) consistently low or undetectable serum iPTH values in all patients with hypoparathyroidism, (2) ability to measure iPTH in greater than 95% of the normal sera and to demonstrate a negative relationship between serum iPTH and total serum calcium over the normal range of the serum calcium (Figure 8), (3) low or undetectable serum iPTH in all patients with hypercalcemia of nonparathyroid origin, other than cancer, and (4) ability to demonstrate iPTH values greater than the upper limits of normal in 90% of the patients with surgically proved primary hyperparathyroidism.

Until recently, bioassays for PTH were not sufficiently sensitive to be used to study circulating PTH. However, this detection limit obstacle has been overcome by two novel approaches. In the first of these, a cytochemical bioassay, the PTH-specific stimulation of the enzyme glucose 6-phosphate dehydrogenase in guinea pig renal slices is measured. This assay is extremely sensitive, allowing femtogram amounts of PTH to be quantitated. However, its considerable technical complexity seems likely to prevent its routine application as a clinical diagnostic tool. A second more convenient approach is based on the use of a GTP analogue, 5'-guanyl-imidodiphosphate, [Gpp(NH)p] to greatly augment PTH sensitivity of a canine renal adenylate cyclase assay system in vitro. In the presence of Gpp(NH)p, as little as 10 pg PTH(1–84)/ml elicits significant enzyme stimulation. Although this bioassay is not as sensitive as the cytochemical bioassay for PTH, it is rapid and simple to perform, and it may utlimately prove to be a useful biologic adjunct to the PTH radioimmunoassay in diagnostic situations.

Although radioligand-binding studies have been used to probe the PTH receptor, no clinically useful radioreceptor assay for PTH has been reported.

II. Calcitonin

A. Structure and Biosynthesis

Calcitonin, a 32-amino-acid 3700-molecular-weight polypeptide with a 1–7 disulfide bridge (Figure 9) is biosynthesized and secreted by the ultimobranchial (parafollicular or "C") cells. It is probably synthesized as a large-molecular-weight precursor, but at this writing, this precursor is not characterized.

The ultimobranchial cells develop from neural crest tissue in the ultimobranchial cleft during embryonic life. They form a discrete organ in submammalian vertebrates called the ultimobranchial gland. In the mammal, the anlage of the cells merge with the embryonic thyroid gland utlimately becoming dispersed in a central region adjacent to the follicular cells.

The amino acid sequences of the calcitonins from many species (i.e., pig, cattle, rat, human, salmon, and eel) have been determined. They differ sufficiently that immunologic crossreactivity between them is incomplete. In fact, radioimmunoassay of human calcitonin was not possible until synthetic human calcitonin and specific antibodies directed against it were available. The entire calcitonin molecule (Figure 9) is required for full biologic activity as is the intact 1–7 disulfide bridge. Salmon calcitonin has a markedly different amino acid sequence from mammalian calcitonins and, interestingly, is approximately 30 times more potent. Synthetic human and salmon calcitonin are commercially available and can be used for the treatment of metabolic bone disease.

B. Control of Secretion

Calcitonin is rapidly released by the "C" cells in response to small increases in plasma ionic calcium. It acts on kidney and bone to restore the level of this cation to just below a normal set point which in turn inhibits the secretion of the hormone (Figure 1). Calcitonin thus is a physiologic antagonist of parathyroid hormone. The hormones act in concert to maintain the normal concentration of extracellular fluid ionic calcium (Figure 2).

Induced hypercalcemia causes an increase in plasma immunoreactive calcitonin (iCT) in approximately 40–60% of normal women and 70–80% of normal men. There is a positive relationship between levels of plasma calcium above the normal range (induced by oral calcium ingestion) and plasma iCT in normal subjects (Figure 10). Pentagastrin injection also causes increased

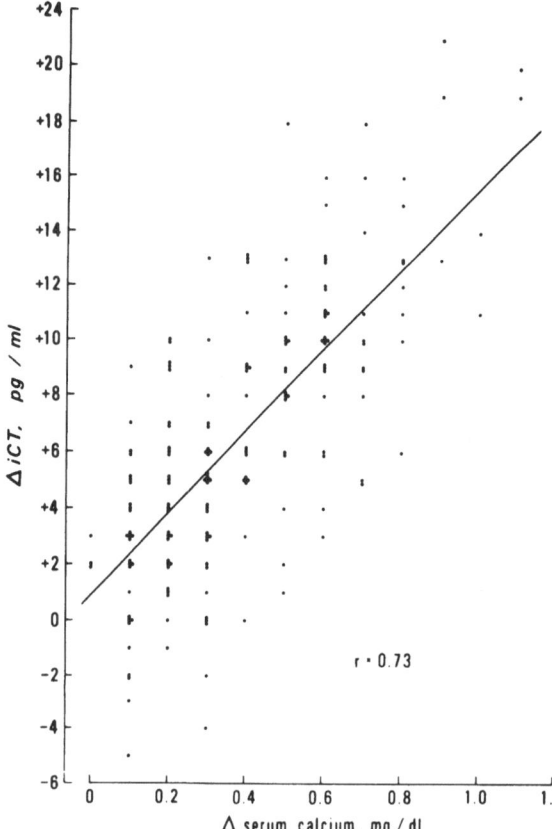

Figure 10. Changes in iCT plotted against the corresponding changes in serum Ca after Ca ingestion. The data represent the responses of 10 subjects to each of the three oral Ca doses at all time intervals studied. r = 0.73. (From Austin LA et al., J Clin Invest 64:1721, 1979.)

plasma iCT. It is unlikely that gastrin is a physiologic calcitonin secretagogue. Other calcitonin secretagogues of unproved physiologic significance include glucagon, β-adrenergic agonists, and alcohol.

C. Metabolism and Circulating Forms

Calcitonin exists in ultimobranchial tissue and in plasma in multiple molecular forms. In contrast to parathyroid hormone, however, the major circulating species are not hormone fragments, but consist of as many as four or five immunoreactive forms with molecular weights larger than the 32-amino-acid calcitonin. Some of these forms are probably polymers of calcitonin with interchain disulfide molecular linkages. Different radioimmunoassays measure these different forms of calcitonin variably, and, therefore, the normal range for plasma calcitonin must be established for each assay. As measured by the most specific sensitive assays, plasma concentrations of iCT are extremely low (less than 100 pg/ml).

D. Actions of Calcitonin

The actual importance of calcitonin in calcium homeostasis of adult humans is not established. Excess or deficiency of parathyroid hormone or vitamin D produce dramatic clinical disorders. In contrast, excess (medullary carcinoma of the thyroid gland) or deficient (postthyroidectomy) calcitonin levels produce no discernable or serious abnormalities of mineral metabolism. The basal plasma levels of calcitonin are lower in women than in men. In adults, calcitonin may function primarily to restrain the bone resorptive effects of parathyroid hormone. If so, the long-term decrease in calcitonin secretion in women and increased parathyroid hormone secretion with aging may constitute one set of etiologic factors in postmenopausal osteoporosis.

When bone turnover rates are high, calcitonin administration produces rapid hypocalcemia and hypophosphatemia. These effects are largely the result of the hormone acting to decrease bone resorption by inhibiting the activities of bone-resorbing cells (osteoclasts). Calcitonin also increases urinary calcium and phosphate, but its action on the kidney is transient and variable. Current evidence does not support a direct effect of calcitonin on the intestinal absorption of calcium. However, several recent reports suggest that the hormone may have a negative or positive influence on the renal production of $1\alpha,25\text{-}(OH)_2D_3$, depending on experimental protocol. Such effects are not completely compatible with the idea that calcitonin is a physiologic antagonist of parathyroid hormone.

E. Mechanism of Action

Calcitonin receptors have been demonstrated in the plasma membranes of target cells (kidney). The hormone stimulates adenylate cyclase in bone and kidney in some species (human and rat), but whether cyclic AMP is a major intracellular mediator of calcitonin action has not been established.

F. Assay in Biological Fluids

The major tool for the assay of calcitonin in biological fluids has been radioimmunologic. Assays for human calcitonin generally crossreact with rat calcitonin, but assays for porcine and bovine calcitonin crossreact poorly with human and rat calcitonin. Although iCT in human plasma is heterogenous, synthetic human calcitonin is generally used as an assay standard. As with PTH, most antisera to human calcitonin are multivalent, and assays developed with them measure the different circulating forms of calcitonin variably. Thus, as noted above, the normal ranges for plasma iCT differ from laboratory to laboratory.

The single most important factor determining the clinical utility of a calcitonin radioimmunoassay is its sensitivity. This is because many patients with calcitonin-secreting medullary cancers of the thyroid gland have normal levels of basal plasma iCT and can be identified only with provocative tests (i.e., calcium infusion, pentagastrin injection; see below). An assay that cannot measure basal levels of iCT cannot reliably demonstrate an abnormal stimulated increase in plasma iCT above basal levels.

Since the development of the original rat hypocalcemic assay for calcitonin, there has been little progress in the development of bioassays for this hormone. Although calcitonin receptors have been demonstrated, radioreceptor assays of calcitonin are too insensitive for the measurement of calcitonin in biological fluids.

III. Vitamin D

A. Physiology, Metabolism, and Action

It is now firmly established that vitamin D and its metabolites are sterol hormones and that their metabolism and mechanism of action have much in common with that of the other steroid hormones (Figure 11). Vitamin D_3 (cholecalciferol) is primarily derived from the skin by ultraviolet (UV) irradiation of 7-dehydrocholesterol; vitamin D_2 (ergocalciferol) is produced by UV irradiation of the plant sterol ergosterol and is used to fortify dairy products. The parent compound, vitamin D, essentially lacks biological activity and requires metabolic transformation to attain potency. The first step in this activation pathway (Figure 11) involves enzymatic 25-hydroxylation, a step predominantly confined to the liver. This conversion does not appear to be tightly regulated, and the circulating concentration of 25-hydroxyvitamin D [25-OHD$_3$], estimated to be approximately 30 ng/ml in normal individuals, is primarily a function of the bioavailability of vitamin D. Hence, it is subject to both seasonal and regional variation due to differences in both diet and sunlight expo-

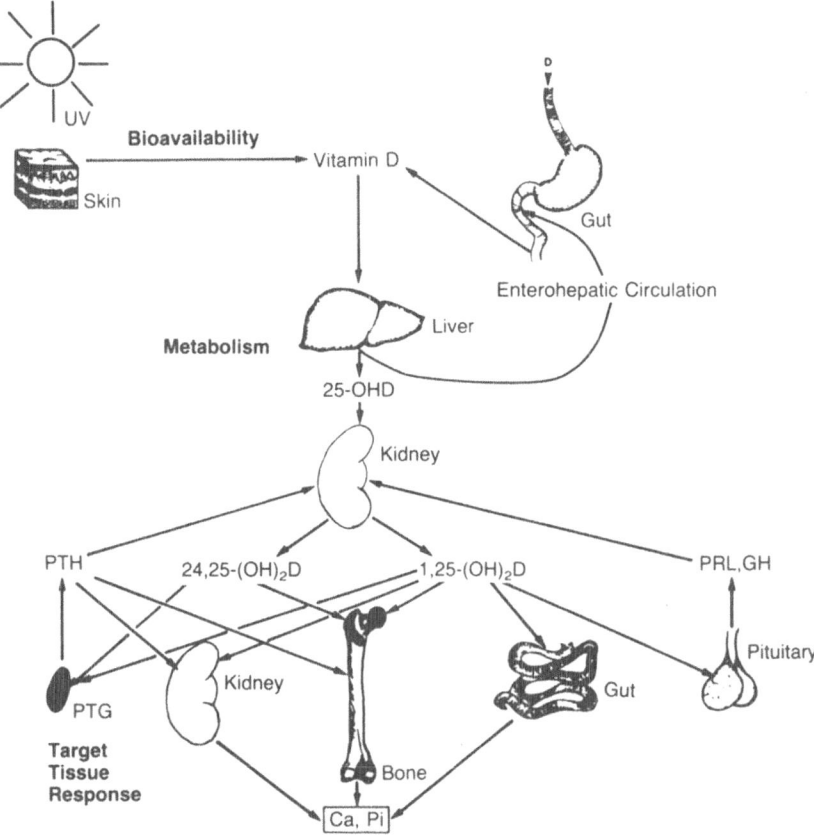

Figure 11. The vitamin D endocrine system. **Bioavailability:** Vitamin D is made available to the body by photogenesis in the skin and absorption from the intestine. Part of the intestinal absorption involves endogenous vitamin D products secreted in bile (enterohepatic circulation). **Metabolism:** The vitamin D must be hydroxylated first in the liver to 25OHD then in the kidney to 1,25(OH)$_2$D and 24,25(OH)$_2$D. Other metabolites and other tissues capable of metabolizing the vitamin D metabolites are known. **Target tissue response:** The three principal target tissues are kidney, bone, and intestine *(gut)*. Parathyroid gland *(PTG)* and anterior pituitary may also be target tissues. Their hormone products (parathyroid hormone, prolactin, and growth hormone) help regulate vitamin D metabolism in the kidney, and PTH, at least, also has independent effects on bone and kidney regulation of calcium and phosphate homeostasis. (From Bikle DD, *Advances in Internal Medicine,* Year Book Medical Publishers, 1982, p. 46.)

sure. 25-hydroxyvitamin D is the most abundant circulating form of the hormone, and is transported in serum bound to a specific globulin-carrier protein (vitamin-D-binding protein), as are the other vitamin D metabolites.

The next step in the bioactivation of vitamin D, 1α-hydroxylation of 25-OHD$_3$, occurs in the renal tubular mitochondria. This hydroxylation is tightly regulated and constitutes the rate-limiting step in the production of the active metabolite, $1\alpha,25\text{-}(OH)_2D_3$. This important regulatory step has been the subject of intensive study. Production of $1\alpha,25\text{-}(OH)_2D_3$ is controlled by the interplay of a number of regulators in accordance with the body's mineral requirements. Hypophosphatemia and hypocalcemia appear to constitute the primary stimuli for renal 1α-hydroxylase. The low calcium stimulus is probably mediated via PTH which stimulates 1-hydroxylase activity either directly or by its hypophosphatemic effects (Figures 1 and 11). The evidence in favor of a negative trophic effect of calcitonin on the activity of the 1α-hydroxylase is not as strong, and recent reports show a stimulatory effect. Additionally, $1\alpha,25\text{-}(OH)_2D_3$ may feed back to regulate its own synthesis either directly or through a suppressive effect on the parathyroid glands where receptors for this D metabolite have been reported.

In normal individuals, $1\alpha,25\text{-}(OH)_2D_3$ is estimated to circulate at concentrations around 30 pg/ml. This metabolite is more rapidly cleared from plasma than its more abundant precursor, $25\text{-}OHD_3$, and it is about 1000 times more potent in stimulating intestinal calcium absorption. It is thus generally considered to be the major biological effector of the vitamin D endocrine system. However, the fact that $25\text{-}OHD_3$ circulates in concentrations 1000 times greater than $1\alpha,25\text{-}(OH)_2D_3$ suggests the real possibility that $25\text{-}OHD_3$ may have intrinsic biological importance as well.

Hydroxylation of $25\text{-}OHD_3$ to $24,25\text{-}(OH)_2D_3$ occurs mainly in the kidney and represents an alternative metabolic fate for $25\text{-}OHD_3$. The regulation of this pathway generally appears to be reciprocal to that leading to $1\alpha,25\text{-}(OH)_2D_3$; for example, PTH stimulates production of $1\alpha,25\text{-}(OH)_2D_3$ but suppresses that of $24,25\text{-}(OH)_2D_3$. At present, the physiologic importance of $24,25\text{-}(OH)_2D_3$ is unclear. It may be involved in normal bone formation, but it has also been proposed that 24-hydroxylation represents the initial step in the metabolic elimination of $25\text{-}OHD_3$. Additional vitamin D metabolites have been identified including $25,26\text{-}(OH)_2D_3$, $1\alpha,24,25\text{-}(OH)_3D_3$, $1\alpha,25,26\text{-}(OH)_3D_3$, and $25\text{-}OHD_3\text{-}26,23$ lactone. Excretion of the vitamin D metabolites occurs primarily in the bile, and evidence has been presented for enterohepatic circulation of both $25\text{-}OHD_3$ and $1\alpha,25\text{-}(OH)_2D_3$.

It is generally accepted that the principal physiologic role of vitamin D is to increase plasma levels of calcium and phosphate to maintain conditions favorable to bone mineralization. The most extensively studied physiologic action of $1\alpha,25\text{-}(OH)_2D_3$, however, is its role in facilitating the absorption of calcium (and phosphate) by the intestine. Much of our knowledge concerning the molecular mechanism of action of vitamin D has been derived from studies of the intestinal cell. Intestinal cytosolic receptors for $1\alpha,25\text{-}(OH)_2D_3$ have been demonstrated. These function to translocate the hormone into the cell nucleus, where their association with chromatin influences transcription and promotes the production of at least one new protein (calcium-binding protein) presumed to facilitate transcellular calcium transport. In this view, vitamin D acts as a classic steroid hormone, but it has also been suggested that the hormone may have cellular actions independent of de novo protein synthesis.

In addition to intestine, $1\alpha,25\text{-}(OH)_2D_3$ receptors have been demonstrated in bone, parathyroid glands, placenta, pancreas, pituitary, and skin. It is possible, therefore, that vitamin D acts directly on these tissues. Vitamin D may also have a role, as yet poorly defined, in muscle function.

B. Assay in Biological Fluids

Using radiolabeled vitamin D derivatives, two general approaches have been developed for the measurement of the various vitamin D sterols in vitro: competitive protein-binding analysis and radioimmunoassay. The sensitive technique of competitive protein-binding analysis takes advantage of naturally occurring specific proteins having high affinity for the vitamin D metabolites, present either in serum (vitamin D-binding protein) or in intestinal target cells (cytosolic receptors). Both of these techniques involve a competition reaction in vitro between a radioligand and the vitamin D metabolite sterol of interest for available sites on the binding protein. Antisera for use in radioimmunoassays are developed against immunogenic derivatives of the vitamin D metabolite sterol molecule.

Existing competitive protein-binding assays and radioimmunoassays are not inherently specific for the measurement of a single circulating vitamin D metabolite because, in addition to the particular vitamin D sterol of interest, the closely related vitamin D metabolites invariably compete to some extent for available binding sites on the binding protein or antibody. Furthermore, other components of blood, including plasma proteins and lipids, may cause nonspecific interference in these assays. To increase the specificities and sensitivies of the assays, it is necessary first to extract the vitamin D sterols from the plasma using a suitable lipophilic solvent and then isolate the particular metabolite of interest by chromatography.

A sensitive bioassay for $1\alpha,25\text{-}(OH)_2D_3$ has been developed using the calcium mobilizing effect of the hormone on embryonic bone organ cultures in vitro as an index of activity. This assay has comparable sensitivity to competitive protein-binding assays and radioimmunoassays but is tedious to do and may require state-of-the-art experience in its performance.

Chapter 2

Resolution and Quantitation of Vitamin D and Vitamin D Metabolites

Ronald L. Horst

For approximately 36 years (1922–1958), the only official method for determining vitamin D in biologic fluids, tissues, and feed-stuffs was the rat line test described by McCollum et al.[1] This test was based on the curative effects of the treatment, with the supposition that after rickets was established in rats they all were cured of the lesion by the addition of vitamin D to the rachitogenic diet.

An alternative approach to the curative method was the preventive method described by Schultzer[2] which is based on the absence or presence of rickets in young rats kept on a rachitogenic diet containing varying amounts of vitamin D.

Both assays are very time-consuming and generally required a 7–14-day assay period. In addition, the inherent variability in response of different animals to the same treatment, as well as errors in the subjective judgment of estimation of curative or preventive effects, make absolute quantitation very difficult. One advantage the line test offered to early scientists was its sensitivity. As little as 25 ng vitamin D_2 or vitamin D_3 could be detected by this method. Early alternatives to these rat bioassays included the measurement of fecal pH changes,[3] growth stimulation test,[4] and colorimetric determinations.[5]

In 1958, a chemical method was adopted as the second official alternative to the rat line test. This method was utilized for pharmaceutical preparations containing large quantities (>250 μg) of vitamin D.[6] In this assay, the "purified" vitamin D preparation was reacted with antimony trichloride in ethylene dichloride. The resulting color could be read at 500 nm, and was linear over a wide range of vitamin D concentrations.

Since 1958, the knowledge of vitamin D physiology and metabolism has progressed rapidly. We now know that vitamin D is metabolized to several compounds but ultimately to 1,25-dihydroxyvitamin D, which is thought to be the active form of vitamin D. This metabolite circulates normally in the plasma at concentrations of 30–40 pg/ml. In addition, other metabolites [24,25-$(OH)_2D$ and 25,26-$(OH)_2D$], which may have biologic roles, also circulate at very low concentrations (0.5–2.0 ng/ml). The need to routinely monitor the plasma concentration of these biologically-active forms of vitamin D has precluded the use of the bioassays, and necessitated the development of more sensitive and rapid assay methods capable of distinguishing between the multiple

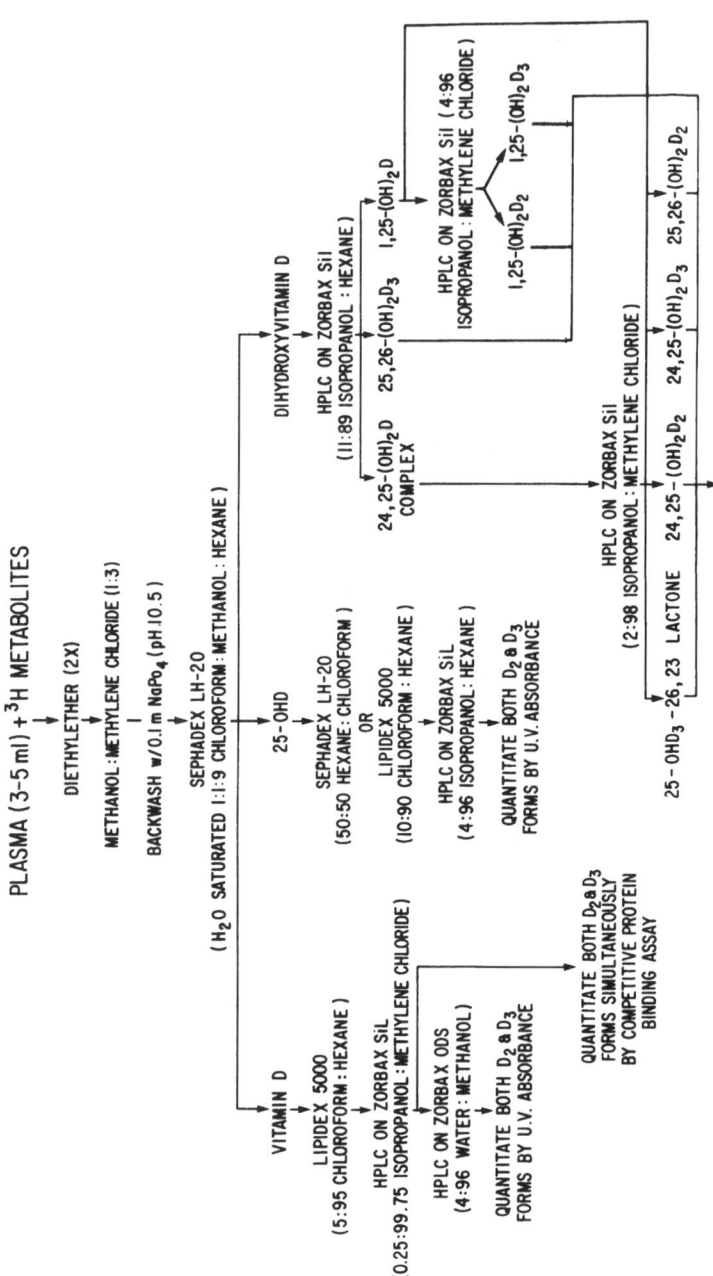

Figure 1. Flow diagram of the purification and ultimate quantitative steps used for the assay of vitamin D and its metabolites.

forms of vitamin D in plasma. The introduction of high-pressure liquid chromatography (HPLC) and the discovery of plasma and intestinal vitamin D and vitamin D metabolite-binding proteins were paramount in the quest for a multiple vitamin D metabolite assay system.

This chapter describes methods currently used in my laboratory for the isolation and ultimate quantitation of vitamin D_2, vitamin D_3, and their metabolites. Figure 1 is the flow diagram used for the purification and ultimate quantitative steps used for the assay of these compounds.

I. Extraction of Vitamin D and Vitamin D Metabolites from Serum or Plasma

A. Diethylether, Methylene Chloride (MeCl₂)/Methanol (MeOH) Procedure

General

1. Thaw sample. Place 3–5 ml of sample into a 50-ml screw-top centrifuge tube (3.0 cm × 10 cm). If the sample consists of less than 5 ml, bring it up to 5 ml with saline.
2. Add 1000 cpm of labeled vitamin D metabolites in 50 µl of ethanol to each plasma sample and to 2 scintillation vials (to estimate recovery). Vortex the mixture and let stand for 15 min.
3. Add 15 ml of ether to the mixture and vent the tube 2–3 times. Secure the cap and shake 5 min on a horizontal shaker at ∼120 opm.
4. After shaking, the samples should be placed into a test tube rack and allowed to settle. After 1–2 min, freeze the bottom layer in a dry ice-acetone bath and pour off the top layer into a 25 × 150 cm tube. Thaw sample.
5. Repeat ether extract: add 15 ml ether, shake for 5 min, freeze, pour off top layer, and thaw the bottom layer.
6. Add 20 ml of 3:1 MeCl₂/MeOH to the remaining bottom layer. Secure the cap and shake 2 min on shaker.
7. Add 5 ml MeOH, cap and shake by hand for a few seconds. Allow to phase.
8. Suction off top layer containing the precipitated proteins. Add 5 ml 0.1 M NaPO₄ buffer (pH 10.5) to the bottom layer, vortex, and allow to phase. Suction off top layer.
9. Repeat NaPO₄ buffer wash of Step 8.
10. Combine the MeCl₂ and ether extracts, place the samples in a heated (∼35°C) water bath, and dry samples down using filtered air in preparation for column chromatography purification. (If water remains in extracts, add a small amount of ethanol to facilitate drying of samples.)

Results

The above procedure generally results in the extraction of 90–100% of the ^{3}H-vitamin D metabolites added to the plasma sample. Extraction of vitamin D metabolites from most species does not require the inclusion of steps 6 through 9; however, if bovine or ovine plasma is being extracted, or if 25-OHD$_3$-26,23 lactone is to be measured in the plasma of any species, it is recommended that these steps be included. As indicated in step 10, it is not necessary to dry down the samples under N$_2$ at this point; air is much cheaper.

Precautions

All the ^{3}H-metabolites used in step 2 should be purified by HPLC before using (Section III). The diethylether used in steps 3 through 5 should be peroxide-free. A simple check involves dissolving 2–3 crystals of ferrous sulfate in 2 ml distilled water. Add 2 ml diethylether and vortex. Allow the sample to set for 4–5 min and, if a yellow color appears in the bottom (water) layer, the diethylether contains peroxides. Also, the use of EM Quant Ether Peroxide Test purchased from E. Merck, Darmstadt, Germany, offers a simpler alternative. We have found that diethylether purchased from Baker (product No. 9248) is generally peroxide-free. The other solvents used were purchased from Burdick and Jackson (Muskegon, Michigan).

Step 6 usually results in the formation of an emulsion; the addition of MeOH (step 7) usually results in breaking the emulsion and ultimate phase separation.

Additional Information

Some of the ^{3}H-vitamin D metabolites are commercially available. 25-OH[^{3}H]D$_3$, 24,25-(OH)$_2$[^{3}H]D$_3$; 1,25-(OH)$_2$[^{3}H]D$_3$; and 1,24,25-(OH)$_3$[^{3}H]D$_3$ are available from Amersham and New England Nuclear. 25,26-(OH)$_2$[^{3}H]D$_3$; and 25-OH[^{3}H]D$_3$-26,23 lactone can be synthesized according to procedures described by Horst et al.[7] Step 3 requires capping the 50-ml tube. We have found that the tin-foil-lined caps for large-mouth scintillation vials work very well and are relatively inexpensive. These caps can also be washed once or twice.

B. Methanol/Methylene Chloride Procedure

General

1. Use 5 ml serum in 25 × 150 tubes (if not 5 ml, bring up to 5 ml with 0.85% NaCl solution).
2. Add radioactive vitamin D metabolites in 50 μl of ethanol for each of the metabolites being analyzed.

3. Add 18.75 ml of cold 2:1 MeOH/MeCl$_2$. Vortex. Let sit 1 h in cooler. Vortex twice during the hour.
4. At the end of the hour, add 6.25 ml cold MeCl$_2$. Vortex. Centrifuge 10 min at ~1000g.
5. Pull out the bottom layer and place into clean tube, using 20-ml glass syringe and long needle. Wash the syringe and needle with ~5 ml MeCl$_2$ after each sample.
6. Add another 6.25 ml MeCl$_2$ to each tube and centrifuge again for 5 min. Pull out bottom layer and combine with first extract.
7. Dry down samples to ~10-ml volume and then add 10 ml of 0.1 M NaPO$_4$ buffer (pH 10.5). Vortex twice and let set until separation of layers is complete.
8. Suction off the top layer with vacuum hose and add another 10 ml of buffer.
9. Vortex twice and suction off the top layer.
10. Dry down sample under filtered air.

Results

The MeOH/MeCl$_2$ extraction is generally more time-consuming and is less effective than the procedure described in Section I.A for vitamin D metabolite extraction. This procedure, however, does not require special items such as screw-capped tubes, shaker, etc. It also offers an advantage over MeOH/chloroform (CHCl$_3$) in that less lipid is extracted from the sample.

II. Primary Low-pressure Liquid Column Chromatography

A. Overview

These represent a series of low-pressure liquid chromatographic steps leading to the ultimate purification of several metabolites for quantitation. The successful completion of these steps, therefore, is paramount to the effective completion of the assays.

B. Separation of Vitamin D and Vitamin D Metabolites into Four Fractions

General

The vitamin D metabolites contained in the lipid extracts (from Section I.A or I.B) are separated by chromatography on a 0.6 × 15.5 cm Sephadex LH-20 column. The column is swelled in 0.015:1:1:9 water/MeOH/CHCl$_3$/hexane and packed in the same solvent system. The lipid extract is applied to the column in 0.5 ml of the same solvent and allowed to run into the column bed. This procedure is repeated with another 0.5 ml. Thereafter, the various vitamin D

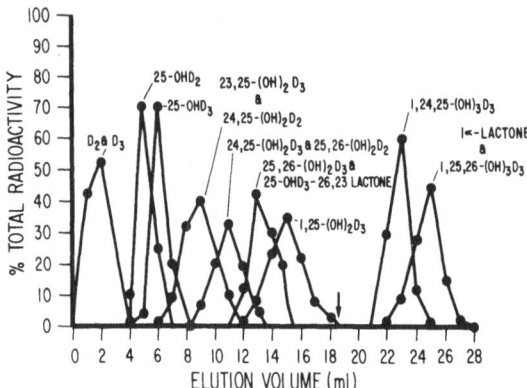

Figure 2. Elution of vitamin D and vitamin D metabolites from the 0.6 × 15.5-cm column of Sephadex LH-20. The first 19 ml of eluent consisted of 0.015:1:1:9 water/$CHCl_3$/MeOH/hexane, and the next 10 ml (solvent change indicated by *arrow*) consisted of 0.015:2:2:8 water/$CHCl_3$/MeOH/hexane.

metabolites can be collected from the column as shown in Figure 2 and described below.

From this column, vitamin D_2 and vitamin D_3 (Fraction 1) elute in the first 2.5 ml, 25-OHD_2 and 25-OHD_3 (Fraction 2) in the next 5.0 ml, and the dihydroxyvitamin D metabolites, including 25-OHD_3-26,23 lactone (Fraction 3), in the next 13.5 ml. If trihydroxyvitamin D metabolites (Fraction 4) are desired, the column solvent can be switched to 0.015:2:2:8 water/MeOH/$CHCl_3$/hexane. The trihydroxyvitamin D metabolites and 1,25-$(OH)_2D_3$-26,23 lactone will elute in 10 ml.

Results

Collection of the four fractions allows for individual handling of each in subsequent purification steps to be described later. The process of eluting the metabolites from application to final collection takes approximately 3 h.

Precautions

The addition of the water to the solvent system is crucial, especially if humidity is poorly controlled in the laboratory. If saturation of the 1:1:9 solvent system occurs during the process of eluting the vitamin D metabolites, the column will begin to swell and the flow rate will fall dramatically. Also, the metabolites will migrate much later than indicated in the normal elution profile (Figure 2). One must also allow several days accompanied by several solvent changes for adequate "swelling" of the Sephadex LH-20. Failure to do so will result in lots of bubbles in the column and erratic migration of metabolites. Usually, 3–4 days of swelling with 2–3 changes of solvent daily is satisfactory. Finally, if the trihydroxy metabolites are of interest, the addition of the 0.015:2:2:8 solvent system to the column will also cause slight swelling and a slower flow rate.

Additional Information

The glass columns used for the Sephadex LH-20 support are made by a local glass blower for $3.50. The column contains a reservoir at the top which will hold ~18 ml. The interface between the column packing and collection point is flat to enable the placement of a small glass fiber disc used for bed support. The discs are generally <0.5 mm in diameter and, therefore, do not affect the bed volume. The glass fiber discs are listed as P.O. No. F2832-9 in the American Scientific Products Catalog. The columns used in this step are also used in subsequent steps. They do not require stopcocks as there is adequate surface tension to avoid the column bed from drying and cracking, that is, the column can sit without any solvent in the reservoir for up to 30 min without any noticeable drying or evaporation of the solvent from the Sephadex LH-20 resin.

III. Secondary Low-pressure Liquid Column Chromatography

A. Further Purification of the Vitamin D Fraction (Fraction 1, Section II.B)

General

The fraction from the $0.015:1:1:9$ $H_2O/CHCl_3/MeOH/hexane$ column containing the vitamin D (first 2.5 ml) is dried under N_2 and chromatographed on a Lipidex 5000 column (0.6 × 14.5 cm) developed in 5:95 $CHCl_3$/hexane. When drying is complete, the sample is dissolved in 0.5 ml of the 5:95 mixture. Let the sample sit for 5–10 min to assure adequate dissolution. The sample is then applied to the column and allowed to run into the column bed. An additional 0.5 ml is used for washing the sample tube and is applied to the column. Thereafter, the first 6.0 ml of eluent is discarded and the next 4.0 ml collected in a 6-ml conical tube. This fraction contains both vitamin D_2 and vitamin D_3 (Figure 3)

Results

This step adequately purifies the sample to allow further purification by high-pressure liquid chromatography (Section IV.C). This Lipidex 5000 column can be run in 60–90 min.

Precautions

Lipidex 5000 arrives suspended in MeOH. To get the resin reequilibrated in $CHCl_3$/hexane, first pour off the excess MeOH followed by 1–2 washings of

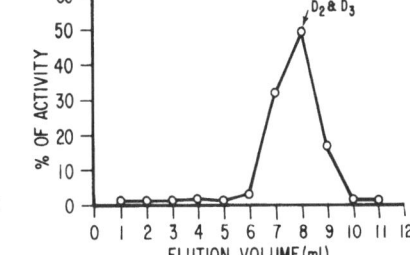

Figure 3. Elution of vitamin D_2 and vitamin D_3 from the 0.6 × 14.5-cm Lipidex 5000 column developed in 5:95 $CHCl_3$/hexane.

1:1 $CHCl_3$/hexane and finally 1–2 washings with 100% hexane. *Reequilibrate only the amount to be used over a period of 4–6 weeks!* After the excess hexane is removed, the resin can be washed with 5:95 $CHCl_3$/hexane and allowed to set overnight to assure equilibration with the new solvent system. The resin should be washed an additional one or two times with column solvent just prior to use. The $CHCl_3$ used to equilibrate the column is purchased from Burdick and Jackson, and is preserved with 1% EtOH. $CHCl_3$ purchased from other vendors may not contain EtOH and therefore the migration pattern of vitamin D may be altered.

Additional Information

Lipidex 5000 can be purchased from Packard Instruments Co., Downers Grove, Illinois (P.O. No. 6008304) at \$150/100 g. This material can be reused after washing the resin with 1:1 $CHCl_3$/hexane followed by MeOH. If the Lipidex 5000 is not frequently used, it is best to store the resin in MeOH rather than solvent systems containing $CHCl_3$. Lipidex 5000 becomes unstable when stored in $CHCl_3$ for long periods.

B. Further Purification of the 25-OHD Fraction (Fraction 2, Section II.B)

General

The fraction from the 0.015:1:1:9 H_2O/$CHCl_3$/MeOH/hexane column containing the 25-OHD fraction (Fraction 2) is dried under N_2 and resuspended in 0.5 ml of 1:1 $CHCl_3$/hexane. Let the sample set 5–10 min and apply it to a 0.6 × 14.5 cm Sephadex LH-20 column developed in 1:1 $CHCl_3$/hexane. An additional 0.5 ml is used for washing the sample tube. After the sample and wash have been applied and allowed to run into the column bed, the first 2.5 ml of column eluent is discarded and the next 6.0 ml collected as 25-OHD$_2$ and 25-OHD$_3$ (Figure 4) in a 6-ml conical tube for further analysis by HPLC (see Section IV.D). Alternatively, the 25-OHD fraction (Fraction 2) from the 0.015:1:1:9 H_2O/$CHCl_3$/MeOH/hexane column can be purified for HPLC analysis using Lipidex 5000 column (0.6 × 14.5 cm) developed in 1:9 $CHCl_3$/hexane. After the sample is applied in column solvent, the first 8.0 ml of eluent is discarded and the second 13.0 ml is collected in a 15-ml conical tube as 25-OHD$_2$ and 25-OHD$_3$.

Results

These columns provide adequate removal of UV-absorbing peak that interferes with the HPLC UV analysis of the 25-OHD$_2$ and 25-OHD$_3$. Generally, the cheaper Sephadex LH-20 column is adequate for the purification of 25-OHD from most species. However, the Lipidex 5000 column is more successful in the cleanup of 25-OHD isolates from human plasma. These columns are also ade-

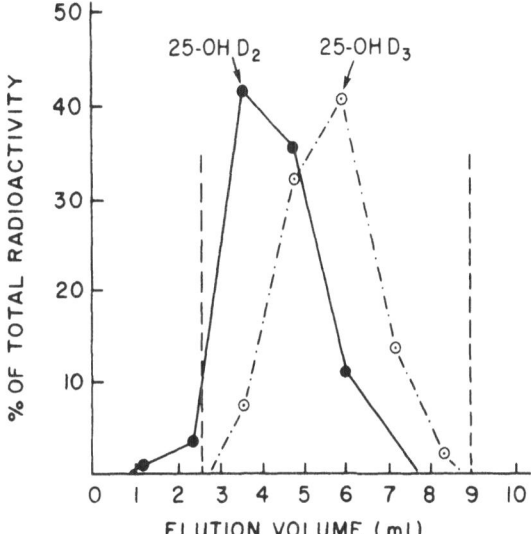

Figure 4. Elution of 25-OHD$_2$ and 25-OHD$_3$ from the 0.6×14.5-cm column of Sephadex LH-20 developed in 1:1 CHCl$_3$/hexane.

quate for the purification of the sample in preparation for competitive protein-binding analysis (Section V.C). These columns can be run in 60–90 min.

Precautions (see Sections II.A and II.B)

When using the Sephadex LH-20 or the Lipidex 5000 columns for purification of the 25-OHD, one must be aware that 25-OHD$_2$ is partially resolved from 25-OHD$_3$ as is shown in Figure 4. Both columns give similar resolution. Therefore, if ^{3}H-25-OHD$_3$ is used for standardizing the columns, adjustments in the collection must be made to include the 25-OHD$_2$.

IV. High-pressure Liquid Column Chromatography (HPLC)

A. Overview

All HPLC is carried out with a Waters Associates Model LC-204 chromatograph fitted with a Model 6000-A pumping system and a Model 440 UV-fixed wavelength (254 nm) detector. The detector response is measured with a Houston Instruments dual pen recorder. The samples and standards are introduced onto the HPLC columns automatically with a No. 710A Waters "intelligent" sampling processor (WISP 710A). The procedure involves transferring the samples into a limited-volume insert (Waters Associates, part No. 72704) with a final volume of 150 μl. Introduction of the samples onto the column with a total volume of 150 μl results in minimal peak spreading (compared to smaller volumes) with negligible sample losses (dead volume of insert was 6 μl). For collection of the individual metabolites from the HPLC columns, a fraction collector is used that requires moving collection tubes at specified times. There-

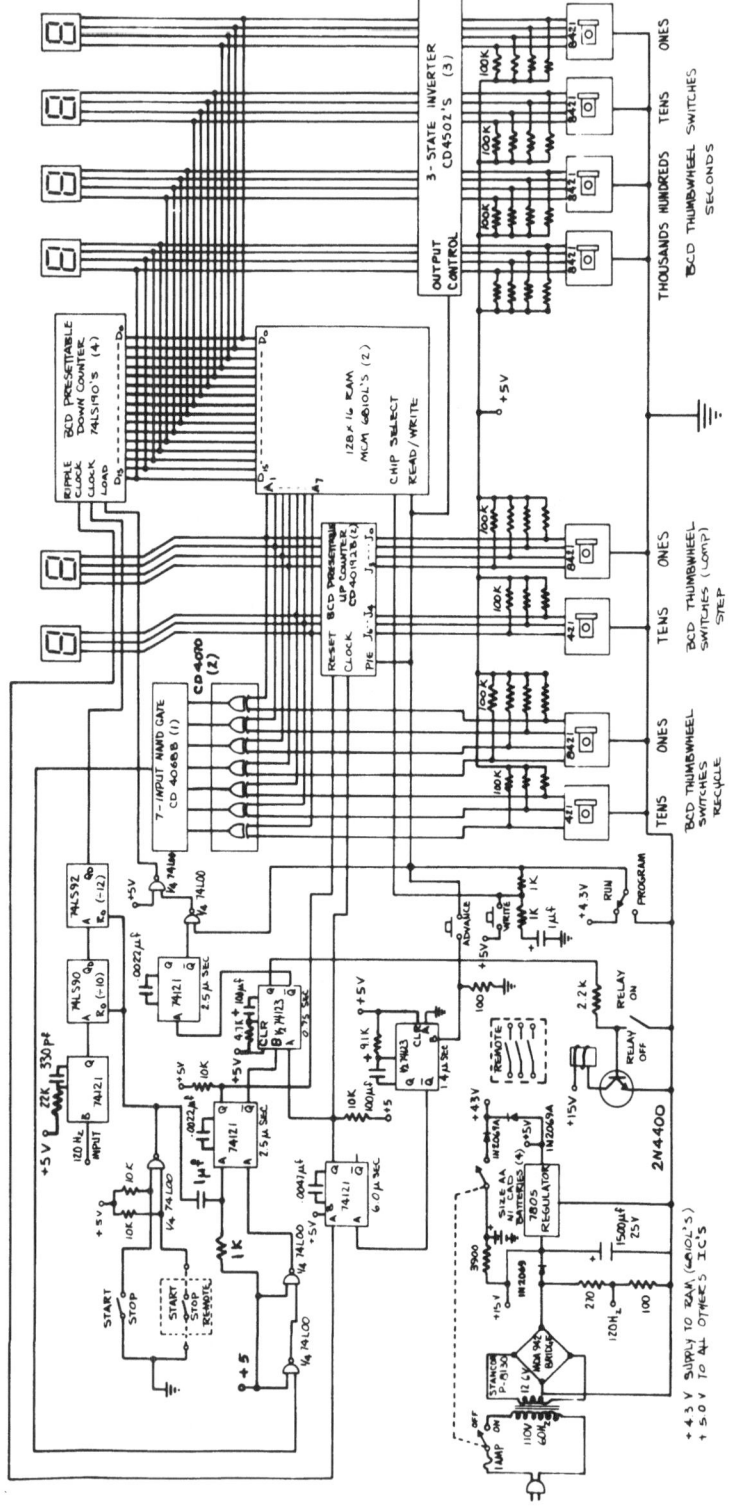

Figure 5. Schematic diagram of the programmable timer.

fore, a programmable timer (Figure 5) was built which would sequence through a set of programmed time intervals, causing relay closure and movement of the fraction collector at the end of each time interval. The timer would continuously repeat this sequence and could be set to initial conditions by a relay closure originating from the WISP 710A. Each time interval could be easily and accurately programmed to the nearest second by placing a switch in the program mode. A display of the step and time interval was provided so the operator could observe where the timer was in sequence. The availability of integrated circuits and integrated circuit memories made the design of this timer straightforward and inexpensive. There is also a commercially available timer from Encoder Products Corp. (Sand Point, Idaho). The solvents used for HPLC are UV spectral grade and are purchased from Burdick and Jackson (Muskegon, Michigan). Prior to use, the solvent mixtures should be degassed to prevent air from entering the HPLC pumps. In our laboratory, the solvent mixtures are stored in the 3.75-l bottles used for shipping by Burdick and Jackson. Each bottle contains a magnetic stir bar. A vacuum is applied to the bottle using a sink aspirator. The solvent is degassed for 5–10 min while the mixture is being continuously stirred. The columns used for HPLC are purchased from DuPont (Wilmington, Delaware). The columns are listed under the names, Zorbax Sil, which is a silicic acid column, and Zorbax ODS, which is a bonded phase-silicic acid column. The Zorbax Sil column is equivalent to the Waters μPorasil, and the Zorbax ODS to the μBondapak C-18. The dimensions of the Zorbax columns used in these assays are 0.45×25 cm unless otherwise stated.

B. Purification of the Vitamin D Fraction

General

The vitamin D fraction (Section III.A) is dried under N_2 and resuspended in 75 μl of isopropanol/$MeCl_2$, 0.25:99.75. After 15–20 minutes, the solution is transferred to the limited volume insert. The conical tube is washed with an additional 75 μl which is combined with the first 75 μl. The vitamin D sample containing $\sim$150 μl is chromatographed on a Zorbax Sil column (0.45×25 cm) developed in 0.25:99.75 isopropanol/$MeCl_2$. On this column, using this solvent system at a flow rate of 2 ml/min, vitamin D_2 and vitamin D_3 comigrate and elute at 7–9 min (Figure 6). The sample can be collected in a 13×100-mm test tube or 6-ml conical tube.

Results

The purification of vitamin D_2 and vitamin D_3 by this method allows for the analysis of the vitamin D by competitive protein-binding analysis (Section V.B and Ref. 7) or by UV absorbance following reverse phase chromatography (Section IV.C). Generally, if the samples are anticipated to have <1–2 ng/ml of either vitamin D_2 or vitamin D_3, the sample should be measured by competitive protein-binding analysis.

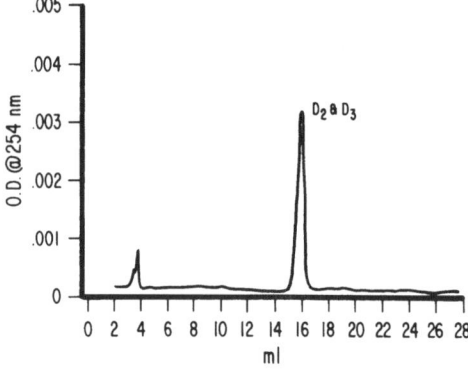

Figure 6. Elution of vitamin D_2 and vitamin D_3 from the Zorbax Sil column developed in 0.25:99.75 isopropanol/MeCl$_2$.

Precautions

As a general rule, all columns should be "stripped" with a polar solvent mixture prior to their use. In our laboratory, the silicic acid columns developed in isopropanol/MeCl$_2$ receive a 15-min strip of 1:1 isopropanol/MeCl$_2$ just prior to use. Likewise, those columns developed in isopropanol/hexane receive a strip of 1:1 isopropanol/hexane just prior to use. Sometimes, the use of solvent systems with MeCl$_2$ as the primary solvent have a continuously increasing baseline when the UV absorbance is monitored, especially when isopropanol is < 5%. This increasing baseline problem is more frequent when isopropanol/hexane solvent systems are immediately followed by isopropanol/MeCl$_2$ solvent systems containing <5% isopropanol. Generally, the 1:1 isopropanol/MeCl$_2$ strip minimized the problems. Sometimes, however, the increasing baseline persists and repeating the 1:1 isopropanol/MeCl$_2$ strip is required. Transferring the solvent system from one containing high concentrations of isopropanol (~50%) to one containing low concentrations (1–2%) of isopropanol in MeCl$_2$ may be required several times before the baseline settles down.

The use of the isopropanol/MeCl$_2$ mixture is necessary if the samples are going to be measured by competitive protein binding. This mixture allows for the adequate resolution of the vitamin D from a compound(s) that enhances the binding of 25-OH[^{3}H]D$_3$ to the binding protein (Figure 7). The substitution of MeCl$_2$ by other primary solvents such as hexane will not allow for adequate resolution of this material from the vitamin D peak.

The HPLC silicic acid columns are very sensitive to H$_2$O. Small amounts of H$_2$O will "deactivate" the column, causing peak spreading and lowered retention times. The columns can be reactivated by pumping a solution of 0.25:1:9 2,2 dimethoxypropane/acetic acid/hexane or MeCl$_2$ across the column for 20–30 min at 2 ml/min. This mixture converts the H$_2$O to acetone and elutes it from the column. The reactivation mixture should be followed by copious amounts of 1:1 hexane/isopropanol. The 2,2 dimethoxypropane can be purchased from Aldrich Chemical Co., Inc. (Cat. No. D13,680-8). The theory behind the procedure for reactivating HPLC silicic acid columns is described in detail by Bredewege et al.[8]

Figure 7. High-pressure liquid chromatography-binding analysis of the semipurified plasma lipid extract taken from the Lipidex 5000 column. The vitamin D fraction was taken from the Lipidex 5000 column and applied to the Zorbax Sil column developed in 0.25:99.75 isopropanol/hexane. The 1-ml fractions were collected in culture tubes, and each tube was analyzed for vitamin D-like activity. B_0 is the total amount of $25\text{-OH}[^3\text{H}]D_3$ bound in the absence of vitamin D_3; B is the amount of $25\text{-OH}[^3\text{H}]D_3$ bound in the presence of vitamin D_3. The ratio B/B_0 is expressed as a percentage.

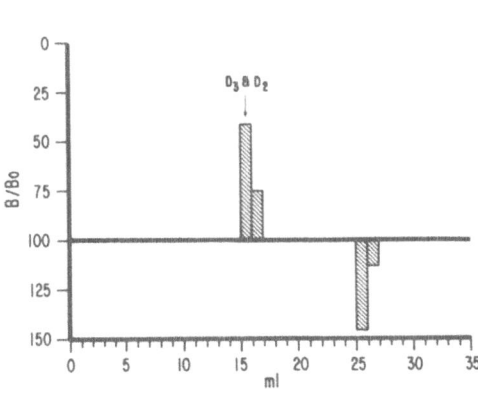

Additional Information

The Zorbax Sil columns used in this and subsequent assays are purchased from DuPont Instruments (Wilmington, Delaware, part No. 850952701). These columns should receive the treatment for silicic acid column reactivation prior to use, as DePont uses a mixture of $MeCl_2/MeOH/H_2O$ to standardize their columns prior to shipping. The columns are, therefore, in a somewhat deactivated state upon arrival.

C. Quantitation of Vitamin D_2 and Vitamin D_3 by HPLC UV Analysis

General

The vitamin D fraction isolated from the first HPLC column (Section IV.B) can be reapplied to a Zorbax ODS reverse phase column for UV quantitation of the vitamin D_2 and vitamin D_3 peaks. The sample is transferred to the limited volume inserts using 4:94 $H_2O/MeOH$. The Zorbax ODS column is developed in the same solvent. The column and detector are standardized over a wide range of vitamin D_2 or vitamin D_3 quantities, usually ranging from 10 to 200 ng. The detector and chart recorder response is linear over this range of standards at settings of 0.005 or 0.01 absorbance units full scale (AUFS). The UV peak height of the unknowns can be compared with those of standard preparations for determining the amount of vitamin D_2 or vitamin D_3 in the sample.

Results

The Zorbax ODS column developed in 4:96 $H_2O/MeOH$ offers the advantage of being able to separate vitamin D_2 and vitamin D_3 (Figure 8). The recorder

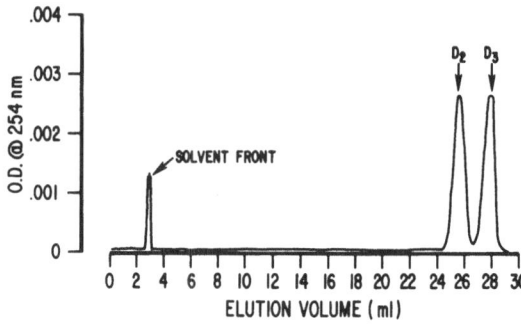

Figure 8. Elution of vitamin D_2 and vitamin D_3 from the Zorbax ODS column developed in 4:96 water/MeOH.

and detector response is linear over a wide range of vitamin D_2 or vitamin D_3 standards (Figure 9). Vitamin D_2 does appear to give a slightly higher peak height per given mass of sterol. ^{3}H-Vitamin D_3 final recovery usually ranges between 50 and 70%.

Precautions

Following the use of reverse-phase solvent systems (i.e., those containing H_2O), the solvent delivery system should be flushed to remove the H_2O-containing solvents. Failure to do so may result in the inadvertent pumping of H_2O/MeOH onto a silicic acid column. The procedure involves first flushing the pump plus the injector or WISP with MeOH followed by a MeOH-miscible solvent, usually $CHCl_3$ or $MeCl_2$. Either of these solvents is miscible with hexane-based solvent systems. The same procedure in reverse should be used for going from hexane-based solvent systems to H_2O-containing solvent systems.

Additional Information

ODS columns can generally be stripped using 100% MeOH. However, if more rigorous stripping is required, the solvent can be changed to $CHCl_3$ and eventually to hexane.

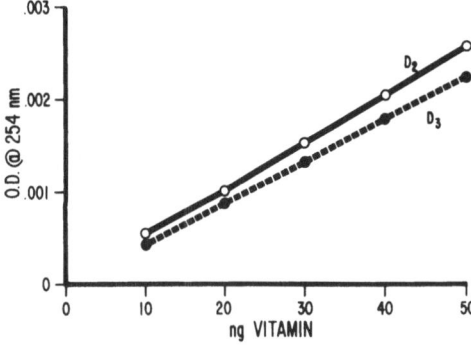

Figure 9. Standard curves for vitamin D_2 and vitamin D_3 using UV absorption (254 nm) peak following HPLC of the standards.

D. Quantitation of 25-OHD₂ and 25-OHD₃ by HPLC UV Analysis

General

The 6-ml conical tube containing the 25-OHD fraction (Section III.B) is dried under N_2, and 75 μl of 1:99 isopropanol/MeCl$_2$ is added. After 15–20 min, the contents are transferred to the limited-volume insert. The tube is washed with an additional 75 μl which is combined with the first 75 μl. The 25-OHD sample is then chromatographed on a Zorbax Sil column developed in 1:99 isopropanol/MeCl$_2$. On this column, using this solvent system at a flow rate of 2 ml/min, the 25-OHD$_2$ elutes in the 10–12-ml region, and the 25-OHD$_3$ elutes in the 12–14-ml region (Figure 10). The samples can be collected directly into scintillation vials which are used for the estimation of 25-OH[^{3}H]D$_3$ recovery.

Results

In combination with the Sephadex LH-20 or Lipidex 5000 column (Section II.D), the 1:99 isopropanol/MeCl$_2$ Zorbax Sil column allows for the uncomplicated quantitation of 25-OHD$_2$ and 25-OHD$_3$ by UV analysis. Both 25-OHD$_2$ and 25-OHD$_3$ give a linear detector and recorder response over a wide range of mass (Figure 11). The 25-OHD$_2$, however, gives a 25% higher peak height than the 25-OHD$_3$. Therefore, if 25-OHD$_3$ standards are used for 25-

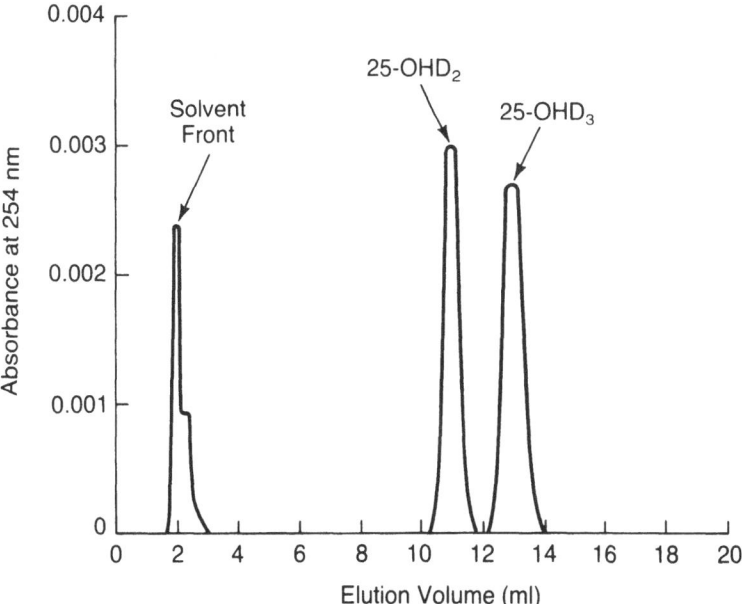

Figure 10. Elution of 25-OHD$_2$ and 25-OHD$_3$ from the Zorbax Sil column developed in 1:99 isopropanol/MeCl$_2$.

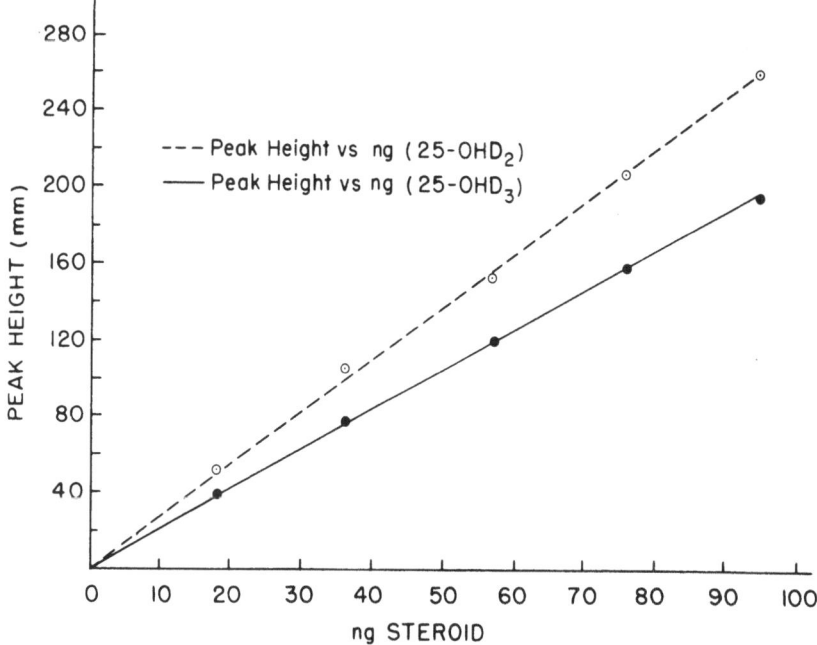

Figure 11. Standard curves for 25-OHD$_2$ and 25-OHD$_3$ using UV absorption (254 nm) peak following HPLC of the standards.

OHD$_2$ quantitation, the values must be adjusted accordingly. The recovery of 25-OH[^{3}H]D$_3$ is 55–75% following this step.

Precautions

It has been our experience that substitution of MeCl$_2$ with hexane is not always successful. Analysis of pig plasma for 25-OHD$_3$ as described above, using the hexane-based solvent 4:96 isopropanol/hexane, results in an interfering UV peak which migrates near the 25-OHD$_3$ peak and interferes with its (25-OHD$_3$) quantitation. The 1:99 isopropanol/MeCl$_2$ solvent system adequately removes this compound from the 25-OHD$_3$ region.

E. The HPLC Resolution of Metabolites in the Dihydroxyvitamin D Fraction (Fraction 3, Section II.B)

General

This fraction represents the most difficult to purify because of the numerous vitamin D$_2$ and vitamin D$_3$ metabolites present. As in Sections IV.C and IV.D, the sample is transferred to the limited-volume insert in preparation for introduction onto the column. The solvent system used is 11:89 isopropanol/hexane. The metabolites are eluted from the Zorbax Sil column in the same solvent.

Using this solvent system at a flow rate of 2 ml/min, the 24,25-(OH)$_2$D complex [consisting of 24,25-(OH)$_2$D$_2$; 24,25-(OH)$_2$D$_3$; 25,26-(OH)$_2$D$_2$; and 25-OHD$_3$-26,23 lactone], migrates in the 7–12-ml region, whereas the 25,26-(OH)$_2$D$_3$ and 1,25-(OH)$_2$D [1,25-(OH)$_2$D$_2$ and 1,25-(OH)$_2$D$_3$] migrate in the 15–18-ml and 26–30-ml regions, respectively (Figure 12). The 24,25-(OH)$_2$D complex should be collected in a 6-ml conical tube and the 25,26-(OH)$_2$D$_3$ and 1,25-(OH)$_2$D$_3$ should be collected in 13 × 100-mm test tubes.

Results

The 25,26-(OH)$_2$D$_3$ and 1,25-(OH)$_2$D are adequately prepared for competitive protein-binding analysis (Section V.C and Section V.D) following chromatography on this column. The 24,25-(OH)$_2$D complex must be further purified (Section IV.F) for analysis of its individual components. The recovery of 25,26-(OH)$_2$[^{3}H]D$_3$ and 1,25-(OH)$_2$[^{3}H]D$_3$ ranges between 50 and 70%.

Precautions

Attempts have been made to analyze 24,25-(OH)$_2$D directly from this column. It has been our experience that 24,25-(OH)$_2$D is rarely prepared adequately for competitive protein-binding analysis at this point. Some species (chick, rat, and pig) have 25-OHD$_3$-26,23 lactone circulating normally, which, if measured with the 24,25-(OH)$_2$D, would result in significant overestimation of the plasma 24,25-(OH)$_2$D status. Also, experimental conditions in which vitamin D$_2$ is fed at high levels can result in the presence of significant amounts of 25,26-(OH)$_2$D$_2$.

F. The HPLC Resolution of the 24,25-(OH)$_2$D Complex

General

The conical tube containing the 24,25-(OH)$_2$D complex is dried under N$_2$ and transferred to the limited-volume insert in 75 μl of 2:98 isopropanol/MeCl$_2$ using a 75 μl wash. The sample is then applied to a Zorbax Sil column eluted with the same solvent. In this system, 25-OHD$_3$-26,23 lactone migrates

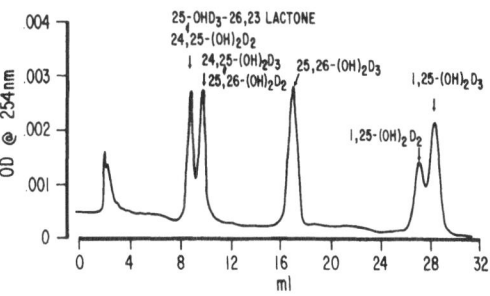

Figure 12. Elution of 25-OHD$_3$-26,23 lactone, 24,25-(OH)$_2$D$_2$, 24,25-(OH)$_2$D$_3$, 25,26-(OH)$_2$D$_2$, 25,26-(OH)$_2$D$_3$, 1,25-(OH)$_2$D$_2$, and 1,25-(OH)$_2$D$_3$ from the Zorbax Sil column developed in 11:89 isopropanol/hexane.

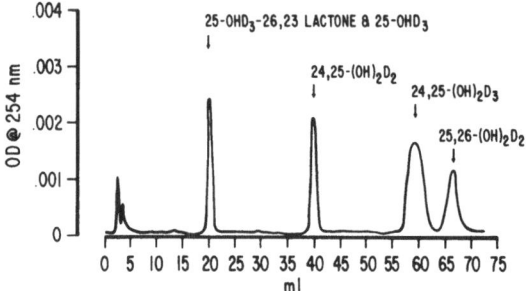

Figure 13. Elution of 25-OHD$_3$-26,23 lactone, 24,25-(OH)$_2$D$_2$, 24,25-(OH)$_2$D$_3$, and 25,26-(OH)$_2$D$_2$ from the Zorbax Sil column developed in 2:98 isopropanol/MeCl$_2$.

between 18 and 22 ml, 24,25-(OH)$_2$D$_2$ between 38 and 43 ml. 24,25-(OH)$_2$D$_3$ between 57 and 63 ml, and 25,26-(OH)$_2$D$_2$ between 64 and 70 ml (Figure 13). The metabolites should be collected in 13 × 100-mm test tubes in preparation for binding analysis. Alternatively, the 24,25-(OH)$_2$D complex can be separated by using a Zorbax NH$_2$ column developed in 1:9 isopropanol/hexane. In this system, 24,25-(OH)$_2$D$_2$ and 24,25-(OH)$_2$D$_3$ comigrate and elute in the 13–15-ml region and 25-OHD$_3$-26,23 lactone migrates in the 19–21-ml region.

Results

When the 2:98 isopropanol/MeCl$_2$ Zorbax Sil column is used, each metabolite can be analyzed individually in a competitive protein-binding assay (Section IV.C). The final recovery of these metabolites generally ranges between 40 and 60%. Frequently, in the case of 24,25-(OH)$_2$D$_2$ and 24,25-(OH)$_2$D$_3$, quantitation can be achieved by UV absorption similar to the UV assays of vitamin D (Section IV.C) and 25-OHD (Section IV.D). The 25-OHD$_3$-26,23 lactone region, however, is frequently contaminated with background UV material and is more difficult to quantitate by UV absorption. In the case of vitamin D$_3$ toxicity or high vitamin D$_3$ intake, the 25-OHD$_3$-26,23 lactone can usually be measured by UV absorbance since the ratio of 25-OHD$_3$-26,23 lactone to background UV is high.

The 24,25-(OH)$_2$D$_2$ and 24,25-(OH)$_2$D$_3$ comigrate on the Zorbax NH$_2$ and therefore can be measured simultaneously. The 24,25-(OH)$_2$D$_2$, however, does not compete as well as 24,25-(OH)$_2$D$_3$ in the rat plasma competitive protein-binding assay (Section V.C); thus the plasma 24,25-(OH)$_2$D may be underestimated if 24,25-(OH)$_2$D$_2$ is the major form.

G. HPLC Resolution of the Trihydroxyvitamin D Fraction (Fraction 4, Section II.B)

General

The samples from this fraction are dried and transferred to the limited-volume inserts using 14:86 isopropanol/hexane. The metabolites are eluted from the column in the same solvent. Using this solvent system at a flow rate of 2 ml/

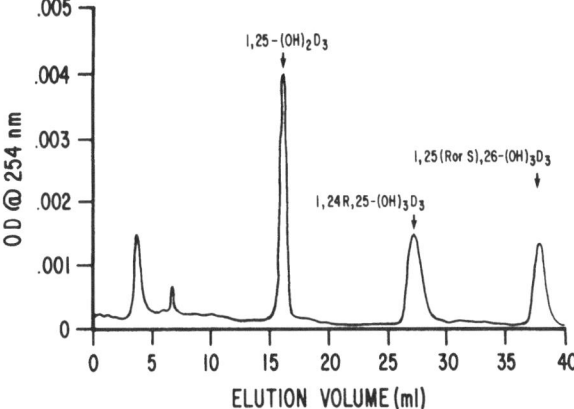

Figure 14. Elution of 1,25-(OH)$_2$D$_3$, 1,24R,25-(OH)$_3$D$_3$, and 1,25,26-(OH)$_3$D$_3$ from the Zorbax Sil column developed in 14:86 isopropanol/hexane.

min, the 1,24,25-(OH)$_3$D$_3$ and 1,25,26-(OH)$_3$D$_3$ elute at 26–30 ml and 37–40 ml, respectively (Figure 14).

Results

The 1,24,25-(OH)$_3$D$_3$ and 1,25,26-(OH)$_3$D$_3$ are adequately prepared for binding analysis (Section V.E). The recovery of the 2 metabolites ranges from 50 to 70%.

Precautions

The compound, 1,25-(OH)$_2$D$_3$-26,23 lactone, comigrates with 1,24,25-(OH)$_3$D$_3$ in this system. If 1,25-(OH)$_2$D$_3$-26,23 lactone is a suspected contaminant, it can be separated by reapplying the 1,24,25-(OH)$_3$D$_3$ region to a Zorbax Sil column developed in 7:93 isopropanol/MeCl$_2$. In this system, the 1,25-(OH)$_2$D$_3$-26,23 lactone migrates well ahead of 1,24,25-(OH)$_3$D$_3$. We have found that the 1,25-(OH)$_2$D$_3$-26,23 lactone is a minor contaminant of 1,24,25-(OH)$_2$D$_3$ isolated from plasma of 1,25-(OH)$_2$D$_3$-treated animals.

V. Competitive Protein-binding Analysis of Vitamin D and Vitamin D Metabolites

A. Overview

The use of competitive protein-binding assays in the quantitation of steroids has offered a fast and reliable method for the quantitation of vitamin D-related compounds. When used properly, the results obtained from these assays are quite reproducible and agree well with direct physical chemical assays. The assays described in this section are used routinely in our laboratory and have been validated for use in the quantitation of vitamin D and vitamin D metabolites in several species. In our competitive protein-binding assay, two pieces of equipment are very helpful for saving time and labor. These are the Micrometic High Speed Automatic Pipette (Model 25004) and the Hamilton PB-600-1 and PB-600-10 Repeating Dispenser. The automatic pipette aspirates a consistent volume of unknown and delivers it along with the diluted binding protein to the incubation tube. The Hamilton repeating dispenser allows for the rapid dispensing of radioactive material or solvent.

B. Competitive Protein-binding Analysis of Vitamin D_2 and Vitamin D_3

General Step Instructions

1. After the vitamin D is purified by HPLC (Section IV.B), the solvent is removed under N_2 and 200 μl of ethanol is added to each sample tube using the Hamilton repeating dispenser. The samples are corked and allowed to set ~20 min in a covered pan containing an ethanol-saturated atmosphere (i.e., add 50–100 ml of ethanol to the covered pan and let it set a few minutes. The atmosphere will become ethanol-saturated if the pan is covered).

2. While the samples are setting, 12 × 75-mm test tubes can be labeled in triplicate for the standard curve and for each unknown. Each 12 × 75-mm test tube then receives ~8000 cpm of 25-OH[^{3}H]D_3 in 20 μl of ethanol. The radioactivity can be easily dispensed using the repeating dispenser.

3. Set up the automatic pipette to aspirate 25 μl of sample or standard in ethanol and deliver 500 μl of a 1:50,000 dilution of sheep plasma in 0.05 M sodium phosphate buffer containing 0.01% gelatin (PBG). (See Appendix for PBG recipe.)

4. Dispense the sample or standard into the appropriately labeled 12 × 75-mm tubes. At the same time, two 25-μl aliquots of sample should be pipetted into two scintillation vials for estimation of ^{3}H-vitamin D_3 recovery. Each 12 × 75 tube then contains 25-OH[^{3}H]D_3 in 20 μl of ethanol, unknown or standard in 25 μl of ethanol, and 500 μl of a 1:50,000 sheep plasma in buffer.

   The vitamin D_3 standards should consist of 0.0, 0.1, 0.2, 0.4, 0.8, 1.6, 3.2, 6.4, and 50 ng of vitamin $D_3/25$ μl of ethanol. Nonspecific binding is estimated by using the 50 ng/25 μl standard. The standard curve should also contain tubes with no vitamin D and no added charcoal to estimate the total 25-OH[^{3}H]D_3 added to the assay. The 0.0 ng/25 μl standard should be pipetted first and the 50 ng/25 μl (nonspecific binding) should be pipetted last.

5. Place the tubes onto a continuous shaker and allow them to incubate and shake in a cold room for 1 h.

6. Remove the tubes from the cold room and add 0.2 ml of a mixture of cold

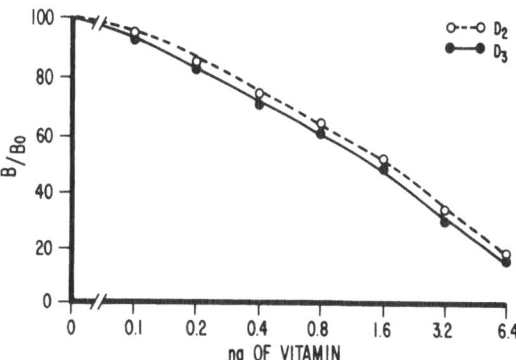

Figure 15. Displacement potency of vitamin D_2 and vitamin D_3 in the competitive binding assay for vitamin D when diluted sheep plasma was used as the source of vitamin D-binding protein. The ratio B/B_0 is expressed as a percentage.

1% Norite neutral charcoal (Fisher Scientific Products) and 0.1% dextran T-70 (Pharmacia Fine Products) suspended in PBG.
7. Allow to set 30–45 min at 4°C, then centrifuge at 1000 g for 10 min at 4°C.
8. Using the automatic pipetter, aspirate 500 μl of supernatant and dispense along with 5 ml of scintillation cocktail into a scintillation vial for quantitation of radioactivity.

Results

This assay will measure as little as 0.1 ng of vitamin D_2 or vitamin D_3 (Figure 15) and is adequate for estimating the total plasma vitamin D status of several species.

Precautions

Diluted plasma from most species discriminates against vitamin D_2 in this assay. An example is given in Figure 16 when using a 1:50,000 dilution of rat

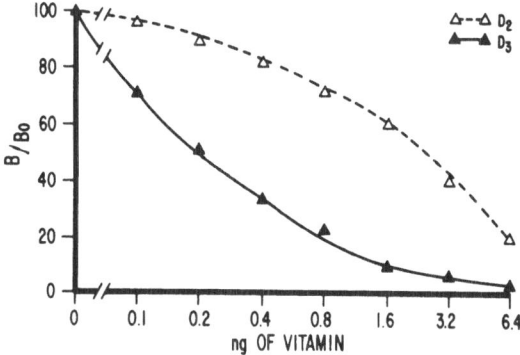

Figure 16. Displacement potency of vitamin D_2 and vitamin D_3 in the vitamin D-binding assay when diluted rat plasma was used as the source of binding protein. The ratio B/B_0 is expressed as a percentage.

plasma. Therefore, care must be taken in choosing the appropriate plasma to avoid this discrimination.

Additional Information

The scintillation cocktail used in our laboratory for step 8 is purchased from Beckman, Cat. No. 566436.

C. The Competitive Protein-binding Assay Used to Measure 24,25-(OH)$_2$D$_2$, 24,25-(OH)$_2$D$_3$, 25-OHD$_3$-26,23 Lactone, 25,26-(OH)$_2$D$_2$, and 25,26-(OH)$_2$D$_3$

General

The isolation of these metabolites from plasma is described in Section IV.E. The assays for these metabolites are conducted exactly as described in Section V.B with the exception that *1:5,000* dilution of *rat* plasma in PBG is substituted for the *1:50,000* dilution of *sheep* plasma in PBG (Section IV.B, Step 3). Also, the standards are made up of the appropriate metabolites being assayed. The same range of standard concentrations as described in Section IV.B, Step 4, can be used for all these metabolites.

Results

The behavior of these metabolites in this binding assay is shown in Figure 17. This dilution of rat plasma in PBG is most sensitive to the compound, 25-OHD$_3$-26,23 lactone. The compounds 25,26-(OH)$_2$D$_3$ and 24,25-(OH)$_2$D$_3$ are indistinguishable, and 24,25-(OH)$_2$D$_2$ and 25,26-(OH)$_2$D$_3$ are 2–3 times less competitive than 24,25-(OH)$_2$D$_3$ or 25,26-(OH)$_2$D$_3$. The data in Figure 17 will offer a reasonable guideline for estimating a standard curve for 25-OHD$_3$-26,23 lactone, 24,25-(OH)$_2$D$_2$, or 25,26-(OH)$_2$D$_2$ if these compounds are unavailable for constructing standard curves.

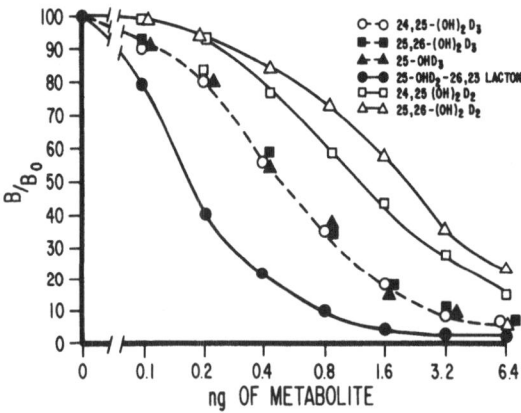

Figure 17. Displacement potency of vitamin D$_2$ and vitamin D$_3$ metabolites in the rat plasma competitive protein-binding assay. The ratio B/B$_0$ is expressed as a percentage.

D. The Competitive Protein-binding Assay for 1,25-(OH)₂D Using Chick Intestinal Cytosol as the Source of Binding Protein

General

This assay is conducted similarly to those described in Sections IV.B and IV.C.

1. Samples obtained from the HPLC purification step (Section III.E) are dried under N_2 and suspended in 200 μl of ethanol as described above (Section IV.B).
2. 5000 cpm of 1,25-(OH)$_2$[^{3}H]D$_3$ is added to each 12 × 75-mm test tube used for the 1,25-(OH)$_2$D assay of the unknowns and standards.
3. Unknowns and each standard in the standard curve (consisting of 0, 5, 10, 20, 40, 80, 160, 520, 640, and 8000 pg/25 μl) are done in triplicate.
4. The unknowns and standards are aspirated and dispensed with 500 μl of chick cytosol-binding protein in dissolved chick intestinal cytosol buffer. (See Appendix for buffer recipe and procedure for making chick intestinal cytosol.) Vortex to assure adequate mixing and incubate the tube for 60 min at 25°C in a heated water bath.
5. Take the samples from the water bath and cool them to 4°C for 15 min. All subsequent steps are carried out at 4°C.
6. All 0.2 ml of charcoal solution containing 0.5% charcoal and 0.05% dextran T-70 suspended in chick intestinal cytosol buffer.
7. After 10 min, centrifuge the tube at 1000 g for 10 min.
8. Aspirate 500 μl of the supernatant and dispense with 5 ml of scintillation cocktail into a scintillation vial. Cap and shake.

Results

When ~0.4 mg of cytosol protein is used per assay tube, this assay is sensitive to 5 pg of 1,25-(OH)$_2$D (Figure 18). Higher sensitivity can be achieved by diluting the cytosol resulting in <0.4 mg of protein per assay tube.

Precautions

The cytosol preparations generally do not last more than 3 months. After this time, the nonspecific binding increases and the total specific binding decreases.

E. Competitive Protein-Binding Assay for 1,25-(OH)₂D and the Trihydroxyvitamin D Metabolites Using the Calf Thymus Cytosol-Binding Assay

General

Calf thymus has recently been demonstrated to have a significant amount of 1,25-dihydroxyvitamin D receptor.[9] The method described for preparing the binding protein (Appendix) results in a very stable preparation which can be stored for several months.

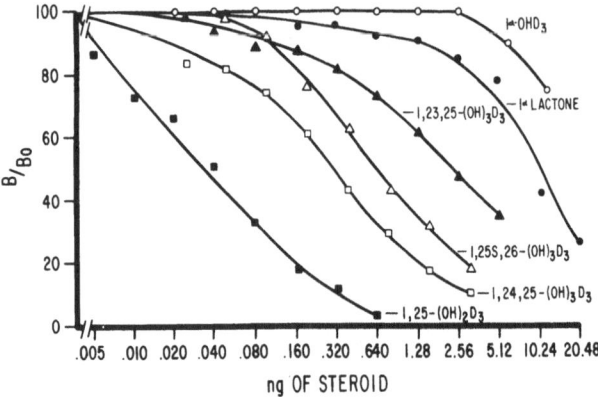

Figure 18. Displacement potency of 1α-hydroxylated vitamin D_3 metabolites in the chick intestinal cytosol-binding assay for 1,25-$(OH)_2D$. The ratio B_0 is expressed as a percentage.

1. Samples obtained from the HPLC purification step (Section IV.E and/or Section IV.G) are dried under N_2 and suspended in 200 μl ethanol.

2,3. Standards for the metabolites to be measured and the incubation tubes are prepared as described in Section V.D, steps 2 and 3.

4. The unknown and standards are aspirated and dispensed with 500 μl calf thymus cytosol-binding protein in buffer (see Appendix for calf thymus cytosol buffer). The samples are vortexed and incubated for 90 min at 25°C in a heated water bath.

5. Take the samples from the water bath and cool them to 4°C for 15 min. All subsequent steps are carried out at 4°C.

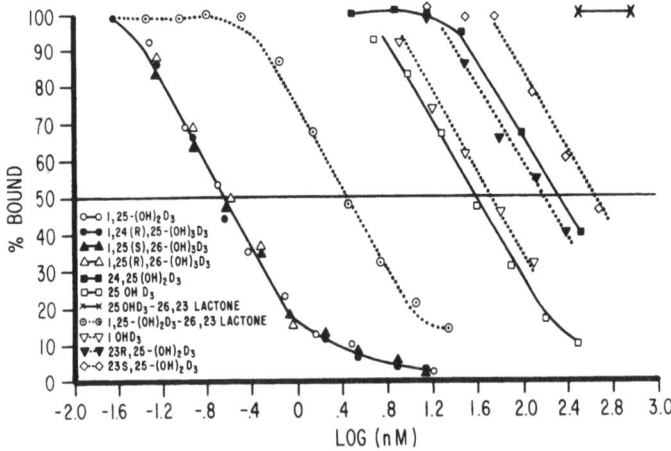

Figure 19. Displacement potency of vitamin D_3 metabolites in the calf thymus cytosol-binding assay for 1,25-$(OH)_2D$, 1,24,25-$(OH)_3D_3$, and 1,25,26-$(OH)_3D_3$.

6. Add 0.2 ml of a charcoal solution containing 1.0% charcoal and 0.1% dextran T-70 suspended in calf thymus cytosol buffer.
7. After 15 min, centrifuge the tubes at 1000 g for 10 min.
8. Aspirate 500 μl of the supernatant and dispense with 5 ml of scintillation cocktail into a scintillation vial. Cap and shake.

Results

The use of calf thymus cytosol provides a sensitive assay for 1,25-(OH)$_2$D as well as 1,24,25-(OH)$_3$D$_3$ and 1,25,26-(OH)$_3$D$_3$. As little as 5 pg of each can be detected (Figure 19). The assay conducted in the above manner offers the advantage of being cospecific for several 1α-hydroxylated vitamin D metabolites, and enough binding protein can be collected from one calf to do $\sim$5000 assay tubes.

References

1. McCollum EV, Simmonds N, Shipley PG, Park EA: Studies on experimental rickets. XVI. A delicate biological test for calcium depositing substances. J Biol Chem 54:41, 1922.
2. Shultzer P: Investigation on the determination of vitamin D: Comparison between the preventive and curative methods. Biochem J 25:1745, 1931.
3. Poulsson E: XIX: The quantitative determination of vitamin D. Biochem J 22:135, 1928.
4. Coward KH, Key KM, Morgan BGE: CLXXXVIII: The quantitative determination of vitamin D by means of its growth-promoting property. Biochem J 26:1588, 1932.
5. Brockman H, Chen Y: Über eine Methode zur quantitativen Bestimmung von Vitamin D. Hoppe Seylers Z Physiol Chem 241:129, 1936.
6. Vitamin D Assay. U.S. Pharmacopeia, XVI ed, p. 910. Mack Printing Co, Yorke, PA, 1960.
7. Horst RL, Littledike ET, Riley JL, Napoli JL: Quantitation of vitamin D and its metabolites and their plasma concentrations in five species of animals. Anal Biochem 116:189, 1981.
8. Bredewege RA, Rothman LD, Pfeiffer CD: Chemical reactivation of silica columns. Anal Chem 51:2061, 1979.
9. Reinhardt TA, Horst RL, Littledike ET, Beitz DC: 1,25-dihydroxyvitamin D$_3$ receptor in bovine thymus gland. Biochem Biophys Res Commun 106:1012, 1982.

Appendix

Preparation of Calf Thymus Cytosol

1. Collect thymus glands from 5–12-week-old calves and wash with cold calf thymus cytosol buffer. All steps are carried out at 4°C.
2. The glands can be frozen in several individual pieces in liquid N$_2$ and stored in

the freezer, or can be homogenized immediately in the calf thymus cytosol buffer (25%, w/v) using four 20-s bursts of a Polytron Pt-20 tissue disruptor at a setting of 7.

3. Centrifuge the homogenates for 1 h at 300,000 g.
4. Aspirate the lipid layer and pour the cytosol into an Erlenmeyer flask.
5. Add 22 g of $(NH_4)_2SO_4$ for every 100 ml of cytosol; be sure pH is maintained at 7.4.
6. Stir the contents in the cold room for 30 min.
7. Add 10 ml of the $(NH_4)_2SO_4$-treated cytosol to each polypropylene tube and centrifuge at 20,000 g for 10 min. Discard the supernatant.
8. Gas the tubes with N_2. *Cap* and store the $(NH_4)_2SO_4$-precipitated pellets at $-20°C$.
9. The amount of buffer the pellet is resuspended in should be adjusted according to the results desired.
10. Prior to the binding assay, resuspend the pellet in calf thymus cytosol buffer, usually around 100 ml.
11. Use 0.5 ml of the resuspended pellet per assay tube.

Chick Intestinal Cytosol Preparation

Note: Always keep cytosol on ice.

1. Weigh ten 50-ml polyethylene tubes with 10 ml cold buffer in each.
2. Kill 40 chicks, 4–8 weeks old, that are being fed a normal stock diet.
3. Remove the duodenum from the stomach to the gallbladder, and remove the pancreas from the duodenum.
4. Clean the duodenum.
   a. Squeeze out the contents and wash the outside with cold tap water.
   b. Wash the inside with 4 ml ice cold 1,25-$(OH)_2D$ buffer.
   c. Squeeze out the remaining buffer.
5. Cut open the duodenum longitudinally by inserting a grooved glass rod and making an incision with a razor guided by the groove.
6. Place the duodenum, mucosa side up, on a glass plate lying on ice, and scrape off the mucosa with a glass slide.
7. Place the mucosa in the iced polyethylene tubes (use the mucosa scraped from four duodena per tube).
8. Reweigh the tubes to determine the grams of mucosa present.
9. Wash the mucosa twice using approximately 5 vol cold buffer.
   a. Fill tubes with buffer.
   b. Vortex tubes covered with parafilm.
   c. Centrifuge first wash for 10 min at 2000 g (4000 rpm, Beckman JA-20).
   d. Discard supernatant.
   e. Resuspend pellet in cold buffer by vortex mixing.
   f. Centrifuge second wash for 5 min at 2000 g.
   g. Discard supernatant.
10. Homogenize pellet in 2 vol cold buffer using three 10-s bursts of a Polytron Pt-20 tissue disruptor at a setting of 7.
11. Balance the tubes and centrifuge at 48,000 g for 1 h (20,000 rpm, Beckman JA-20).

12. Aspirate off the lipid layer (<0.1 ml). Pour the supernatant into an Erlenmeyer flask and mix thoroughly.
13. Aliquot 6 ml of the supernatant to test tubes (20 × 150 cm).
14. Freeze supernatant on the side of the test tubes in dry ice-acetone, vortexing frequently during the freezing process.
15. Lyophilize frozen cytosol overnight.
16. Perform a Biuret assay on the protein to determine the mg protein per ml (optional).
17. Run a set of standards (0.4–0.8 mg cytosol protein/0.5 ml buffer) to determine the best concentration of the cytosol.
18. Select dilution giving best curve.

Chick Intestinal Cytosol Buffer

6.8 g KH_2PO_4 (50 mM)
7.45 g KCl (100 mM)
0.154 g DTT (dithiothreitol) (Cleland's Reagent) (1.2 mM)
800 ml of deionized, distilled water
Adjust to pH 7.4 with 1.0 N KOH
qs to 1000 ml

Calf Thymus Cytosol Buffer

6.06 g Tris-HCl (50 mM)
0.60 g EDTA (1.5 mM)
0.62 g DTT (5 mM)
2.42 g $Na_2MoO_4 \cdot 2 H_2O$ (10 mM)
22.37 g KCl (500 mM)
800 ml deionized, distilled water
Adjust to pH 7.4 with HCl
qs to 1000 ml

Phosphate Buffer with Gelatin (PBG) Used in the Binding Assay for 24,25-(OH)₂D, 25,26-(OH)₂D, and 25-OHD₃-26,23 Lactone

6.9 g NaH_2PO_4 (0.05 mM)
0.1 g gelatin*
10 ml 1% merthiolate solution
800 ml deionized, distilled water
Adjust to pH 7.4 with NaOH
qs to 1000 ml

*The gelatin is dissolved in the buffer by heating the suspension to 40°C for 20–30 min with intermittent shaking. The solution is then cooled to room temperature prior to adjusting the pH.

Chapter 3

Survey of Competition Assays for the Vitamin D Metabolites

John G. Haddad

Much of the literature concerned with clinical disorders of mineral homeostatis contains speculations regarding vitamin D "sensitivity" and vitamin D "resistance." Whereas vitamin D depletion and excess can be confirmed by bioassay of the antirachitic activity in serum or tissue extracts,[1-4] the bioassay techniques are tedious, costly, and imprecise. The introduction of radioactive antirachitic sterols provided new tools that allowed further understanding of the metabolism of these substances.[5,6] Since the parent vitamin is now recognized to be metabolized to more potent compounds in the liver and kidney, the lack of specificity of the bioassay of simple serum or tissue extracts is easily understood (Table 1).

In recent years, assays of the vitamin D metabolites in serum have been developed or modified in several laboratories in order to more precisely evaluate the influences of environment, diet, disease, and drugs on vitamin D status in humans. Most recently, additional techniques have been developed to quantitate the potent renal metabolite, 1,25-dihydroxyvitamin D [1,25-(OH)$_2$D] as well. In both instances, these assays were made possible by the recognition of high-affinity binding proteins in serum and tissues, by the availability of suitably high specific activity radioactive sterols, the adaptation of chromatographic separation steps in refining serum lipid extracts, and by the utilization of adsorbent techniques to separate specifically bound sterol. This chapter reviews the development and modifications of these assays.

I. Vitamin D Assays

Although somewhat overlooked during the period of emphasis on metabolites, vitamin D itself has received attention in recent years. In much earlier days, monumental extraction and chromatographic efforts were required to provide an isolate from which a spectral quantitation of vitamin D could be made. Also, large amounts of starting material were required. The recent availability of high-performance liquid chromatography (HPLC) technology and high specific-activity sterols has permitted their application to saturation analyses of vitamin D.

A huge problem with vitamin D, compared to its metabolites, is that it is

Table 1. Vitamin D Metabolites in Human Plasma

	Vitamin D	25-OHD	1,25-(OH)$_2$D	24,25-(OH)$_2$D
Total plasma concentration (range)	2.6–13 nmol/L (1–5 ng/ml)	26–78 nmol/L (10–30 ng/ml)	52–104 pmol/L (20–40 pg/ml)	2.5–10 nmol/L (1–4 ng/ml)
Plasma half-life	1 day	12–20 days	Hours	2–3 weeks
Relative biopotency[a]	Inactive	1	50–1000	<1
Plasma concentration as an index of vitamin D stores	Not reliable	Excellent	Not reliable	Not reliable
Plasma concentration as an index of calcium economy[b]	Not reliable	Not reliable	Excellent	Not reliable

[a] Activity in stimulating ^{45}Ca release from fetal rat bones in organ culture, or occupancy of 1,25-(OH)$_2$D receptor.
[b] Alternatively, index of body's need to acquire calcium.

poorly soluble in assay media. Another problem is that no known binding proteins display a high affinity for vitamin D. Although some workers tried to sidestep these problems by carrying out equilibrium incubations of vitamin D with plasma vitamin D-binding protein (DBP) for extended periods,[7] some success has only recently been gained by employing dysequilibrium techniques.[8]

Suitable extraction and chromatographic procedures for vitamin D usually include mixing serum with chloroform–methanol or dichloromethane–methanol after addition of [^{3}H]-vitamin D$_3$ to monitor the recovery of the sterol. The vitamin can be isolated from other D sterols by its early elution from silicic acid columns in ether/n-hexane (1:2, v/v) or early elution from Sephadex LH-20 in CHCl$_3$/n-hexane (1:1, v/v). Additional purification steps can be carried out on Lipidex 5000, straight-phase (silica) or reverse-phase (octadecyl-silica) HPLC columns or cartridges. It is important to remember that these steps do not separate cholesterol from D metabolites. These isolates are lipid rich, and not easily solubilized in aqueous media. Therefore, a prolonged incubation of these isolates (with ethanol and gelatin) with plasma DBP *prior* to the addition of 25-OH[^{3}H]D$_3$ has been shown to result in the displacement of the tracer in the 100 pg range.[8] The method seems to do the best with the difficulties inherent in the assay: (1) solubility of sterol, and (2) relatively poor affinity for vitamin D by plasma DBP.

Currently, direct estimation of vitamin D by its absorption characteristics during elution from HPLC appears to be a secure and reproducible method. The saturation dysequilibrium technique, however, does permit the estimation of smaller amounts of the vitamin. The reproducibility of the latter technique awaits better definition. Other modifications may prove to be helpful, since there is movement of vitamin from chylomicrons to the α-globulin at 37° over 1 h. Others[9] have employed ion-exchange filtration to adsorb DBP, and this technique sidesteps some of the difficulties seen with the coated charcoal method of separating bound and free sterol. Finally, it seems likely that suitable antisera to vitamin D itself will provide a higher-affinity system than plasma DBP. Such reagents should permit reliable and sensitive assays of the parent vitamin.

II. 25-Hydroxyvitamin D Assays

A. Binding Proteins

Serum

Bioassay recognition of the nature of antirachitic sterol transport in the blood[10,11] has been vigorously examined with the use of labeled sterols,[12-19] and revealed that mammals possess an α-globulin that preferentially binds 25-OHD with high affinity. In contrast, avian sera contains less anodic transport proteins with different binding recognition[20-22] of vitamin D_3 (D_3) and 25-OHD. Further, the affinities of binding of the cholecalciferol and ergocalciferol metabolites are different in avian sera[21,22] and complement the recognized lower potency of the ergocalciferol compounds in this species.[23] Purification and partial characterization of the human serum-binding protein have been performed in several laboratories[19,24,25] and revealed a single-chain polypeptide that sediments at 3.46S in the analytical ultracentrifuge. Saturation analyses, with sucrose gradient ultracentrifugation separation of bound sterols, revealed this human serum inter-α-globulin to bind D_3, 25-OHD$_3$, or 1,25-(OH)$_2$D$_3$ mole for mole, and have highest affinity for 25-OHD$_3$ (K_d 6.8 $\times$ 10^{-8} M, at 4°C). Whereas 24,25-(OH)$_2$D$_3$ competes equally with 25-OHD$_3$ in the displacement of [^{3}H]25-OHD$_3$ from this protein, D_3 and 1,25-(OH)$_2$D$_3$ are five to ten times less competitive.

Radioimmunoassays of the human serum antirachitic sterol-binding protein reveal that it circulates at ~10^{-6} M concentration and normally is only 3 to 5% occupied by these sterols.[24] The application of pregnancy human serum as a binding protein source is well-supported by the recognized increase of the binding protein level in pregnant and estrogen-treated subjects.[26,27] Since the human serum-binding protein recognizes 25-hydroxycholecalciferol (25-OHD$_3$) and 25-hydroxyergocalciferol (25-OHD$_2$) equally[26] (Figure 1), it is a suitable binding protein to measure total 25-OHD (25-OHD$_3$ and 25-OHD$_2$) in blood, which contains both moieties.[28] Although disputed,[29] rat tissue and serum-binding protein have also been shown not to distinguish between 25-OHD$_3$ and 25-OHD$_2$.[30] Although antisera to the human binding protein do not crossreact with rat sera,[24,27] available studies indicate these serum-binding proteins share similar physicochemical and ligand-recognition properties.[15,16,20,31]

Tissue

Soluble, trypsin-sensitive molecules capable of high-affinity, specific binding of [^{3}H]25-OHD$_3$ were discovered in rat kidney and skeletal muscle extracts,[31]

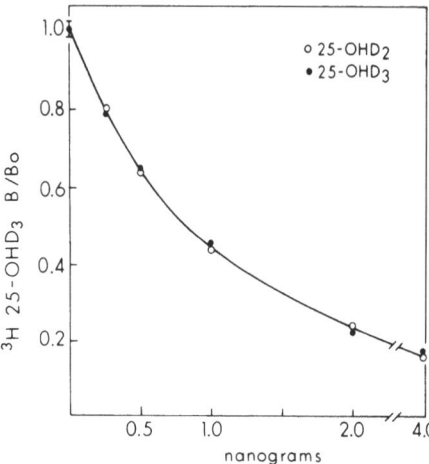

Figure 1. Equipotency of 25-OHD$_3$ ($\bullet$) and 25-OHD$_2$ ($\circ$) in the displacement of 25-OH[^{3}H]D$_3$ from human serum 25-OHD-binding protein. (Reprinted with permission from Ref. 26.)

and also were found to be useful in competitive binding assays.[32,33] These proteins sediment at approximately 6S, and are not altered by high salt or dithiothreitol treatment.[34] Thus far, they have been demonstrated in all nucleated tissues examined.[34–39] Similar large 25-OHD-binding proteins have been observed in nucleated tissues of humans.[40] This tissue-binding protein binds 25-OHD with higher affinity than either D or 1,25-(OH)$_2$D. In the mammal, the tissue protein recognizes 25-OHD$_3$ and 25-OHD$_2$ equally.[30] In the chicken, however, 25-OHD$_3$ is bound preferentially by both serum- and tissue-binding proteins.[21,22,41] The function of the tissue 25-OHD-binding protein is not known at present. It is clear from saturation analyses, however, that it normally exists largely in an apoprotein form, since sera or tissue extracts from vitamin-D-repleted animals serve very well as binding-protein sources for radioligand assay of 25-OHD.[16,26,42,43] Recently, the tissue protein has been demonstrated to be a complex of serum DBP and monomeric actin.

High-speed supernatants from muscle and kidney extracts bind 25-OHD$_3$ with slightly higher affinity than the corresponding serum-binding protein.[34] In practice, however, the relatively large amounts of 25-OHD in serum (10^{-8}–10^{-7} M) and availability of high specific-activity [^{3}H]25-OHD$_3$ enables one to use either serum or tissue protein in the 25-OHD assay. Increased or decreased sensitivity can be obtained by simply decreasing or increasing the amount of binding protein and tracer sterol. Since human sera is most cheaply and easily obtained, does not distinguish between 25-OHD$_3$ and 25-OHD$_2$, and can be used at 1:5000 or greater dilution, we currently use it routinely as our source of binding protein in the 25-OHD assay.

Antiserum

Suitable, specific antisera may permit easier estimation of 25-OHD in serum extracts. Several laboratories are working on this possibility.

B. Extraction Procedures

Whereas total lipid extraction[45] was consistently applied in the metabolic studies leading to our present understanding of vitamin D metabolism, other techniques for the purpose of assay have been successfully employed. Although vigorous extraction with absolute ether fails to extract the parent vitamin or its less polar esters with certainty, this is an advantage if one is interested only in the more polar derivatives. Separation of phases is quicker with ether than with the chloroform method, and ether is more rapidly (and less expensively) evaporated under a nitrogen stream.[32] Fresh ether must be used to avoid peroxides that cause oxidation of 25-OHD. The trace HCl content of some batches of chloroform poses a similar threat. More recently, dichloroethane has been reported to be another attractive option.[43]

C. Chromatography

Both 25-OHD$_3$ and 25-OHD$_2$ conveniently elute after D$_3$ and before 24,25-dihydroxyvitamin D [24,25-(OH)$_2$D] on small silicic acid columns.[22,32,46] Cholesterol, D$_3$, and other less polar compounds are discarded in an ether-n-hexane (1:2, v/v) mixture, whereas 25-OHD elutes at 70–80% ether concentration. Since 24,25-(OH)$_2$D does not elute until CH$_3$OH is added, and ether is most easily dried, 100% ether can be used to elute the 25-OHD from the micro- (4 cm in transfer pipette) silicic acid columns.

Sephadex LH-20 liquid–gel partition chromatography has also been successfully employed to isolate 25-OHD from lipid extracts of serum.[43,47] Since the elution pattern of 25-OHD is influenced by the solvent mixture employed,[48] the column length should ensure the timed or volume-determined collection of 25-OHD away from vitamin D and, more importantly, 24,25-(OH)$_2$D. Preparative thin-layer chromatographic systems can also be used and have been described.[49,50] Since the mentioned extraction and chromatographic procedures are all suitable, and monitored by the addition of recovery tracer [^{3}H]25-OHD$_3$ to serum, their selection depends on the necessity for high recoveries, convenience, and preference.

Recently, a nonchromatographic system has been advocated in which ethanol extracts of serum are used directly in the competitive binding assay.[29,51] Although the potential competition by vitamin D is controversial,[29,43,51] these authors failed to recognize the equipotent competitive displacement and sizable amount of 24,25-(OH)$_2$D in human serum[52] (Figure 2). This fact is underscored by the higher "25-OHD" values observed in the nonchromatographic assay,[29,51] despite some evidence to the contrary.[53] In addition, the utilization of lipoprotein buffers in competitive assay of ethanol extracts of serum yields data considerably different from that derived in its absence.[48] At present, therefore, it is reasonable to consider the valid competitive binding assays of 25-OHD in serum to be those preceded by chromatographic isolation of 25-OHD.[32,43,46,47,49,54] Additional studies of the serum content of 24,25-(OH)$_2$D$_3$

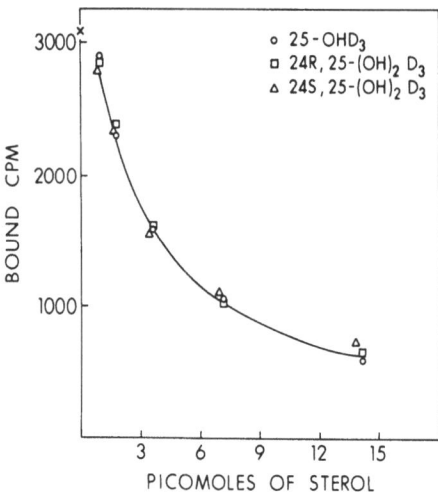

Figure 2. Equipotency of 25-OHD₃ (○); 24R,25-(OH)₂D₃ (□); and 24S, 25-(OH)₂D₃ (△) in the displacement of 25-OH[³H]D₃ from rat serum 25-OHD-binding protein. (Reprinted with permission from Ref. 51.)

are necessary to determine the magnitude of its interference in the "direct" serum 25-OHD assays. Finally, chromatography eliminates interfering lipid from the assay systems.

D. Assay Conditions and Characteristics

Although original reports[46] included protracted incubation times to facilitate the quantitation of vitamin D in addition to 25-OHD, the recognized lower-affinity binding of vitamin D by serum[16,29,47] and tissue[31] binding proteins, and the conspicuous absence of vitamin D assay data by such a method, this assay does not appear to be reliable for vitamin D measurements.[47] Because the association time of 25-OHD₃ and binding protein is very rapid,[33,43] incubation time can easily be shortened to 1 h or less.[33,43,47]

Since the specific-binding of 25-OHD by human serum is optimal at neutral to alkaline pH,[16,43] it follows that most investigators have selected buffers in this pH range. Successful displacement curves can be obtained at room temperature, but enhanced binding, greater binding protein dilution, and increased sensitivity, as well as more consistent results, are realized at 0–4°C. The transfer of 25-OHD₃ requires its solubilization in organic solvents, and ethanol is often used. Although specific binding occurs after addition of binding proteins to sterol dried in the assay tube, enhanced binding is observed when small amounts of ethanol (<10%) are included.

Separation of specifically bound sterol is usually achieved with dextran-coated charcoal, although Florisil has been used. The adsorbent is a convenient method, but its use must be carefully monitored since contact time with incubation mixture must be uniform for all assay tubes, and the amount of adsorbent dispensed should be uniform. Bound radioactivity is usually assayed in scintillation fluids designed for aqueous samples.

The specific activity of [^{3}H]25-OHD$_3$ currently available (1 to 12 Ci/mmol) allows the production of standard displacement curves sensitive to picogram and nanogram amounts of 25-OHD. Higher-specific-activity [^{3}H]25-OHD$_3$ is sold, but these preparations may degrade rapidly. Methods to handle such materials and preserve the integrity of the sterol would permit even greater sensitivity. The realization of high-activity γ-emitting isotope substitution in a corticosteroid protein-binding assay is provocative and encouraging, but similar results in vitamin D work have not presently been achieved.

Major problems in recovery or assay consistency usually involve deterioration of [^{3}H]25-OHD$_3$ or reference standard sterols. Preservation of these sterols in benzene, benzene-ethanol, or dried under N_2 in a dark, cold environment is important to avoid degradation. It is also helpful to purify the radioactive sterol on short silicic acid or Sephadex LH20 columns just prior to its use as recovery or assay tracer. Unlabeled sterol can be stored under similar conditions, and a series of identical "stock" aliquots can be used once as sources of standard solutions for the assay. We have also found it useful to analyze routinely the concentrated 25-OHD$_3$ standard (2–10 μg/ml of ethanol) at 263 nm in the spectrophotometer, and adjust dilutions proportional to the sterol's extinction coefficient.[26] In general, the serum-binding protein is stable, and can be kept in laboratory freezers for extended periods without loss of specific binding activity. Tissue-binding proteins, as the neat cytosol, also store well frozen, but seem to last even better in the lyophilized state.[32]

Plasma 25-OHD concentrations, as determined by competitive protein-binding assay, approximate levels derived by bioassay[4] and gas–liquid chromatography[50] of chromatographed serum extracts. Since a variety of successful 25-OHD assay techniques have been developed, a partial list of their characteristics has been collated and listed in Table 2.

III. 24,25-Dihydroxyvitamin D Assays

A. Binding Proteins

In contrast to the assays previously mentioned, the assay of 24,25-(OH)$_2$D$_3$ has been described only recently. The important observations which made these assays possible were the recognition that rat serum diluted 1:3500 and rat kidney cytosol were capable of binding 24,25-(OH)$_2$D$_3$ with high affinity. In fact, 25-OHD$_3$ and 24,25-(OH)$_2$D$_3$ are equipotent in their ability to displace 25-OH[^{3}H]D$_3$ from these binding proteins (Figure 1). Avian sera do not bind either of these sterols as tightly as sera from mammals, and avian sera bind the ergocalciferol compounds less tightly than the cholecalciferol metabolites.

It is clear that both 25-OHD$_3$ and 25-OHD$_2$ are measured equally in mammalian DBP assays, but the recognition of 24,25-(OH)$_2$D$_3$ is presumed now to be equal to 24,25-(OH)$_2$D$_2$ recognition, since reference 24,25-(OH)$_2$D$_2$ is not currently available. In biologic fluids, it is recognized that the hydroxyl group

John G. Haddad

Table 2. Features of Serum 25-OHD Assay Methods

Country (reference)	Extraction	Chromatography	Tracer Recovery (%)	Binding Protein	Incubation Time (temperature, °C)	Separation	Normal Range (ng/ml)
USA (32)	Ether	Silicic acid	64	Rat kidney cytosol or serum	1 hr (25)	Dextran/charcoal	11–55
USA (46)	Chloroform methanol	Silicic acid columns	80	Rat serum	10 days (4)	Heparin–MnCl$_2$	18–36
France (49)	Dichloromethane methanol	Silica gel thin-layer	25	Human serum	20 hr (4)	Florisil	10–23
UK (47)	Chloroform methanol	Sephadex LH-20 columns	77	Rat serum	0.5 hr (25)	Dextran/charcoal	8–30
UK (54)	Chloroform–methanol–water	Silicic acid columns	87	Rat serum	15 hr (4)	Dextran/charcoal	4–33
Belgium (43)	Dichloromethane methanol–water	Sephadex LH-20 columns	82	Rat or human serum	1 hr (0)	Dextran/charcoal	5–22
USA (29)	Ethanol	None	95	Rat serum	2 hr (4)	Dextran/charcoal	20–100
Switzerland (71)	Ethanol	None	91	Rat serum	2 hr (4)	Dextran/charcoal	20–61

at the C-24 position is the R isomer, and 24R,25-(OH)$_2$D$_3$ is equal to 25-OHD$_3$ in its ability to displace 25-OH[^{3}H]D$_3$ from mammalian DBP.

B. Extraction and Chromatography

The procedures are similar to those previously described for extraction of vitamin D metabolites from serum. Thus far, Sephadex LH-20 chromatography seems to offer the most convenient isolation technique. In one technique[52] 0.9 × 30-cm columns are eluted with CHCl$_3$-n-hexane, 64:35 (v/v) as a single step.[53] 24,25-(OH)$_2$D elutes between 33 and 45 ml.

Validation of the isolation performed on small Sephadex LH-20 columns has been done with longer columns and in a high-speed liquid chromatographic system. Although 24,25-(OH)$_2$[^{3}H]D$_3$ can be used as the assay tracer,[52] 25-OH-[^{3}H]D$_3$ serves as well, and is less expensive. It now appears, however, that 24,25(OH)$_2$D peaks include other metabolites unless HPLC or multiple chromatographic systems are used.

C. Assay Procedures

The methods employed for the assay of 24,25-(OH)$_2$D are very similar to those described for the 25-OHD assay. After isolation, aliquots are used in the assay (50 μl) and for estimation of recovery. Incubation conditions and adsorbent separation steps are identical to those used in the 25-OHD assay. To date, few measurements have been reported, but the levels of 24,25-(OH)$_2$D in human sera are apparently in the 1.0–5.0 ng/ml range. Of considerable interest is the presence of 24,25-(OH)$_2$D in sera from patients who have had total nephrectomy, indicating extrarenal 24-hydroxylase activity.[53] The availability of agents and the relative ease of this assay ensure its future application to a variety of disorders of mineral homeostasis.

D. Comment

Although several groups have applied assay techniques to the estimation of 24,25-(OH)$_2$D in human plasma, there has been considerable disagreement about the absolute values found. The potential for the coelution of 25,26-(OH)$_2$D$_2$ with 24,25-(OH)$_2$D has been cited as a factor contributing to higher values in non-HPLC systems. For example, the presence of 24,25-(OH)$_2$D in anephric sera has been reported,[53,54] but others do not find it in similar sera.[55,56] It is not at all certain, in the author's opinion, that 24,25-(OH)$_2$D has a biological role other than being a metabolite along the catabolism pathway. Until its role is defined, measurements of this sterol would likely attract less attention.

IV. 1,25-Dihydroxyvitamin D Assay

A. Binding Proteins

In contrast to serum and tissue DBP's lower-affinity binding of 1,25-(OH)$_2$D, extracts from avian and rat intestinal mucosa have been shown to contain a protein that binds 1,25-(OH)$_2$D$_3$ with high affinity.[57-59] Recently, competitive radioassays for 1,25-(OH)$_2$D$_3$ have been developed on the basis of the ability of avian intestinal mucosal extract to bind this sterol with high affinity. In contrast to the tissue cytosol 5 S–6 S binding protein, the 1,25-(OH)$_2$D$_3$ binding protein is apparently in lower concentration, and is more labile.

Binding protein can be prepared from vitamin D-deficient white leghorn chickens. Duodena are removed and rinsed with 5.0 ml 0.05 M phosphate and 0.05 M KCl, pH 7.4. The mucosa is scraped off, minced in 5 volumes of buffer, and centrifuged at 2000 g for 10 min. The material is washed twice with 5 volumes buffer, and the supernatants are discarded. The mucosal pellet is resuspended in 2 volumes buffer and homogenized at 0°C. The homogenate is centrifuged at 50,000 rpm for 45 min; the lipid layer is aspirated; and the clear cytosol is lyophilized.[60]

Several factors contribute to the difficulty of the 1,25-(OH)$_2$D$_3$ assay: (1) serum content of 1,25-(OH)$_2$D is much lower than serum 25-OHD; (2) 1,25-(OH)$_2$D$_3$ cochromatographs with 25,26-(OH)$_2$D$_3$ in the silicic acid and Sephadex LH-20 chromatographic procedures; (3) very-high-specific-activity 1,25-(OH)$_2$-[^{3}H]D$_3$ has not, until recently, been available except by in vitro biosynthesis from 25-OH[^{3}H]D$_3$ by avian kidney; and (d) the 1,25-(OH)$_2$D$_3$-binding protein is labile. Depite these obstacles, workers have developed techniques which enable them to isolate and assay this important metabolite.

B. Extraction and Chromatographic Techniques

Following the addition of 3500 dpm 1,25(OH)$_2$-[^{3}H]D$_3$ to serum to monitor recovery, large volumes (5.0–20 ml) of serum are extracted with chloroform/ methanol, 1:2, or CH$_2$Cl$_2$/CH$_3$OH mixtures. The chloroform or dichloromethane layer is dried under N$_2$ prior to chromatography.

Initially, an arduous series of chromatographic steps was employed to isolate the 1,25-(OH)$_2$D. Silicic acid columns were used but have recently been deleted in favor of columns of Sephadex LH-20, eluting with 65% chloroform in hexane or hexane:chloroform:methanol 9:1:1. Final purification originally relied on the liquid–liquid partition of these eluates on columns of Celite.[61] Recently, Sep-PAK cartridges of ODS-silica have been employed. Also, diatomaceous earth columns have been used as a convenient extraction and chromatographic step.

Superior resolution of the metabolites has been reported with high-pressure liquid chromatography (HPLC) on a 0.4 × 30-cm column of 10 μm silicic acid in 10% isopropanol in hexane at 1.8 ml/min and 800 psi. The resolution of D metabolites is excellent in high-pressure systems, although very recent observations indicate that even the HPLC separation of 1,25-(OH)$_2$D from all other metabolites may not be complete as currently performed.

C. Assay Procedures

The two reported techniques vary in the preparation of the avian intestinal mucosal-binding protein. In one instance, an intestinal cytosol–chromatin complex is utilized by preparing 100,000 g supernatant and detergent-extracted chromatin from 6.0 gm intestinal mucosa from vitamin-D-depleted chickens. In the other method, only the high-speed supernatant of extracts of mucosa is used and is prepared in the presence of 0.05 M KCl. Optimal pH for either method is in the 7–8 range.

The cytosol–chromatin assay utilizes incubations of 200 μl of binding protein solution, 1,25-(OH)$_2$[^{3}H]D$_3$, and either standards or unknowns in 20 μl ethanol at 25°C for 20 min, followed by the addition of 1.0 ml cold 1% Triton-X-100 in 0.01 M Tris, pH 7.5, and filtration of the suspension on glass fiber filters. Following rinsing with detergent buffer, the filters containing bound 1,25-(OH)$_2$[^{3}H]D$_3$ are heated in counting vials containing 5 ml CH$_3$OH-CHCl$_3$ (2:1, v/v), the solvent is dried, and radioactivity is assayed in toluene-based scintillation fluid.

The cytosol assay utilizes 1.0-ml dilutions (1.0 mg protein/tube) of binding protein, 70 pg 1,25-(OH)$_2$[^{3}H]D, and either standards (10–140 pg) or unknowns, and is conducted at 25°C for 1 h. As with the cytosol–chromatin assay, radioactive and nonradioactive sterols are introduced in small amounts (50 μl) of ethanol. After 1 h, tubes are placed at 0°C, and 1.0 ml of cold 40% polyethylene glycol is added. The tube contents are mixed and centrifuged at 4800 g for 30 min at 4°C, and the supernatants are discarded. The precipitates are solubilized and counted in a dioxane-based scintillation solution.

Variable sensitivity has been reported using 6–12 Ci/mmol tracer 1,25-(OH)$_2$-[^{3}H]D$_3$ synthesized from 25-OH[^{3}H]D$_3$ by chicken kidney homogenates. Recently, high-specific-activity 1,25-(OH)$_2$[^{3}H]D$_3$ has been synthesized. This will permit greater sensitivity and the processing of smaller serum samples. Although the source of the binding protein used is identical in these assays, the cytosol–chromatin assay apparently recognizes 1,25-(OH)$_2$D$_2$ less well than the cytosol assay which is reported not to distinguish between the ergo and chole forms of this dihydroxy metabolite. The apparent contradiction demands additional study in order to resolve this issue.

At present, the 1,25-(OH)$_2$D assays are difficult because of the necessity for at least two chromatographic procedures. Either Celite columns or high-pressure chromatographic systems are required to isolate 1,25-(OH)$_2$D$_3$ from other metabolites. The use of cytosol-binding protein seems to represent an improvement over the preparation of subcellular fraction mixtures, and the widespread availability of high-specific-activity 1,25-(OH)$_2$[^{3}H]D$_3$ should attract more workers into this area of investigation. The features of these assays are depicted in Table 3. It is now recognized that receptor stability can be enhanced in

Table 3. Serum 1,25-$(OH)_2D$ Assay Methods

Step	Method 1	Method 2
Serum extraction	20 ml of CH_3OH-$CHCl_3$ (2:1)	5.0 ml dichloromethane
Chromatography	a. Two columns of Sephadex LH-20,1.0 × 15 cm[$CHCl_3$-hexane (65:35)] b. Celite 0.8 × 8.5 cm (aqueous ethanol and ethyl acetate-hexane)	a. Sephadex LH-20(0.7 × 9 cm)[hexane-$CHCl_3$-CH_3OH (9:1:1)] b. High-pressure (silica 10 μm) (isopropanol-hexane)
Recovery	40–75%	68%
Binding protein	Avian small intestine; cytosol–chromatin suspension	Avian small intestine; cytosol preparation
Assay	25°C for 20 min	25°C for 60 minutes
Harvest	Glass filters with Triton-X-100 rinse	PEG 6000; count precipitate
Sensitivity	20 pg/tube	10 pg/tube
Normal level (human blood)	64 ± 12 pg/ml (n = 20)	29 ± 2.0 pg/ml (n = 20)
Later studies	33 ± 6 pg/ml (n = 29)	CH_2Cl_2-CH_3OH extraction; dextran-coated charcoal harvest

Note: a = Brumbaugh et al. (1974a,b), Haussler et al. (1976), and Hughes et al. (1976). b = Eisman et al. (1976a,b).

buffers containing 0.1–0.3 M KCl, molybdate, and sulfhydryl-reducing agents. Very recently, the receptor protein in avian intestinal cytosol was purified to homogeneity.[62]

Recently, groups of workers have reported success in raising antisera to vitamin sterols. Suitable radioimmunoassay (RIA) procedures have been reported in very recent papers, and are discussed elsewhere in this volume. Similarly, the cytoreceptor assay, reflecting a competitive assay utilizing entire cells as binding components, is described in another section.

V. Comments

It is clear that methodology employing competition assays for vitamin D sterols has undergone considerable investigation over the past 11 years. Analogous to the situation with steroid competitive protein-binding (CPB) assays a few years ago, it is apparent that reliable, stereospecific antisera will be provided for the D assays. Several workers have developed useful antisera employing sterol–protein immunogens.[63-66] More specific reagents may lead to less cumbersome procedures for presenting sterols to the antibody without loss of specificity. More simple preparative chromatography on cartridges of silica[67] is an attractive alternative. The recent reports that vitamin-D-sufficient rabbit[68] or pig[69] intestinal cytosol can serve as a receptor source suitable for 1,25-(OH)$_2$D assay is encouraging.

Although multiple assay schemes have been developed, the availability and maintenance of extraction and chromatographic systems are limiting features. Additional improvements seem to be needed in the preparative phases (simplified extraction or no extraction in the presence of highly specific antisera or cell-receptor assays; simple isolation steps prior to competitive assay). It is noteworthy that monoclonal antibodies to 1,25-(OH)$_2$D$_3$ have been produced;[70] this augurs well for the development and availability of suitable assay reagents.

References

1. Warkany J: Am J Dis Child 52:831–847, 1936.
2. Warkany J, Mabon HE: Am J Dis Child 60:606–613, 1940.
3. Warkany J, Guest GM, Grabill FJ: J Lab Clin Med 27:557–565, 1941.
4. Haddock L, Vazquez M: J Clin Endocrinol Metab 26:859–866, 1966.
5. DeLuca HF: J Lab Clin Med 87:7–26, 1976.
6. Norman AW, Henry HL: Rec Prog Horm Res 30:431–480, 1974.
7. Belsey R, DeLuca H, Potts J: J Clin Endocrinol Metab 33:554–557, 1971.
8. Hollis B, Roos B, Lambert P: Steroids 37:609–619, 1981.
9. Kawakami M, Imawari M, Goodman DS: Biochem J 179:413–423, 1979.
10. Thomas WC Jr, Morgan HG, Connor TB, Haddock L, Bills CE, Howard JE: J Clin Invest 38:1078–1085, 1959.
11. de Crousaz P, Blanc B, Antener I: Helv Odontol Acta 9:151–155, 1965.
12. Chalk KJI, Kodicek E: Biochem J 79:1–7, 1961.
13. Chen PS, Bosman HB: J Nutr 83:133–139, 1964.
14. Rikkers H, DeLuca HF: J Phys 213:380–386, 1967.
15. Rikkers H, Kletziens R, DeLuca HF: Proc Soc Exp Biol Med 130:1321–1324, 1969.
16. Haddad JG, Chyu KJ: Biochim Biophys Acta 248:471–481, 1971.
17. Rosenstreich SJ, Volwiler W, Rich C: J Clin Nutr 24:895–905, 1971.
18. Smith JE, Goodman DS: J Clin Invest 50:2159–2167, 1971.
19. Peterson PA: J Biol Chem 246:7748–7754, 1981.
20. Edelstein S, Lawson DEM, Kodicek E: Biochem J 135:417–426, 1973.
21. Belsey RE, DeLuca HF, Potts JT Jr: Nature 247:208–209, 1974.
22. Bouillon R, Van Baelen H, Tan BW, De Moor P: J Biol Chem 255:10925–10930, 1980.
23. Steenbock H, Kletzein SWF, Halpin JC: J Biol Chem 97:249–264, 1932.
24. Haddad JG, Walgate J: J Biol Chem 251:4803–4809, 1976.
25. Imawari M, Kida, K, Goodman DS: Clin Res 24:458A, 1976.
26. Haddad JG, Hillman L, Rojanasathit S: J Clin Endocrinol Metab 43:80–91, 1976.
27. Haddad JG, Walgate J: J Clin Invest 58:1217–1222, 1976.
28. Haddad JG, Hahn TJ: Nature 244:515–517, 1973.
29. Belsey RE, DeLuca HF, Potts JT Jr: J Clin Endocrinol Metab 38:1046–1051, 1974.
30. Haddad JG, Chyu KJ, Hahn TJ, Stamp TCB: J Lab Clin Med 81:22–27, 1973.

31. Haddad JG, Birge SJ: Biochem Biophys Res Commun 45:829–834, 1971.
32. Haddad JG, Chuy KJ: J Clin Endocrinol Metab 33:992–995, 1971.
33. Haddad JG: In Norman AW, Schaefer K, Grigoleit HG, von Herrath D, Ritz E (Eds.): Vitamin D and Problems Related to Uremic Bone Disease. deGruyter, Berlin, 1975, pp. 285–293.
34. Haddad JG, Birge SJ: J Biol Chem 250:299–303, 1975.
35. Haddad JG, Hahn TJ, Birge SJ: Biochim Biophys Acta 329:93–97, 1973.
36. Oku T, Ooizumi K, Hosoya N: J Nutr Sci Vitaminol 20:9–25, 1974.
37. Lawson DEM, Wilson PW: Biochem J 144:573–583, 1974.
38. Brumbaugh PF, Hughes MR, Haussler MR: Proc Natl Acad Sci USA 72:4871–4875, 1975.
39. Reynolds RD, Knutson JC, DeLuca HF: Fed Proc 34:893(A), 1975.
40. Haddad JG, Walgate J, Min C, Hahn TJ: Biochim Biophys Acta 444:921–925, 1976.
41. Haddad JG, unpublished observations.
42. Haddad JG, Stamp TCB: Am J Med 57:57–62, 1974.
43. Bouillon R, Van Kerkhove P, DeMoor P: Clin Chem 22:364–368, 1976.
44. Jones G, DeLuca HF: J Lipid Res 16:448–453, 1975.
45. Bligh EG, Dyer WJ: Can J Biochem 37:911–917, 1959.
46. Belsey RE, DeLuca HF, Potts JT Jr: J Clin Endocrinol Metab 33:554–557, 1971.
47. Edelstein S, Charman M, Lawson DEM, Kodicek E: Clin Sci Mol Med 46:231–240, 1974.
48. Holick, M, DeLuca HF: J Lipid Res 12:460–464, 1971.
49. Bayard F, Vec P, and Louvet JP: Eur J Clin Invest 2:195–198, 1972.
50. Sklan D, Budowski P, Katz M: Anal Biochem 56:606–609, 1973.
51. Haddad JG, Min C, Walgate J, and Hahn TJ: J Clin Endocrinol Metab 43:712–715, 1976.
52. Taylor CM, Hughes SE, DeSilva P: Biochem Biophys Res Commun 70:1243–1249, 1976.
53. Haddad JG, Mendelsohn M, Min C, Slatopolsky E, Hahn TJ: Arch Biochem Biophys 182:390–395, 1977.
54. Lambert P, Stern P, Avioli R, Brackett N, Turner R, Greene A, Fu I, Bell N: J Clin Invest 69:722–725, 1982.
55. Horst R, Littledike E, Gray R, Napoli J: J Clin Invest 67:274–280, 1981.
56. Taylor C, Mawer E, Wallace J, St. John J, Cochran M, Russell RG, Kanis J: Clin Sci Mol Med 55:541–547, 1978.
57. Brumbaugh PF, Haussler MR: J Biol Chem 249:1258–1262, 1974.
58. Frolik CA, DeLuca HF: Steroids 27:433–440, 1976.
59. Lawson DEM, Wilson PW: Biochem J 144:573–583, 1974.
60. Eisman JA, Hamstra AJ, Kream BE, DeLuca HF: Arch Biochem Biophys 176:235–243, 1976.
61. Haussler MR, Baylink DJ, Hughes MR, Brumbaugh PF, Wergedal JE, Shen FH, Nielsen RL, Counts SJ, Bursac KM, McCain TA: Clin Endocrinol 5:151s–165s, 1976.
62. Simpson R, DeLuca H: Proc Natl Acad Sci USA 79:16–20, 1982.
63. Peacock M, Taylor G, Brown W: Clin Chim Acta 101:93–101, 1980.
64. Bouillon R, DeMoor P, Baggiolini E, et al. Clin Chem 26:562–567, 1980.
65. Clemens T, Hendy G, Graham R, Baggiolini E, Uskokovic M, O'Riordan, J: Clin Sci Mol Med 54:329–332, 1978.

66. Gray T, McAdoo T, Pool D: Clin Chem 27:458–463, 1981.
67. Adams J, Clemens T, Holick M: J Chromatogr 226:198–201, 1981.
68. Duncan W, Walsh P, Haddad J: Am Fed Clin Res National Meeting Abstracts, Clin Res 30:390A, 1982.
69. Clayton J, Guillard-Cumming D, Kanis J, Russell RG: Fifth Workshop on Vitamin D, Abstracts p. 299, 1982.
70. Perry M, Chappel J, Clevinger B, Haddad J, Teitelbaum S: Fifth Workshop on Vitamin D, Abstracts p. 70, 1982.
71. De Hertogh R, Van der Heyden I, Ekka E: J Steroid Biochem 6:1333–1337, 1975.

Chapter 4

Microassay for 25-Hydroxyvitamin D: Method and Interpretation

Sara B. Arnaud and
Juliann Meger

The most widely used measurement of "vitamin D" is the assay of 25-hydroxy-vitamin D. It provides an objective assessment of the nutritional vitamin D status of patients with disorders of calcium and phosphorus metabolism which is useful in their diagnosis and management. General indications for this assay are given in Table 1. Osteomalacia due to vitamin D deficiency is an occult disease whose diagnosis is aided by measurements of the endocrine response to prolonged calcium deficiency (serum parathyroid hormone) associated with reduced levels of serum 25-hydroxyvitamin D. In later stages of vitamin D deficiency, manifested by either hypocalcemia and/or hypophosphatemia, the measurement is confirmatory. Patients with unexplained muscle weakness may have hypovitaminosis D, and those with disturbances in gastrointestinal function (e.g., peptic ulcer disease, diarrhea, or steatorrhea) are likely candidates for hypovitaminosis D. The assay is a useful guide in monitoring vitamin D or 25-hydroxyvitamin D_3 therapy, and increased values are indicative of hypervitaminosis D in hypercalcemic patients.

25-hydroxyvitamin D is an intermediate in the metabolic pathway of vitamin D. The prohormone, vitamin D, either synthesized in skin or absorbed from the intestine from food, undergoes 25-hydroxylation primarily in the liver. 25-hydroxyvitamin D circulates in plasma in microgram quantities, bound to a highly specific binding protein. This protein stores and transports 25-hydroxyvitamin D to tissues which further metabolize this compound to forms which carry out the actions of vitamin D in calcium and phosphorus metabolism. In this scheme, the assay of 25-hydroxyvitamin D provides information on the quantity of substrate for biologically active derivatives of vitamin D. Even though 25-hydroxyvitamin D, at physiological concentrations, may have little direct biological effect itself, its measurement has more clinical application than that of any of the other vitamin D metabolites at this time.

I. Principle of the Assay of 25-Hydroxyvitamin D

The metabolite is extracted from serum or plasma with organic solvents, isolated from other lipids and vitamin D sterols by chromatography, dissolved in

Table 1. Indications for the Assay of 25-Hydroxyvitamin D

Deficiency	Excess	Metabolic Abnormalities
Hypocalcemia	Vitamin D therapy	Anticonvulsant medication
Hypophosphatemia	Hypercalcemia	Corticosteroid treatment
Prematurity		
Gastrointestinal disease		
Sprue		
Crohn's disease		
Gastrointestinal surgery		
Ileal resection		
Bypass for obesity		
Hepatobiliary disease		
Cholestasis		
Renal disease		
Nephrotic syndrome		
Hypocalcemia		
Myopathy		

ethanol, and quantitated by a competitive protein binding assay.[1,2] An explanation and description of each of the steps illustrated in Figure 1 follows.

II. Extraction of 25-Hydroxyvitamin D from Serum

The classic lipid extraction of Bligh and Dyer, or a modification of it, is the most commonly used method for the extraction of 25-hydroxyvitamin D.[3] This extraction usually involves two steps. The first uses a mixture (2:1) of methanol and chloroform or methanol and methylene chloride to precipitate proteins and solubilize free sterol. After centrifugation and transfer of the supernatant to a vessel for evaporation, additional chloroform or methylene chloride is added to separate the organic solvents containing 25-hydroxyvitamin D from an aqueous methanolic phase. The upper aqueous phase is removed by aspiration or the lower organic solvent phase by drip, if a separatory funnel is used. Recovery of the sterol, monitored by [³H]-25 hydroxyvitamin D, is invariably better than 90% with this classic method. Equally effective extraction of 25-hydroxyvitamin D is accomplished with ether or ethanol.

The selection of the extraction method is based on a number of considerations, most of which have little to do specifically with the assay of 25-hydroxyvitamin D. Methylene chloride forms an azeotrope with water which is removed from the lipids during evaporation with nitrogen, yielding drier sediments than extraction with chloroform. Ether is not used in our laboratory because of considerations of safety. Its use is preferred by others who are doing multiple assays of the vitamin D sterols since it extracts all of them with nearly equal efficiency, can be done in one step, and evaporates rapidly. In this procedure, the aqueous phase is the lower one. It can be frozen rapidly with dry ice and acetone, effectively separating the upper organic phase containing the vitamin D sterols which can then be poured off into a vessel for evaporation. Ten volumes of methanol or ethanol alone is also a convenient extraction method, resulting in good recovery of 25-hydroxyvitamin D. However, evaporation of alcoholic extracts takes more time and uses more nitrogen gas than evaporation of ether, chloroform, or methylene chloride.

The detailed procedure we use for the extraction of 25-hydroxyvitamin D follows.

1. Pipette 0.2 ml serum into the bottom of a 16 × 125-mm disposable tube. Add 1000 cpm [^{3}H]-25-hydroxyvitamin D_3 to each unknown and to 2 scintillation counting vials. Allow 15 min for the labeled sterol to bind to DBP. Then add 0.8 ml saline (0.85% NaCl) to improve the efficiency of extraction and to bring the total volume to 1 ml in instances where it may be desirable to vary sample volumes from 0.05 to 1 ml. Vortex.
2. Add 3.75 ml of a mixture of methanol and methylene chloride (2:1, v/v) which has been prepared fresh. Vortex. Proteins are precipitated in this two-step procedure.
4. Centrifuge at 3500 rpm for 10 min at 4°C.

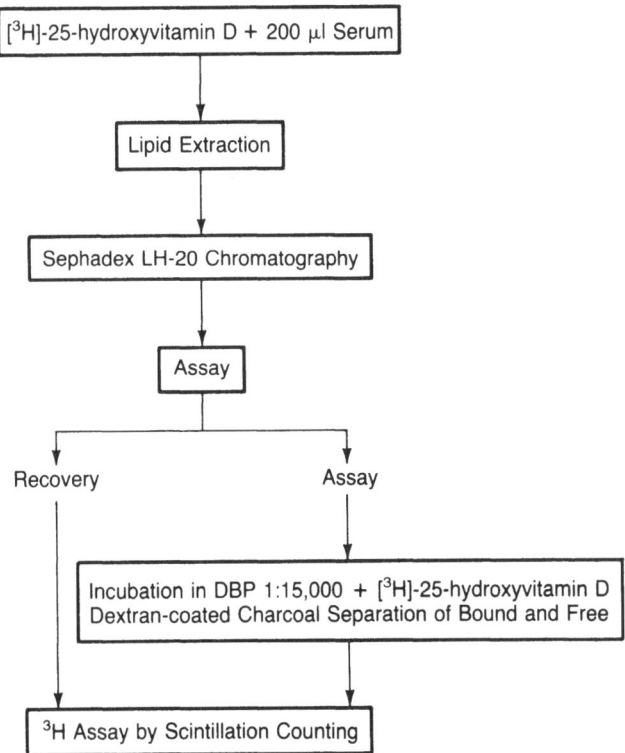

Figure 1. General scheme for the assay of 25-hydroxyvitamin D.

5. Pour off the supernatant into another 16 × 125-mm tube, using a technique which takes advantage of capillary action to ensure complete transfer of the supernatant. Discard tubes containing the protein precipitate.
6. Add 1.75 ml methylene chloride, vortex, and allow phases to separate at 25°C for 1 h or at 4°C overnight after capping each tube.
7. Aspirate the aqueous upper phase with vacuum through a Pasteur pipette which is wiped clean with solvent between each sample.
8. Evaporate the lower methylene chloride phase with nitrogen in a water bath at 37°C.
9. When tubes have evaporated to dryness, add 0.2 ml solvent for chromatography, for example, chloroform/hexane in a 1:1 mixture.

The above procedure requires about 2 h to extract 20 samples and involves the loss of approximately 10% of the 25-hydroxyvitamin D during the transfer. Greater losses than this suggest other problems in the extraction of 25-hydroxyvitamin D.

Failure to obtain clear phase separation may be due to errors in the amounts of solvents added to the volume of serum, the variable water content of methanol (an hydroscopic solvent which must be kept tightly capped), or change in the proportions of methanol and methylene chloride in solutions kept for long periods of time. Icteric serum may require the addition of more water to effect good phase separation.

Other reasons for low recoveries include inadequate time allowed for phase separation and overzealous aspiration of the aqueous phase.

III. Chromatography for the Isolation of 25-Hydroxyvitamin D

Small columns of variable size filled with silicic acid, Sephadex LH 20, or Lipidex 5000 are now used for the isolation of 25-hydroxyvitamin D. In most laboratories, recoveries from any of these systems are better than 80%. Commonly used solvents are chloroform and hexane for Sephadex and ether/hexane mixtures for silicic acid.[4] The solvent used to elute the sterol can be added all at once to single columns of the "tulip" design which originated in Wisconsin, or it is allowed to flow from a single reservoir through a manifold connected to a series of columns. Solvent flow through these columns is sufficiently rapid under gravity that the isolation of the metabolite can be accomplished in 30–60 min. Silicic acid columns are usually discarded after use, whereas the Sephadex can be reused more than 20 times without loss of resolution or carryover of substances in serum, provided it is "stripped" after each use. More polar compounds are removed from Sephadex columns stripped by allowing a 70% chloroform in hexane solution to flow through in amounts equal to at least 4 times the void volume of the column. In carrying out multiple assays of vitamin D derivatives from the same serum, others prefer the purification of 25-hydroxyvitamin D for quantitation by direct UV absorbance with silicic acid

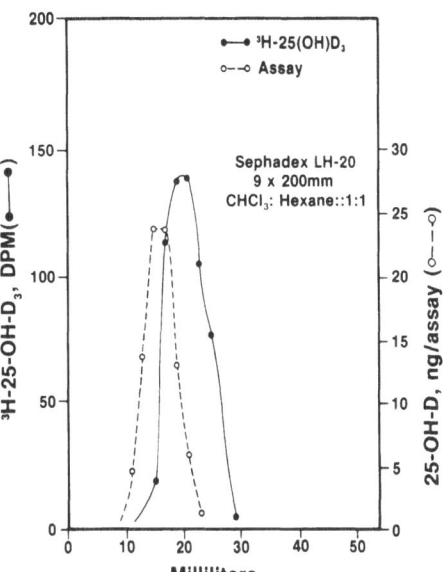

Figure 2. Isolation of 25-hydroxyvitamin D containing both derivative D_2 and D_3 by chromatography which uses "short" 0.9 × 20-cm columns of Sephadex LH 20 in chloroform/hexane, 1:1, v/v, as the solvent system. In this calibration experiment, the source of 25-hydroxyvitamin D_2 was a patient serum extract assayed by competitive protein-binding assay of every 4-ml fraction of effluent, and the source of 25-hydroxyvitamin D_3, the radioactive compound, assayed by scintillation counting of 4-ml fractions.

or Lipidex 5000.[5,6] Two chromatographic steps are required to remove 25-hydroxyvitamin D from the lactone and other more polar derivatives which interfere in the assay if the solvent system, hexane/chloroform/methanol, 9:1:1, v/v/v, used by Eisman et al.[7] for 1,25-dihydroxyvitamin D purification is used for 25-hydroxyvitamin D.

The short columns we use do not separate 25-hydroxyvitamin D_2 from 25-hydroxyvitamin D_3.[8] The peak of the D_2 derivative elutes slightly earlier than the D_3 derivative with a few milliliters separating the 2 peaks in these 15-cm columns as illustrated in Figure 2. This incomplete separation is advantageous if the total 25-hydroxyvitamin D component is to be quantitated. Columns at least 50 cm long (0.9 cm diameter) must be used for complete separation of the 2 compounds when this information is desired. The practice of vitamin supplementation with D_2 and the content of vitamin D_2 in American diets is so variable that assays of 25-hydroxyvitamin D_3 alone will not provide the clinician with an accurate estimation of the status of vitamin D nutrition, especially in children, nor permit its application to the management of patients receiving vitamin D_2, the most commonly used form of vitamin D for treatment of disorders of calcium metabolism.

The step-by-step procedure we use for the chromatography to isolate 25-hydroxyvitamin D follows; 10 samples are chromatographed at one time.

1. The columns are poured with a slurry of Sephadex LH 20 in chloroform/hexane, 65:35, v/v, and filled to 19 cm. The Sephadex shrinks to 15 cm when it is equilibrated with chloroform/hexane, 50:50, v/v, the solvent used for the chromatography. Each set of 10 freshly poured columns is calibrated with 10,000 cpm [^{3}H]-25-hydroxyvitamin D_3 by counting 25 fractions, 2 ml each, of the effluent following application of the radioactivity.

2. Each lipid extract dissolved in 0.2 ml chloroform/hexane, 50:50, v/v, is vortexed, and is applied with a Pasteur pipette to the surface of the Sephadex after the column has been equilibrated with the same solvent. Two rinses of 0.1 ml volume are applied to the surface of the Sephadex and the combined 0.4 ml allowed to run through followed by additions of 0.5–1 ml solvent. When the lipid extract has migrated well below the surface of the Sephadex, the column is filled with solvent and connected to the reservoir.

3. Scintillation vials are used to collect the effluent to avoid transfer of solutions, and premarked to collect two fractions. The first 8-12 ml containing void volume, lipoproteins, and vitamin D is discarded, and the second 22-ml fraction is evaporated to dryness with nitrogen and resolubilized in 0.5 ml ethanol for the assay.

4. The columns are prepared for the next use by changing the solvent in the reservoir to chloroform/hexane, 70:30, v/v and allowing at least 50 ml to flow through each to remove metabolites of vitamin D more polar than 25-hydroxyvitamin D.

Low recoveries from this simple chromatography are most commonly due to failure to equilibrate the columns adequately in the solvent for the isolation of 25-hydroxyvitamin D. The time spent in the two rinses of the tube containing the lipid extract is also important to prevent losses at this step. A saline blank is run with each set of 10 columns, using a different column for each run, to determine the presence of contamination by substances that may interfere in the assay.

IV. Protein-binding Assay

The assay for 25-hydroxyvitamin D is a competitive protein-binding assay which uses naturally occurring vitamin-D-binding protein (DBP). This protein is present in large excess in the serum of vertebrates, but there are species differences in the binding affinity. The unpurified protein from rat serum, diluted 1:15,000, as first described by Belsey,[2] is the most widely and successfully used protein. Its source from vitamin-D-replete animals has proven to be as potent as material from vitamin-D-deficient animals. The protein is diluted in either phosphate or barbital acetate buffer to which is added albumin, gelatin, or other compounds such as β-lipoprotein or polyvinyl alcohol to solubilize the 25-hydroxyvitamin D. In the selection of an albumin preparation, one needs to carry out preliminary tests to check for the presence of DBP or other proteins which result in more than 10% binding of labeled 25-hydroxyvitamin D.

The assay is an equilibrium assay with a very short incubatiom time; maximum binding is greatest in 30–60 minutes at 25° C. Longer incubations at 4°C carried out in some laboratories have no scientific advantage over shorter incubations, but may be a preferred time schedule.

The most widely used techniques for the separation of bound and free 25-

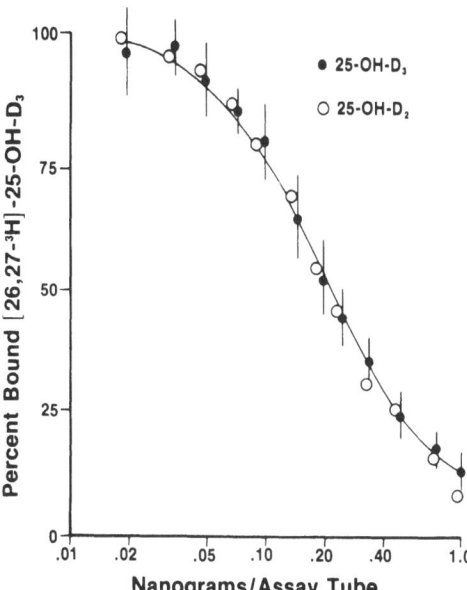

Figure 3. Standard curve of the assay. Data from an assay of 25-hydroxyvitamin D₂ standard are compared to data from 10 consecutive assays ($\pm$SD) of the 25-hydroxyvitamin D₃ standard. (Both standards were generous gifts of Dr. John Babcock of the Upjohn Company).

hydroxyvitamin D use solutions of 5% dextran-coated charcoal. The free 25-hydroxyvitamin D is bound to the charcoal, which is pelleted by centrifugation. Only the supernatant is aliquoted and counted after being mixed with scintillation solution. As in many of the sterol assays, dextran-coating of the charcoal is not necessary to separate free from protein-bound sterol, but is used because it improves the character of the pellet. Dextran-coated charcoal is centrifuged and suspended in dextran-free buffer before use since dextran itself competes with charcoal for free 25-hydroxyvitamin D. Both the amount of charcoal and the timing of the separation procedure are critical since the bond between sterol and transport protein is not as firm as the antigen–antibody bond in radioimmunoassays where an excess of charcoal is desirable to remove free antigen. Relatively small amounts of charcoal in 2 or 5% solutions are added to the incubation mixture, and separated within 1 or 2 h. If it is anticipated that a longer interval will be necessary between the addition of charcoal and the time that aliquots of the supernatant are taken, a second standard curve is added to the end of the assay to check for dissociation of sterol from the charcoal.

A typical standard curve is shown in Figure 3. This curve is constructed from the assay which we routinely perform, the details of which follow.

A. Procedure

Reagents

0.05 *M* barbital acetate buffer, pH 8.6
Albumin (Pentex from Miles laboratories) 0.05 g/100 ml buffer
Charcoal (Norit from Fisher)

Dextran, T 70 (Pharmacia)
Ethanol, absolute

Supplies

13 × 75-mm borosilicate or flint glass test tubes
Scintillation vials: 20-ml capacity for counting the recovery aliquot, 7.5-ml
    capacity for the assay
Automatic repipette (Beckman)
Manual repipette (Labindustries)

1. The dextran-coated charcoal is prepared first to allow sufficient mixing
   time. It is routinely prepared fresh for each assay. However, it can be made
   once a week, in albumin-free buffer, stored at 4°C, and then centrifuged
   and resuspended in albumin-containing buffer 1 h before use. The routine
   preparation involves weighing 2 g charcoal and 0.2 g dextran. These are
   dessolved in 0.05 M barbital acetate buffer separately, and then mixed
   together in 100 ml buffer without albumin. One-half hour before the assay
   incubation is completed, the dextran-coated charcoal is centrifuged at 3000
   rpm for 10 min. The fines and dextran supernatant are poured off and the
   charcoal is resuspended in buffer containing albumin. It is stirred on ice and
   added to the assay tubes with a Rainin pipette while being stirred.
2. Standards are pipetted into empty assay tubes on ice in three 50-μl aliquots.
   In addition, unknowns are pipetted in two 20- and two 10-μl aliquots. The
   smaller amounts are made up to volume with ethanol. Larger aliquots are
   added to the large scintillation vials to monitor recovery. The ethanol in
   these vials is evaporated in a Brinkman concentrator at 37°C while the
   assay is incubating.
3. The DBP buffer is prepared 30 min before use by adding 8 μl rat serum to
   120 ml 5% albumin buffer. Half that amount is needed for an overnight
   incubation. Then 1000 cpm [^{3}H]-25-hydroxyvitamin D$_3$ in 10 μl ethanol is
   added to each assay tube with a Beckman repipette followed by a wash of
   1 ml of albumin buffer (for the first tube containing 50 μl to assess nonspecific binding) or DBP buffer. Total radioactivity added to each tube is
   assessed by adding 1 ml albumin buffer at the same time the DBP buffer is
   added and 0.4 ml albumin buffer in lieu of charcoal at the separation step.
   All tubes are vortexed after the addition of either albumin buffer or DBP
   buffer and incubated for 2 h at 25°C.
4. Separation of bound and free [^{3}H]-25-hydroxyvitamin D is performed by
   the addition of 0.4 ml dextran-coated charcoal in albumin buffer to all tubes
   except those marked "T" for monitoring the total radioactivity added to the
   assay. Assay tubes are then centrifuged at 4°C for 15 min at 4000 rpm.
   Tubes are kept on ice while a 900-μl aliquot is removed with a Labindustries
   repipette and washed out with 6.5 ml scintillation solution into a small scintillation vial for counting for, usually 5 min.

5. After the aliquots for recovery are evaporated to dryness, 10 ml scintillation solution is added and the vials are counted for at least 10 min or for the time needed to obtain at least 5% sigma.

Calculations

Recoveries of the fraction of radioactivity originally added to serum which is finally assayed in the 50-μl ethanolic extract are determined in the following way: DPM, corrected for background, in each 200-μl aliquot is divided by the total amount of radioactivity added to the serum at the start. The fraction of radioactivity in 200-μl is then divided by 4. This represents one tenth of the total recovery since it is a 50-μl aliquot of a 500 μl volume of ethanolic extract and is usually 6–7%.

The assay calculations include correction for nonspecific binding, which is the amount of radioactivity that remains in the supernatant after the addition of charcoal to the assays tubes containing *no* DBP. Ideally, it is a zero; in actuality, it may be as high as 10% of the radioactivity added and shows some variation in each of the unknown extracts of serum. The points on the standard curve and each unknown are determined by calculating the percent bound (B/T) in each sample and then normalizing this value by dividing it by the B/T of the 0 standard and the saline control as follows:

$$\frac{\text{CPM in supernatant} - \text{CPM nonspecific binding}}{\text{Total CPM added to the assay} - \text{CPM nonspecific binding}} = \text{B/T}$$

B/T in 0 concentration standard and/or in saline control = 100%

Ideally the B/T in these two samples agree. Often, they do not, and one then needs to use the 0 concentration standard to determine the standard curve results and the saline control B/T for the sample 100% reference. Since the 0 standard is made up at the same time as the rest of the standards, and the standards can be used for a few weeks, the 0 standard reflects deterioration of the ethanol or contamination of these vials with plasticizers or other unknown substances from plastic pipette tips.

$$\frac{\text{B/T in unknown}}{\text{B/T in 0 standard or saline control}} = \text{"normalized" B/T}$$

The standard curve is plotted on semilogarithm paper as illustrated and values in pg/tube in the unknowns determined from it (see Figure 3).

$$\frac{\text{pg/tube}}{\text{recovery in the same aliquot}} \times 5 \ (\text{if 0.2 ml extracted}) = \frac{\text{pg}}{\text{ml serum}}$$

$$\frac{\text{pg/ml}}{400} = \frac{\text{pmol}}{\text{ml}}$$

B. Sensitivity

Now that the specific activities of radioactive 25-hydroxyvitamin D_3 range from 80 to 160 Ci/mmol, the assays are sensitive enough to determine 25 or 50 pg of the sterol per tube. Allowing for duplication of ethanol extracts in the assay, aliquots to determine recovery, and assuming recoveries of 60% or better from the extraction procedures, a level of sensitivity of less than 1 ng/ml can be achieved. Since 25-hydroxyvitamin D circulates in much greater quantities than this, even in nutritional deficiency states (e.g., less than 5 ng/ml), the amount of serum required for the assay is quite small. This level of sensitivity makes "microassays" quite practical in laboratories doing routine analyses and is of particular value in pediatric patients or in research laboratories using small animals. Microassays require so little sample that it is possible to optimize the conditions and equipment for this assay in laboratories where multiple assays requiring larger volumes of serum or plasma and different chromatographic systems are needed, and perform both single and multiple assays conveniently. Up to 40 samples can be extracted, chromatographed, and assayed in 2.5 days, when 20 columns are set up for the isolation of the metabolite.

C. Precision

The precision of the assay is monitored by noting shifts in the standard curve and by appropriate controls. The initial effort is to determine the reproducibility of the assay itself by determining the degree to which known amounts of 25-hydroxyvitamin D added to a single serum can be accurately quantitated and recovered in 10 or more consecutive assays. These samples do not keep well, however, as continuing controls for the assay for longer than 2 months and are not stable enough specimens for quality control measurements which must be carried out for 12 months. Shifts in the standard curve, the first record of continuing quality control, is monitored by keeping track of the picograms of standard that displace 50% of the added label on the standard curve. Since the straight-line portion of the curve shows the least variation, lack of change in this measure is one of the least difficult standards of reproducibility to meet. Acceptable variation in this control is generally less than 10% or 15–20 pg. More variation in this value serves to alert the assayist to any change in the potency of the standards. The second quality control on the reproducibility of the assay is the results obtained from two pools of low-to-normal- and high-value samples. The source of a low-value pool can be aliquots of a single outdated pint of plasma which has a value of 10–15 ng/ml (25–30 pmol/ml). Extracts of such plasma displace labeled metabolite in the lower ranges of the standard curve, a region which is likely to be more variable than the midregion. This sample is run as an unknown in each assay and monitors the extraction procedure as well as the interassay variation. An acceptable upper limit of variation is 20% or 2 ng/ml. The high-value control (e.g., more than 200 ng/ml or

500 pmol/ml) is a high-value sample that is either subjected to repeated extraction and assay or extracted once and the ethanolic extract run with each assay. Repeated assay of the high control monitors the high portion of the standard curve and should not vary more than 10–15%.

D. Quality Control

In addition to the high- and low-value samples which are used to determine the precision of the assay, as described above, evaluation of reused columns for contamination and poor resolution, and readjustment of standards, are part of the necessary quality control procedures. Contamination of columns is clearly demonstrated by either assaying a blank for each column before the unknown sample is applied, or carrying an extract of saline or water through the assay procedure each time it is performed. Results from these experiments early in the course of developing the assay indicate the procedure of choice. We use a saline extract control in a different column each assay run now since our earlier experience with column blanks proved the adequacy of our procedures for stripping columns of contaminants. The assay of a saline extract should give the same depression of the B/F ratio as an aliquot of ethanol. When they do not agree, values for unknowns are more correctly calculated from the saline control than from the 0 standard which is used in the same assay to calculate the standard curve.

The potency of the standards is kept constant by procedures used to keep any solution of ethanol in good condition as well as a few procedures unique to the storage of vitamin D compounds. Since 25-hydroxyvitamin D is sensitive to oxidation, it is stored in tightly sealed screwcap vials under nitrogen or argon. It is kept at 4° or −20°C to minimize evaporation of ethanol. Newly prepared standards are made from a 1:10 dilution of a stock standard containing 1 ug/ml, a concentration at which the absorbance can be easily monitored. This UV stock solution is adjusted for evaporation of ethanol by the addition of a few drops of ethanol every 4–8 weeks to maintain the selected absorbance at a specific wavelength. There is 0.450 absorbance at a 265 mu in our 25 mM standard, using an extinction coefficient of 18,000 (concentration in molarity = extinction coefficient/absorbance at 265 mu). Comparison of our assay with an HPLC method in another laboratory[10] revealed agreement of each sample-pair to be within 12%.

E. Interpretation

Normal Values

There is seasonal variation in serum 25-hydroxyvitamin D with the highest values in mid and late summer months, and the lowest ones in midwinter and early spring. Typical seasonal variation is illustrated in Figure 4 which shows

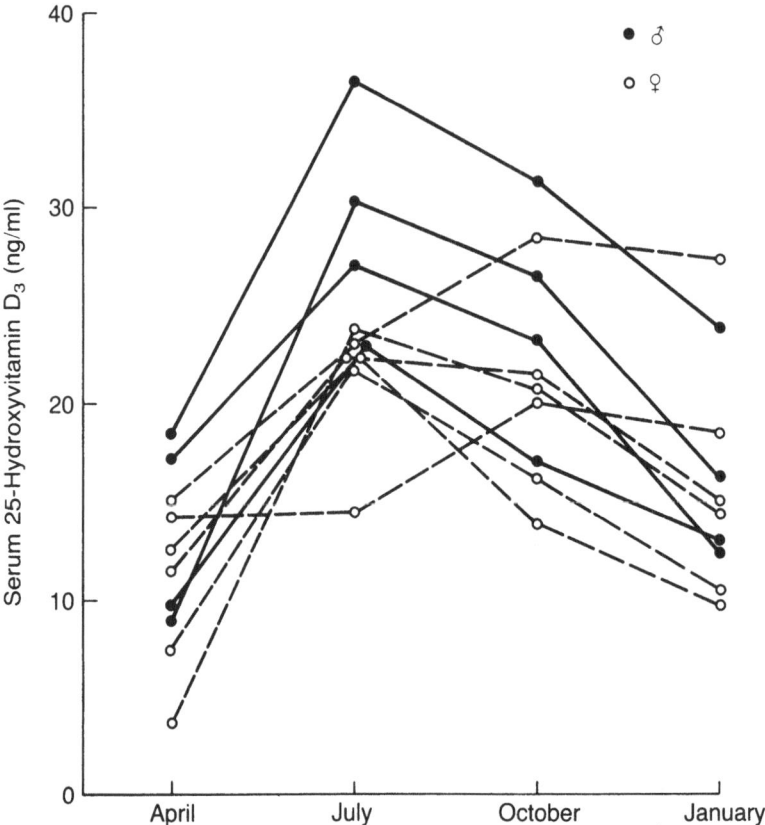

Figure 4. Longitudinal study of the seasonal variation in serum 25-hydroxyvitamin D in 10 healthy adults living at a latitude of 42°.

repeated measurements in 10 healthy adults living at a latitude of 42°. Minimum seasonal change in the concentration of serum 25-hydroxyvitamin D in one subject was 3 ng/ml and maximum 25 ng/ml. The pattern of change with season parallels the change in the intensity of the ultraviolet rays from the sun, the principal source of circulating 25-hydroxyvitamin D.[8] The influence of occupation, as it relates to the environment, is one factor to consider in evaluating results. Sex and age differences are reported, values being slightly higher in children and slightly lower in women.[11] Values in American children are slightly higher than adults because of the vitamin D_2 supplementation of their diets as demonstrated by measurement of the two components of 25-hydroxyvitamin D in children between 6 months and 6 years of age in summer, autumn, and winter months (see Figure 5). The normal range in our laboratory, derived primarily from healthy volunteers in southern Minnesota and northern California is as follows: Adults: 10–50 ng/ml; 25–125 pmol/ml; children: 10–64 ng/ml; 25–160 pmol/ml.

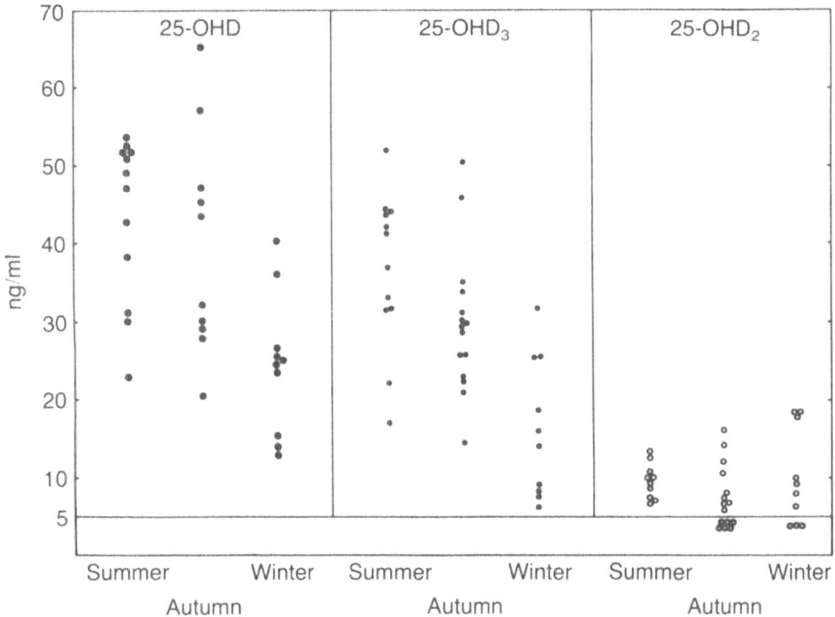

Figure 5. Serum concentrations of total 25-hydroxyvitamin D and its D_3 and D_2 derivatives in healthy children between the ages of 6 months and 6 years. The specific assays were performed by using 1 ml serum and long 0.9 × 50-cm columns to separate the two 25-hydroxyvitamin D compounds. These fractions were then assayed separately in the same manner as in the short-column assay, using radioactive 25-hydroxyvitamin D_3 to estimate the recovery of 25-hydroxyvitamin D_2.

Figure 6. Serum 25-hydroxyvitamin D_3 in patients with clinical, biochemical, and/or radiographic evidence of osteomalacia and rickets. The normal range of the bar graph in adults is derived from age-matched healthy controls who were sampled between October and April.

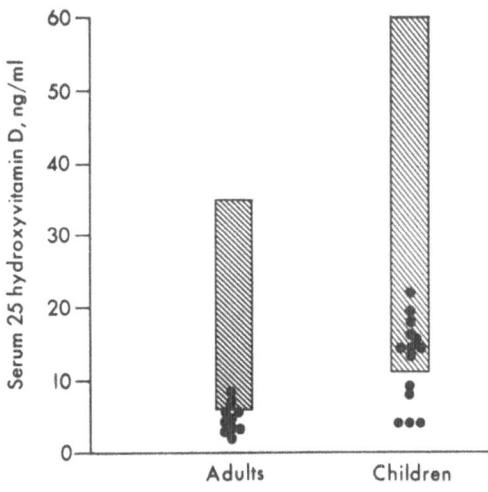

Abnormal Values

Values below the normal range are indicative of a low status of vitamin D nutrition. Low values do *not,* by themselves, indicate the presence of bone disease, calcium malabsorption, or a disturbance in any of the biologic activities of vitamin D carried out by the "hormonal" forms of the vitamin *unless* these reduced values are associated with hypocalcemia and/or hypophosphatemia. The result of the 25-hydroxyvitamin D assay is, most simply, an index of vitamin D reserve. Therefore, in the diagnosis of diseases associated with deficiency or excess of vitamin D, the assay is most importantly used as a confirmatory test to aid in separating the "nutritional" causes of the disorder in mineral metabolism in question from other metabolic abnormalities.

Concentrations in neonates are similiar to those in mothers at birth and are maintained in the first weeks of life except in premature infants who regularly show decreases of 2–5 ng/ml between 1 and 3 weeks of age.[12] At parturition, serum 25-hydroxyvitamin D is lower in the mother of the premature than of the full-term infant.[13] This observation and the high risk of the premature infant for the development of bone disease are sufficient indication for assessing the vitamin D status in mothers and their low-birth-weight infants by microassays geared to measure the metabolite in 1–200 μl serum. There is no evidence that neonatal hypocalcemia is etiologically related to serum 25-hydroxyvitamin D, but evaluation of the status of vitamin D nutrition in this common disorder of calcium homeostasis is helpful in providing a guide to vitamin D supplementation in hypocalcemic and sick newborns.

Assay results consistent with hypovitaminosis-D-induced osteomalacia or rickets are less than 10 ng/ml in adults and 15 ng/ml in children (see Figures 6 and 7). In the presence of hypocalcemia and/or hypophosphatemia, these

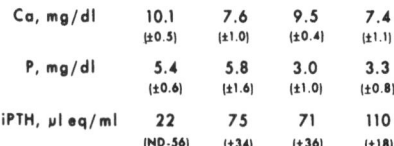

Ca, mg/dl	10.1	7.6	9.5	7.4
	(±0.5)	(±1.0)	(±0.4)	(±1.1)
P, mg/dl	5.4	5.8	3.0	3.3
	(±0.6)	(±1.6)	(±1.0)	(±0.8)
iPTH, μl eq/ml	22	75	71	110
	(ND-56)	(±34)	(±36)	(±18)

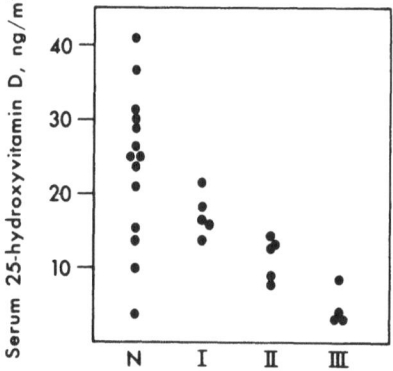

Figure 7. Relationship of 25-hydroxyvitamin D to mild *(stage I),* moderate *(stage II),* and severe *(stage III)* degrees of infantile vitamin D deficiency rickets. Infants were assigned to the different stages of vitamin D deficiency by the radiographic evidence of bone disease *(stages II and III)* or lack of it *(stage I).* Average values for calcium *(Ca),* phosphorus *(P),* and immunoreactive parathyroid hormone *(iPTH)* in the serum of infants in each group are shown above the graph. *N* = normal.

Table 2. Serum 25-Hydroxyvitamin D in Patients with
Gastrointestinal Disease Sampled between October and April

Diagnosis	N	Age in Years (average)	Serum 25-OHD (ng/ml)
Normal	14	50	22 ± 8
Nontropical sprue	15	49	9 ± 4[a]
Postgastrectomy	13	54	19 ± 10
Alcoholic cirrhosis	17	55	20 ± 14
Primary biliary cirrhosis	21	47	16 ± 12
Pancreatic carcinoma	21	61	19 ± 7

[a] $p < .05$ from normal

values are indicative of bone disease and are related to its severity. As illustrated in Figure 7, we found low normal values associated with hypocalcemia in the early stages of infantile rickets, with hypophosphatemia in moderately severe rickets, and below-normal serum 25-hydroxyvitamin D associated with decreases in both mineral concentrations in advanced stages of the disease.[14]

The measurement is of value in the evaluation of osteopenia and/or hypocalcemia in patients with hepatobiliary and gastrointestinal disorders. As illustrated in Table 2, in disorders of altered intestinal transit time, postgastrectomy syndrome and nontropical sprue, both of which are associated with an increased incidence of osteomalacia, only patients with nontropical sprue have reduced levels of serum 25-hydroxyvitamin D. Series of patients with regional enteritis, ileal resection in adults, and jejunoileal bypass are reported to have low values.[15-17] In the latter group, the low serum 25-hydroxyvitamin D was related to the frequency of stools and weight reduction. There is not general agreement about the levels of 25-hydroxyvitamin D in patients with liver disease. Normal and low values are reported in patients with alcoholic cirrhosis.[18,19] We have found reduced values regularly only in icteric patients whose primary liver disease is the result of cholestasis or hepatocellular injury. The greater the degree of hyperbilirubinemia, the lower the concentration of serum 25-hydroxyvitamin D ($r = -0.412$, $p < .05$). Considering the many factors which may contribute to the increased incidence of vitamin D deficiency in these patients (e.g., malabsorption of dietary vitamin D,[20] defective hepatic hydroxlation,[21] disturbance of enterohepatic recycling,[22] and increased urinary and fecal excretion[23] of 25-hydroxyvitamin D compounds), its measurement is an helpful adjunct to both treatment and diagnosis.

Drug therapy with anticonvulsants is commonly associated with reduced serum concentrations of 25-hydroxyvitamin D.[24] There is less evidence that treatment with corticosteroids causes reduced levels of this metabolite.[25]

Concentrations of 25-hydroxyvitamin D in chronic renal failure are usually normal, but may reveal associated nutritional deficiency of vitamin D in patients with hypocalcemia if reduced.[26] Reduced values are reported in patients with the nephrotic syndrome.[27]

Values consistent with treatment with pharmacologic doses of vitamin D not

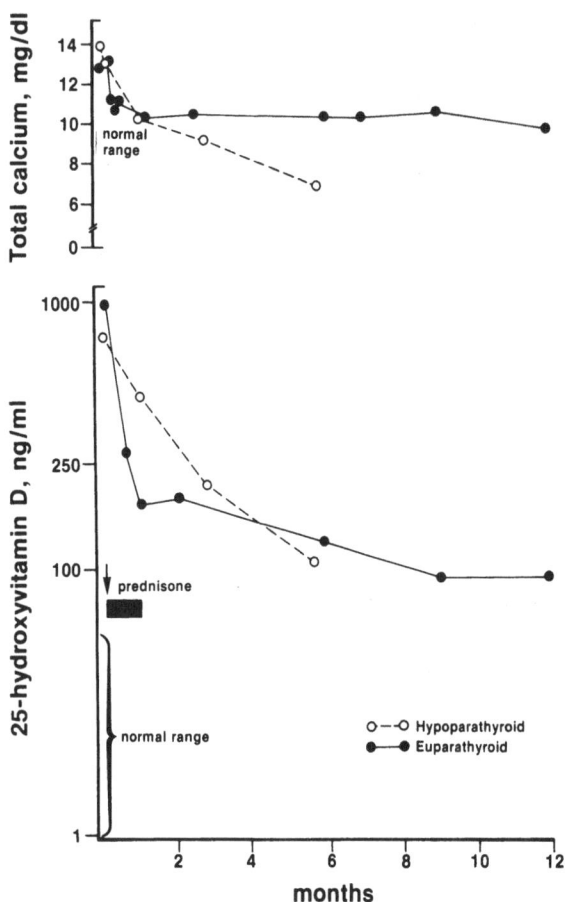

Figure 8. Comparison of the changes in total serum calcium and serum 25-hydroxy-vitamin D in two patients with hypervitaminosis D_2. One had surgical hypoparathyroidism (o) and the other was treated with vitamin D_2 for "rheumatism" (●). Both patients received prednisone treatment for their vitamin D intoxication for 1 month.

usually associated with hypercalcemia range from 100 to 400 ng/ml and are a function of the daily dosage of vitamin D.[28] Circulating values higher than 400 ng/ml without associated hypercalcemia were observed in 20% of Haddad's series of 41 patients receiving pharmacologic doses of vitamin D_2.[28] We have not, however, measured concentrations lower than this in hypercalcemic patients with hypervitaminosis D before treatment or in excess of 400 ng/ml in eucalcemic patients receiving pharmacologic doses of vitamin D. As illustrated in Figure 8, the decreases in serum 25-hydroxyvitamin D and serum calcium parallel one another in patients with hypervitaminosis D treated with prednisone. The toxicity of vitamin D is currently attributed to the increase in 25-hydroxyvitamin D since serum 1,25-dihydroxyvitamin D does not usually increase; however, toxicity may be due to pharmacologic effects of other vitamin D derivatives as well.[29] Normocalcemia occurs in hypoparathyroid patients when serum 25-hydroxyvitamin D levels are in the 200 ng/ml range and in patients with parathyroid glands intact when values reach the 100 ng/ml range. Based on these data, the experience of others, and the relative biologic potency of 25-hydroxyvitamin D, it appears[30] that concentrations ranging from 100 to 250 ng/ml are therapeutic in hypoparathyroid patients treated with pharmacologic doses of vitamin D_2 or 25-hydroxyvitamin D_3.

Measurements of 25-hydroxyvitamin D have been used to monitor the treatment of patients with renal osteodystrophy with 25-hydroxyvitamin D_3.[31] Although it is not yet known whether the effectiveness of treatment was due to the pharmacologic effects of 25-hydroxyvitamin D directly or to other vitamin D compounds, circulating levels from 100 to 250 ng/ml were achieved and maintained without the complications of vitamin D intoxication. Vitamin D therapy, monitored in this fashion, is important not only for adjusting dosage but also for determining patient compliance.

Acknowledgments This study was supported by grants from the Mayo Foundation and the National Institutes of Health AM12302 and AM19525

References

1. Haddad JG, Chyu KJ: Competitive protein-binding radioassay for 25-hydroxycholecaldiferol. J Clin Endocrinol Metab 33:992, 1971.
2. Belsey R, DeLuca HF, Potts JT Jr: Competitive binding assay for vitamin D and 25-OH vitamin D. J Clin Endocrinol Metab 33:554, 1971.
3. Bligh EG, Dyer WJ: A rapid method of total lipid extraction and purification. Can J Biochem 37:911, 1959.
4. Holick MF, DeLuca HF: A new chromatographic system for vitamin D_3 and its metabolites: resolution of a new vitamin D_3 metabolite. J Lipid Res 12:460, 1971.
5. Caldas AE, Gray RW, Lemann J: The simultaneous measurement of vitamin D metabolites in plasma: studies in healthy adults and in patients with calcium nephrolithiasis. J Lab Clin Med 94:840, 1978.
6. Horst RL, Littledike ET, Gray RW, Napoli JL: Impaired 24,25-dihydroxyvitamin D production in anephric human and pig. J Clin Invest 67:274, 1981.
7. Eisman JA, Hamstra AJ, Kream BE, DeLuca HF: A sensitive, precise, and convenient method for determination of 1,25-dihydroxyvitamin D in human plasma. Arch Biochem Biophys 176:235, 1976.
8. Haddad JG, Hahn TJ: Natural and synthetic sources of circulating 25-hydroxyvitamin D in man. Nature 244:515, 1973.
9. Arnaud SB, Matthusen BS, Gilkinson JB, Goldsmith RS: Components of 25-hydroxyvitamin D in serum of young children in upper midwestern states. Am J Clin Nutr 30:1082, 1977.
10. Eisman JA, Shephard RM, DeLuca HF: Determination of 25-hydroxyvitamin D_2

and 25-hydroxyvitamin D₃ in human plasma using high pressure liquid chromatography. Anal Biochem 80:298, 1977.

11. Stryd RP, Gilbertson TJ, Brunden MN: A seasonal variation study of 25-hydroxyvitamin D₃ serum levels in normal humans. J Clin Endocrinol Metab 78:771, 1979.

12. Hillman LS, Haddad JG: Perinatal vitamin D metabolism. II. Serial 25-hydroxyvitamin D concentrations in sera of term and premature infants. J Pediatr 86:928, 1975.

13. Rosen JF, Roginski M, Nathenson G, Finberg L: 25-hydroxyvitamin D: Plasma levels in mothers and their premature infants with neonatal hypocalcemia. Am J Dis Child 127:220, 1974.

14. Arnaud SB, Stickler GB, Haworth JC: Serum 25-hydroxyvitamin D in infantile rickets. Pediatrics 57:221, 1976.

15. Driscoll R, Meredith S, Wagonfeld J, Rosenberg I: Bone histology and vitamin D status in Crohn's Disease: assessment of vitamin D therapy. Gastroenterology 72:1050, 1977.

16. Compston JE, Creamer B: Plasma levels and intestinal absorption of 25-hydroxyvitamin D in patients with small bowel resection. Gut 18:171, 1977.

17. Hey H, Lund B, Sorensen OH, Lund BJ, Christensen MS: Impairment of vitamin D and bone metabolism in patients with bypass operation for obesity. Acta Med Scand [Suppl] 624:73, 1979.

18. Hepner GW, Roginski M, Moo HF: Abnormal vitamin D metabolism in patients with cirrhosis. Dig Dis 21:527, 1976.

19. Long RG, Meinhard E, Skinner RK, Varghese Z, Wills MR, Sherlock S: Clinical biochemical, and histologic studies of osteomalacia, osteoporosis, and parathyroid function in chronic liver disease. Gut 19:85, 1978.

20. Krawitt EL, Grundman MJ, Mawer EB: Absorption, hydroxylation, and excretion of vitamin D₃ in primary biliary cirrhosis. Lancet 2:1246, 1977.

21. Ponchon G, Kennan AL, DeLuca HF: "Activation" of vitamin D by the liver. J Clin Invest 48:2032, 1969.

22. Arnaud SB, Goldsmith RS, Lambert P, Go VLW: 25-hydroxyvitamin D₃: Evidence of an enterohepatic circulation in man. Proc Soc Exp Biol Med 149:570, 1975.

23. Jung RT, Davie M, Siklos P, Chalmers TM, Hunter JO, Lawson DEM: Vitamin D metabolism in acute and chronic cholestasis. Gut 20:840, 1979.

24. Hahn TJ, Hendin BA, Scharp CR, Haddad JG Jr: Effect of chronic anticonvulsant therapy on serum 25-hydroxycalciferol levels in adults. N Engl J Med 287:900, 1972.

25. Klein RG, Arnaud SB, Gallagher JC, DeLuca HF, Riggs BL: Intestinal calcium absorption in exogenous hypercortisolism; role of 25-hydroxyvitamin D and corticosteroid dose. J Clin Invest 60:253, 1977.

26. Eastwood JB, Stamp TCB, Harris E, de Wardener HE: Vitamin D deficiency in the osteomalacia of chronic renal failure. Lancet 2:1209, 1976.

27. Schmidt-Gayk H, Grawunder C, Tschope W, Schmitt W, Ritz E, Pietsch V, Andrassy K, Boullion R: 25-hydroxyvitamin D in nephrotic syndrome. Lancet 1:105, 1977.

28. Haddad JG, Stamp TCB: Circulating 25-hydroxyvitamin D in Man. Am J Med, Third F. Raymond Keating, Jr. Memorial Symposiupm, p. 185, 1974.

29. Hughes MR, Baylink DJ, Jones PG, Haussler MR. Radioligand receptor assay

for 25-hydroxyvitamin D_2/D_3 and la25dihydroxvitamin D_2/D_3. Application of hypervitaminosis D. J Clin Invest 58:61, 1976.

30. Kooh SW, Fraser D, DeLuca HF, Holick MF, Belsey RE, Clark MB, Murray TM: Treatment of hypoparathyroidism and pseudohypoparathyroidism with metabolites of vitamin D_3: evidence for impaired conversion of 25-hydroxyvitamin D to 1 alpha- dihydroxyvitamin D. N Engl J Med 293:840, 1975.

31. Recker R, Schoenfeld P, Letteri J, Slatopolski E, Goldsmith R, Brickman A. The efficacy of caldifediol in renal osteodystrophy. Arch Intern Med 138:857, 1978.

Chapter 5

A Method for Assay of Serum or Plasma Concentrations of 1,25-Dihydroxyvitamin D

Richard W. Gray

The development in Haussler's laboratory of a radioreceptor assay to measure plasma levels of $1,25(OH)_2$-vitamin D was a major advance in vitamin D technology.[1] It made possible not only a study of the regulation of $1,25(OH)_2$-D synthesis in vitamin-D-replete animals,[2] but also the understanding of the pathogenesis of a number of disease states involving altered vitamin D metabolism.[3-5] The observation that the receptor could be extracted from the mucosa under conditions of high ionic strength led to the use of cytosol preparations for the receptor and the use of polyethylene glycol to separate bound from free hormone.[3] This technique along with the advent of high-performance liquid chromatography (HPLC) greatly simplified the original filter assay which employed celite liquid/liquid chromatography for sample purification.[1] More recently the commercial availability of high-specific-activity $[^3H]$-$1,25(OH)_2$-vitamin D_3 and the use of hydroxylapatite or dextran-coated charcoal to separate bound from free hormone has made possible even greater sensitivity and has made it possible for a number of laboratories all over the world to develop assays for plasma levels of $1,25(OH)_2$-vitamin D.[6-11] The assay for plasma $1,25(OH)_2$-vitamin D described below is basically that of Eisman et al.[3] as modified by Caldas et al.[12] using the dextran-charcoal separation developed by Lambert et al.[13]

I. Procedure

A. Extraction

There are two commonly used extraction methods currently in use in our laboratory. The most universally applicable is the extraction with diethyl ether.[12] Aliquots of serum or plasma (1–10 ml) are placed in 50-ml glass centrifuge tubes with teflon-lined screw caps (Corex No. 8422-A). The samples are "spiked" with a small amount (1800–2000 dpm) of $[^3H]$-$1,25(OH)_2$ vitamin D_3 to monitor recovery through the extraction and purification procedure. The radioactivity in six replicate aliquots of the "spike" is averaged for use in calculating recovery. After incubation at room temperature for 10 min, 2 vol peroxide-free ether (Mallinckrodt No. 0848-1) are added. It is very important that the ether be peroxide-free, and the presence of peroxides can be tested for by the use of peroxide test strips (EM lab No. 10061-9G). The tubes are then capped and shaken on a horizontal shaker for 30 min. They are then centrifuged with the caps on for 10 min at 1000–1500 rpm in an IEC size 2 centrifuge to separate the phases. (We use a 16-place rotor for this centrifuge with double trunnion rings so that as many as 32 tubes can be spun at once.) After centrifugation, the upper phase is drawn off with a 25-ml serological pipette using a "propipette" and placed in a 125-ml separatory funnel (Kimax No. 29048F) containing 40 ml 0.1 M $NaPO_4$ buffer, pH 10.5 (alkaline buffer). The lower, aqueous phase remaining in the tube is then reextracted twice with ether as described above, the combined extracts shaken in the glass-stoppered separatory funnels, and the phases allowed to separate (this normally takes only a few minutes). These operations are all done under a fume hood. The separatory funnels are held in a 10-place rack (SP No. S9192-1). After separation of the phases, the lower aqueous phase is discarded and the ether phase is drained into a 250-ml boiling flask (SP F-4020-250).

The ether is evaporated on a rotary evaporator (Buchi Rotovapor-R with "V" stand SP No. E5047-1) using a water aspirator as a source of vacuum. As the ether evaporates, the extract may become cloudy because of small amounts of residual water. This can be removed by the addition of small amounts (5–10 ml) of absolute ethanol and further evaporation of this to dryness using a vacuum pump. (We use a rotary-vane pump with a separate explosion-proof motor which generates about 63.5–76.0 cm of mercury vacuum.) The sample is then quantitatively transferred with peroxide-free ether to a 13 × 100-mm glass tube. The ether is then evaporated under N_2 using a multiple evaporator (Organomation Ass. model No. 112, Northborough, Massachusetts, 01532, N-Evap analytical evaporator) and the extract is dissolved in 0.5 ml of 9:1:1 hexane/chloroform-methanol in preparation for the first chromatography step. The sample can, at this point, be stored overnight at 4°C.

The second extraction procedure involves a solid phase extraction using commercially available extraction columns from Analytichem International (Harbor City, California, Clin Elut). We use the largest size column (20 ml) so that sample sizes up to 10 ml can be extracted. Aliquots of the plasma or serum are "spiked" with ^{3}H-1,25(OH)$_2$-vitamin D_3 as described for the ether extraction. The sample is then loaded onto the extraction column which has already had at least 10 ml of 0.1 M phosphate buffer (pH 10.5) applied to it. The sample is then rinsed with alkaline buffer until the final volume applied to the column (buffer + sample) is 20 ml. The samples are then allowed to "soak" into the column for about 10–15 min, and an 18-gauge needle is placed on the column with a 250-ml boiling flash under it. The sample is then eluted with 50 ml of dichloromethane (DCM). After the solvent has run through, the column is plugged at the bottom and 20 ml of 20% acetone/DCM added. After 1 h, the plug is removed from the column and the solvent is allowed to drain into the flask: The column is then rinsed with two 20 ml-aliquots of the 20% acetone/DCM. The combined extracts are then evaporated to dryness with a rotary evaporator and vacuum pump and prepared for subsequent purification as described for the ether extraction. This method is much quicker than the ether extraction method and involves less glassware, but other vitamin D metabolites such as 25-OH-D may not be completely extracted by this method.

B. Sephadex LH-20 Chromatography

We routinely use Sephadex LH-20 chromatography to purify samples prior to HPLC even though it may be possible to eliminate this step when Clin-Elut columns are used for extraction and 2 ml or less of serum or plasma are extracted.[8,9] We use LH-20 because no two samples are alike with respect to their lipid composition, and this is certainly the most conservative way of sample purification. The solvent system used was developed in DeLuca's laboratory[3] and is a mixture of hexane/chloroform methanol in the ratio 9:1:1 v/v. This solvent mixture, because of its methanol content, imparts some gel filtration characteristics to the system so that more efficient removal of contaminating lipids is possible than with the chloroform/hexane (65:35) mixture formerly used for chromatography of vitamin D metabolites.[14] However, with the 9:1:1 mixture it is necessary to preswell the Sephadex several times before it can be packed into the column. In addition, it is helpful to add 1.5 ml distilled water per 1100 ml of the 9:1:1 mixture to saturate the system with water. This results in more consistent retention times for the vitamin D metabolites under conditions of varying humidity (R. Horst, personal communication). The columns used are 9 mm in diameter and 150 mm high with a 100-ml bulb reservoir at the top and a Teflon step-cock at the bottom (Glenco Scientific, Houston, Texas, No. 3126-093). Several days prior to use, 20-g aliquots of Sephadex LH-20 are swelled in an excess amount of the water-saturated 9:1:1 mixture in a 250-ml glass-stoppered Ehrlenmeyer flask. This mixture is allowed to stand for at least 1 h. The solvent is then decanted, and the LH-20 is swelled again with more solvent and allowed to stand overnight. Periodically on the following day, the LH-20 is again reequilibrated three or four times before being packed into the columns. The columns are placed under a hood, plugged with a small amount of glass wool, and 1–2 ml of the 9:1:1 mixture added to each column. The Sephadex is then added using a 25-ml serologic pipette with a "propipette" control. The LH-20 is packed to a height of 10.5 cm from the stop-cock. If the LH-20 is not adequately equilibrated, bubbles will form in the column, and the column must be repacked. When packed in this way, each column contains about 2.8 gm Sephadex.

The LH-20 column is calibrated with [³H]-25(OH)D₃ and [³H]-1,25(OH)₂D₃. These standards should be applied and eluted exactly as described below for the samples except that 1-ml fractions of the eluate are collected so that the elution volume of each metabolite can be determined. Once the calibration has been established, column calibration need not regularly be performed. Each LH-20 column is used for only a single sample.

The sample is applied in 0.5 ml 9:1:1 hexane/chloroform/methanol and rinsed on with two 0.5-ml portions of the solvent; 60 ml of solvent are then added to the reservoir. The first 20 ml of eluate is discarded. The remaining volume is collected in a 125-ml boiling flask. The sample is evaporated to dryness with the Rotovapor and vacuum pump, transferred to a glass tube with ether, and then evaporated to dryness under N_2. The sample may be stored overnight at this point in 0.5 ml HPLC solvent.

C. HPLC

The final purification of the sample must include HPLC, not so much to remove other vitamin D metabolites (this could be done with LH-20) but to remove small amounts of lipid contaminants which interfere with the receptor assay for 1,25(OH)$_2$D. Sample purification by HPLC is routinely performed using a microparticulate silica column, Dupont Zorbax-Sil, 6.2 mm × 25 cm. Other investigators also use a smaller Zorbax-Sil column, 4.6 mm × 25 cm.[15] The particular type of chromatograph used is not critical. We initially used a Glenco Scientific HPLC System I employing a Milton-Roy pump and Valco loop injector. This simple apparatus is clearly sufficient for purification of small numbers of samples (8–12/day). More recently we have used a Waters Model 204 HPLC in conjunction with an automatic sample injector (WISP) and microprocessor-controlled fraction collector for automatic injection and collection of a larger number of samples (35–40 samples).[9]

The retention time and recovery of 1,25(OH)$_2$D on the HPLC column must be routinely checked before each series of samples using the same batch of eluting solvent mixture.

The polarity of the eluting solvent mixture (10–12% isopropanol in hexane) and the column flow rate (1.5–2 ml/min) are adjusted to obtain a retention time for 1,25(OH)$_2$D of 18–20 minutes. The column is calibrated with 1800–2000 dpm [^{3}H]-1,25(OH)$_2$-D to avoid loading the column with large quantities of unlabeled 1,25(OH)$_2$D. One-half minute fractions of the column eluate are sequentially collected, toluene-counting cocktail added, and radioactivity determined to establish retention time. The recovery of the labeled 1,25(OH)$_2$D should be 90% or better and, if not, it should be repurified using a separate HPLC column.

Using the Glenco HPLC, the samples are dissolved in 100 μl of the HPLC solvent and injected manually into the sample loop. Using the WISP, the samples are dissolved in 100 μl of the HPLC solvent and transferred to microvolume insert vials using a 50-μl rinse. The WISP is programmed to inject a 150-μl sample. For each unknown sample, column eluate from 2 min before to 2 min after the peak [^{3}H]-1,25(OH)$_2$D retention time is collected in a 16 × 125-mm glass tube. Using the Glenco HPLC the time of this fraction is determined with a timer. Since the retention time for 1,25(OH)$_2$D$_3$ in this system is about 18–20 min, 16–18 min of column eluate must be discarded before the 4-min fraction is collected for the sample. Using the WISP, the 1,25(OH)$_2$D fraction is collected automatically using the microprocessor. Since the tubes in the fraction collector will hold only about 8–9 min of column eluate, two fractions must be collected before the 4-min sample collection. In addition, a 6–8-min rinse is discarded after the samples and before injection of the next sample. Therefore, the total run time for each sample is about 26–28 min. This amounts to at least 18 h of HPLC time for an assay involving 36 samples. After the samples have been collected, they can be corked and stored up to 24 h at 4°C.

D. Potential Problems with the WISP

1. It is possible, if left unattended, for the WISP to inject all of the samples into the waste line. This can happen when there is an increase in back pressure due to a plug in the sample line and the pressure limit on the HPLC pump is exceeded, automatically shutting off the pump. When the HPLC pump is shut off, the WISP will continue to run and inject all of the samples into the waste line. This can be prevented by purchasing a System Controller from Waters, Inc. (approximately $7,000) which will interface the HPLC with the WISP. Alternatively, one can purchase a relay from Radio Shack (approximately $4.50) which can be wired into the pump pressure limit circuitry. This relay will then act to shut the system down in case of a buildup in back pressure.
2. It is also possible to use a second relay to shut the pump off when the WISP is finished injecting all of the samples so that solvent will not continue to be pumped through the system after the run is over. Consult your local Waters technical representative for details on how to set up the relays.
3. Because of the relatively large volume injected, the injection syringe on the WISP may need periodic lubrication.
4. Finally, the HPLC pump should also be serviced on a regular basis to insure uniform collection of all the samples during the run.

E. Radioceptor Assay

The HPLC solvent is evaporated under N_2 using the N-Evap and the sample taken up in exactly 1 ml of absolute ethanol. The resulting solution is drawn up into a Hamilton micro-lab-P programmable pipettor. Duplicate 150-μl aliquots are pipetted into scintillation vials to monitor recovery of the added "spike" through the extraction and purification procedure. Triplicate 200-μl aliquots are then pipetted into 12 × 75-mm polystyrene "Lancer" tubes (SP No. T1225-2). The ethanol is evaporated to dryness using the N-Evap and 50 μl of ethanol added to each tube.

F. Standards

Pure, crystalline 1,25(OH)$_2$D suitable for use as a standard can be obtained from Hoffmann-LaRoche. A concentrated 1,25(OH)$_2$D$_3$ solution containing approximately 1 mg/ml is made up in absolute ethanol. This solution can be stored for long periods of time at $-70°C$. A 20-μl aliquot of this solution is diluted to 2 ml in absolute ethanol to obtain a concentrated standard containing about 10 $\mu g/ml$. The exact concentration is determined by measuring the absorption maximum at 264 nm and calculating the concentration from the extinction coefficient using the equation $\mu g/ml = (MW \times Abs)/18.3$ where MW equals the molecular weight of the vitamin D metabolite, Abs equals the maximum absorbance at 264 nm, and 18.3 mM^{-1} cm^{-1} is the extinction coefficient for 1,25(OH)$_2$-vitamin D$_3$[16]. From this standard, a 200-fold dilution is made to obtain a working stock (approximately 50 ng/ml) solution which is further diluted to obtain a series of standard solutions for the standard curve as follows:

Volume of 50 ng/ml Working Stock (ml)	Q.S. to Volume with Abs. ETOH (ml)	pg 1,25(OH)$_2$D$_3$ 50 μl
0.6	10	150
0.4	10	100
0.25	10	62
0.4	25	40
0.2	25	20
0.1	25	10
0.05	25	5

Fifty-microliter aliquots of each of these standards are pipetted into assay tubes ("Lancer" tubes, see above) in triplicate. A 100-fold dilution of the 10 $\mu g/ml$ standard is also made (NSB solution) and a 50-μl aliquots of this solution are pipetted into assay tubes to obtain a measure of nonspecific binding (NSB) in the presence of excess unlabeled 1,25(OH)$_2$D$_3$ (5 ng/tube).

To each assay tube are added 50 μl of absolute ethanol for samples after evaporation as described above, 50 μl of the appropriate standard (or NSB solution), or 50 μl of ethanol alone for maximum binding ("Bo"). In addition, all of the tubes for the standard curve contain column eluate added in 50 μl of ethanol and evaporated to dryness (see below). Radioactive 1,25(OH)$_2$D$_3$ "trace" (about 6000 dpm/tube) is then added to each tube in 40 μl of ethanol. The radioactive 1,25(OH)$_2$D can be obtained from either New England Nuclear or Amersham and should have a radiochemical purity greater than 95% (see below). The assay tubes are then placed in an ice bath and 1 ml of reconstituted chick intestinal cytosol protein (containing about 0.3 mg of protein, see below) is added to each tube. The tubes are then vortexed and placed in a shaker at room temperature for 1 h. They are then returned to the ice bath and 0.3 ml of dextran-coated charcoal suspension (see below) is added to each. After addition of the charcoal, the tubes are vortexed and kept in the ice bath for 30 min, being vortexed at 10-min intervals. The tubes are then centrifuged in a refrigerated centrifuge at 2000 g for 20 min (Beckman J-6B centrifuge with JR-3.2 rotor), after which they are removed and the supernatant poured

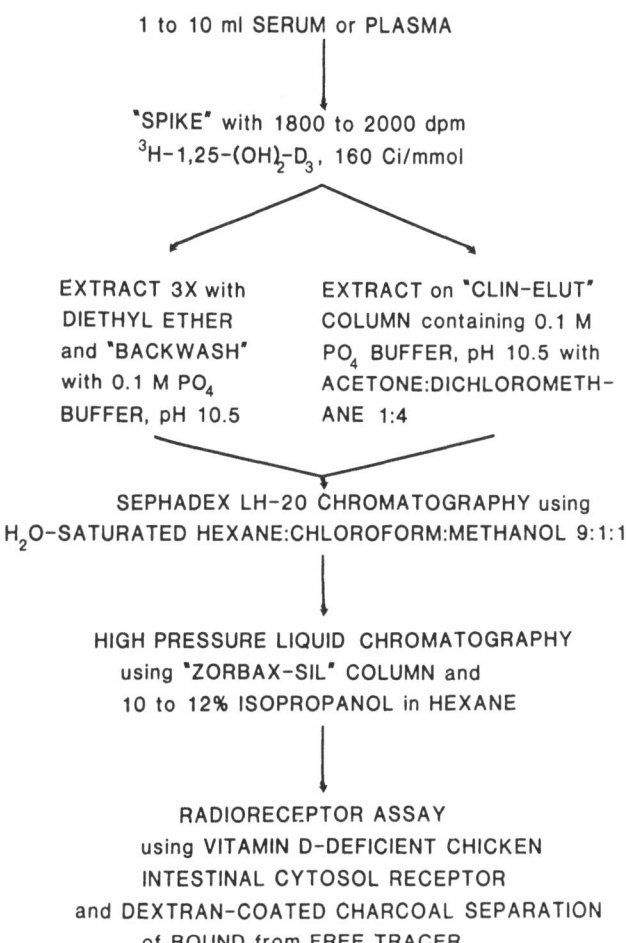

1 to 10 ml SERUM or PLASMA

"SPIKE" with 1800 to 2000 dpm
^{3}H-1,25-(OH)$_2$-D$_3$, 160 Ci/mmol

EXTRACT 3X with
DIETHYL ETHER
and "BACKWASH"
with 0.1 M PO$_4$
BUFFER, pH 10.5

EXTRACT on "CLIN-ELUT"
COLUMN containing 0.1 M
PO$_4$ BUFFER, pH 10.5 with
ACETONE:DICHLOROMETH-
ANE 1:4

SEPHADEX LH-20 CHROMATOGRAPHY using
H$_2$O-SATURATED HEXANE:CHLOROFORM:METHANOL 9:1:1

HIGH PRESSURE LIQUID CHROMATOGRAPHY
using "ZORBAX-SIL" COLUMN and
10 to 12% ISOPROPANOL in HEXANE

RADIORECEPTOR ASSAY
using VITAMIN D-DEFICIENT CHICKEN
INTESTINAL CYTOSOL RECEPTOR
and DEXTRAN-COATED CHARCOAL SEPARATION
of BOUND from FREE TRACER

Figure 1. Summary of technique.

off into 20-ml scintillation vials containing 10 ml of Scintisol (Isolab). Radio-activity is determined in the samples and standards by liquid scintillation counting with external standard correction for efficiency. Toluene cocktail is used to determine radioactivity in the sample aliquots taken for recovery.

The flow diagram in Figure 1 summarizes the basic steps described above.

G. Calculations

Recovery of the added "spike" for each sample is calculated by dividing the total recovered dpm (observed dpm/0.15) by the total dpm initially added (this usually ranges from 50 to 60% for most serum samples).[12] The standard curve is calculated by standard radioimmunoassay (RIA) techniques to obtain "B/Bo," which is then plotted as a function of the pg of 1,25(OH)$_2$D$_3$ per tube. The term "B/Bo" is obtained by dividing the dpm specifically bound for the

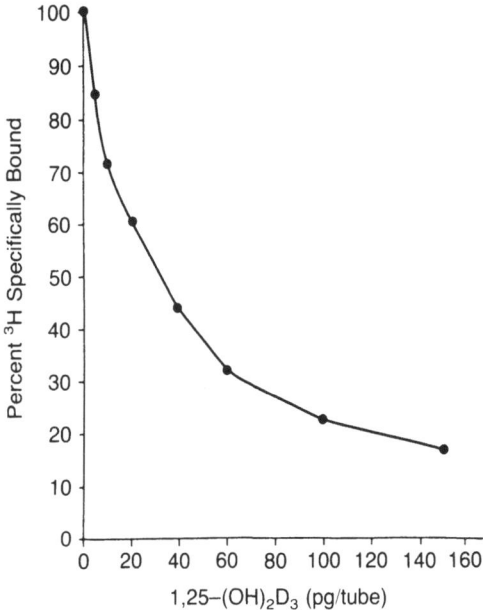

Figure 2. Representative standard curve.

sample (total − NSB) by the dpm specifically bound in the absence of any unlabeled 1,25(OH)$_2$D$_3$ (Bo − NSB) and expressing the result as a percentage. The resulting standard curve can then be fitted using a polynomial regression program or a logit transformation to read the unknowns. A representative standard curve is shown in Figure 2. The final 1,25(OH)$_2$D concentration in pmol/l can thus be calculated from the following equation:

$$pmol/l = \frac{pg/sample \times 5}{\text{fractional recovery} \times \text{ml plasma extracted} \\ \times (416\ pg/pmol/1000\ ml/L)}$$

For human plasma samples, the total 1,25(OH)$_2$D concentration represents both 1,25(OH)$_2$D$_2$ and 1,25(OH)$_2$D$_3$ since both of these metabolites comigrate very closely on HPLC[6,15] and both are equally recognized by the cytosol-binding protein.[3]

H. Effect of Solvents on the Standard Curve

Because large amounts of organic solvents are used in purifying the 1,25(OH)$_2$D$_3$ prior to assay, it is necessary to determine whether these solvents have any effect on the standard curve. This can be accomplished by running standard curves in the presence of an equivalent amount of HPLC column eluate used to purify the samples. For example, if a 4-min fraction is collected from the HPLC for the sample and 25% aliquots of the sample are taken for analysis, then 1 min of column eluate solvent is present in each sample assay tube. An equivalent amount of solvent can also be added to all of the tubes in the standard curve by adding 1 min of column eluate evaporated to dryness and dissolved in 50 μl ethanol to each tube and evaporating the ethanol to dryness before running the standard curve. Although standard curves run in the presence and absence of column eluate are not significantly different from one another (data not shown), small differences are sometimes apparent. For this reason we routinely include column eluate in each standard curve to insure that the standards are determined under exactly the same conditions as the samples.

I. Validating the Assay

Recovery of $1,25(OH)_2D_3$ Applied Directly to the HPLC

Fourteen replicate 332-pg aliquots of $1,25(OH)_2D_3$ were "spiked" with 1800 dpm of $[^3H]$-$1,25(OH)_2D_3$. Each sample was individually loaded into a microinsert vial and chromatographed on HPLC using the WISP to apply the samples. Recovery of radioactivity in the $1,25(OH)_2D_3$ fractions from the HPLC averaged $85.4 \pm 7.5\%$ (n = 14) and radioreceptor assay of each sample resulted in an average recovery of 329 ± 22 pg or $99.0 \pm 6.5\%$ of the $1,25(OH)_2D_3$ applied to the column.

Recovery of $1,25(OH)_2D_3$ Added to Serum and Carried Through the Entire Extraction, Chromatography, and Assay Procedure

Eight replicate 332-pg aliquots of $1,25(OH)_2D_3$ were added to 10-ml aliquots of a human plasma pool and carried through the entire assay procedure. In three separate assays, recovery of the added $1,25(OH)_2D_3$ averaged 342 ± 74 pg or $103 \pm 22\%$.

J. Intra- and Interassay Coefficients of Variation

In two recent assays, replicate measurement of $1,25(OH)_2D_3$ in a human plasma pool showed intrassay coefficients of variation of 11% (n = 7) and 9% (n = 4).

From October 1980 through May 1982, we have evaluated the interassay coefficient of variation (CV) in three human plasma pools. The mean $1,25(OH)_2D$ concentrations were

Pool 1 (17 assays) 84 ± 15 pmol/l, CV = 18%
Pool 2 (18 assays) 74 ± 14 pmol/l, CV = 19%
Pool 3 (19 assays) 326 ± 59 pmol/l, CV = 18%

II. Materials

A. Glassware

All glassware must be scrupulously cleaned. We routinely scrub all glassware in Alconox (SP C-6301-3), rinse in water, soak in Contrad-70 (C & M 117-655) for at least 2 h, rinse in water, and then rinse in deionized water (do not use acid-washed glassware). After drying, all glassware is rinsed in methanol (Mallinkrodt Nanograde).

B. Reagents

For HPLC purification of the sample the highest-purity HPLC-grade solvents must be used, for example, Burdik-Jackson, Fisher HPLC grade, or Mallinkrodt Chrom AR. For LH-20 chromatography slightly cheaper solvents can be used, such as Mallinkrodt Nanograde.

For extraction we use nanograde dichloromethane and AR-grade acetone. As mentioned above, all ether must be anhydrous and peroxide-free. We use Mallinkrodt anhydrous ether with preservative to prevent peroxide formation. Every can must be tested with peroxide strips before use. (We occasionally find a can that is peroxide-positive.)

Radioactive $1,25(OH)_2D_3$ of the highest specific radioactivity available should be used, and its radiochemical purity should be checked before use. We use $[26(27)-methyl-^3H]-1,25(OH)_2$-vitamin D_3 obtained from Amersham or New England Nuclear with a specific activity ranging from 140 to 160 Ci/mmol. It may be necessary after a period of several weeks to repurify the $[^3H]$-$1,25(OH)_2D_3$. This should be done with a separate Zorbax Sil column kept solely for that purpose.

C. Preparation of Chick Intestine Mucosa Cytosol

Chickens

One-day-old white leghorn cockerels can be obtained from a local commercial chicken hatchery. Once received in the animal facility, they can be housed in stainless steel chick brooders and kept in a separate room with an incandescent light bulb as the only source of light. They are fed a commercially available vitamin D-deficient-diet (Rachitogenic Chick Diet, ICN, Cleveland, Ohio) for at least 3–4 weeks.

Preparation of Cytosol

At the time of sacrifice, the duodenal loops are removed and flushed with ice-cold phosphate-KCl buffer (0.5 M PO_4, 0.1 M KC, 1.0 mM dithioerythritol, pH 7.4). All subsequent operations are carried out in a cold room at 4°C. The loops are slit open lengthwise and the mucosa scraped from the serosa using a microscope slide. The mucosal scrapings are collected in a preweighed plastic weighing boat, and when all of the mucosa has been collected, the total wet weight is determined using a top-loading balance. The mucosa is then placed in a 400-ml beaker with 5 vol (v/w) of the PO_4-KCl buffer and is gently stirred with a teflon-coated rod. The suspension is then divided equally into 3–4 50-ml polycarbonate centrifuge tubes and centrifuged in a Sorvall centrifuge (SS-34 rotor) at 1000 rpm for 10 min.

The supernatant is carefully decanted, the tubes are filled with PO_4-KCl buffer, the pellet gently resuspended with the stirring rod, and the tubes centrifuged again as above. After being decanted a second time, the mucosal pellets are combined and homogenized in 1.8 vol PO_4-KCl buffer using a Potter-Elvehjem homogenizer with a teflon pestle. The resulting homogenate is then divided into polycarbonate ultracentrifuge tubes for the Beckman 70.1 Ti rotor. The tubes are then centrifuged at 2–3° C in the 70 Ti rotor at 120,000 g for 1 h. It is very important to keep the mucosa and the mucosal homogenate from warming up at any point during the isolation procedure. On the other hand, the homogenate must not be allowed to freeze. This also destroys the receptor. After centrifugation, the tubes are carefully removed from the rotor, and the lipid floating on top is sucked off with a disposable pipette. The supernatant is then decanted into a glass-stoppered tube kept in an ice bath, and 4-ml aliquots are pipetted into 20 × 185-mm screw cap vials for lyophylization. Aliquots of the cytosol are also taken for Lowry protein determination. (Protein content normally ranges from 10 to 14 mg/ml depending on the amount of buffer used for homogenization and how vigorously the mucosa is washed before it is homogenized.) The cytosol is shell-frozen in the vials using liquid nitrogen and lyophylized for at least 5 h but not longer than 6 h. The vials are then removed, flushed for 30 s with a stream of nitrogen, capped, and stored at −70° C. Cytosol prepared and stored in this way is suitable for use in the radioreceptor assay up to 12 weeks after the initial isolation.

The optimal protein concentration for the radioreceptor assay is determined by running a series of "mini" standard curves at different protein concentrations, for example 0.25, 0.3, and 0.4 mg protein/ml of diluted cytosol. The cytosol is reconstituted by adding 4 ml of deionized water to the vial and diluting to the appropriate volume with PO_4-KCl buffer to obtain the desired protein concentration. The abbreviated standard curve (see above) contains the 5-, 10-, 40-, and 60-pg standards in addition to the "zero-binding" and NSB tubes. A protein concentration is chosen that will distinguish the 5-pg standard from zero and that will bind 40–60% of the added trace $1,25\text{-}(OH)_2D_3$. (If these conditions cannot be met, more chicks must be obtained and another batch of cytosol prepared.) For most cytosol preparations the optimal protein concentration per tube is usually 0.25–0.4 mg protein.

D. Dextran-coated Charcoal

Neutralized activated charcoal is obtained from Sigma, washed several times with deionized water to remove the "fines," and dried. Dextran T-70 is obtained from Pharmacia. A 0.5% charcoal/0.5% dextran (w/v) suspension is made in 0.05 M $NaPO_4$ buffer, pH 7.4. After being stirred for 10–15 min, the mixture is centrifuged at 250 g for 10 min at 0–5°C and resuspended in one-half the original volume of buffer containing 5% human plasma to yield a 1.0% charcoal/0.1% dextran/5% plasma concentration. This mixture is stirred overnight at 4°C, centrifuged at 250 g for 10 min, and resuspended in the original volume of 0.05 M PO_4 buffer to again yield a 0.5% charcoal/0.05% dextran suspension. For the radioreceptor assay, this suspension is continuously stirred in an ice bath, and 0.3-ml aliquots are added to each of the assay tubes.

E. Application of the Assay

Table 1 summarizes some of our measurements of serum or plasma $1,25(OH)_2D$ concentrations in humans and lower animals. In healthy adult humans eating normal self-selected diets, plasma $1,25(OH)_2D$ levels range from 40 to 140 pmol/l (mean $\pm$ 2 SD). These values are similar to those reported from a number of other laboratories.[1,3,6,11] We have been unable to detect either diurnal or long-term variation in serum $1,25(OH)_2D$ levels when repetitive measurements are made in the same subjects.[20,23,30] We have also been unable to detect a significant difference in plasma $1,25(OH)_2D$ levels in

Table 1. Plasma $1,25-(OH)_2D$ Concentrations in Humans and Animals

Subjects	Number	$1,25-(OH)_2D$ pmol/l[a]	References
Humans			
Healthy adults, age 20–50 yrs eating normal diets	97	89 ± 25	17
Healthy adults, age 20–40 yrs, eating diets providing only 4 mmol Ca/day for 4–20 days	16	106 ± 20	18,19
Healthy adults, age 20–40 given $CaCO_3$, 0.85–1.5 mmol/kg/day for 4 days	17	37 ± 23	20
Calcium nephrolithiasis	93	124 ± 47	21
Primary hyperparathyroidism	14	170 ± 60	22
Chronic renal disease creatinine 5–40 ml/min	8	48 ± 10	23
Anephrics	24	< 10	20
Rats			
Males, 6 weeks	40	228 ± 76	24
Females, 6 weeks	17	146 ± 44	24
Males, 6 months	9	105 ± 41	24
Females, 6 months	4	101 ± 11	24
Males, 2 years	10	41 ± 10	25
Mice			
Males and females, 13 weeks	8	70 ± 4	26
Chickens			
Cockerels, 6 weeks old, on normal diet	9	146 ± 41	19
Cockerels, 6 weeks old, on low PO_4 diet for 2–15 days	12	363 ± 74	19
Dogs			
Females: adult mongrels	3	83 ± 14	19

[a]Mean $\pm$ SD

healthy women as compared to men: women (n = 37) 86 ± 29 pmol/l, men (n = 54) 88 ± 26 pmol/l (not significant). Plasma 1,25(OH)$_2$D levels rise by about 20% in response to dietary calcium restriction (p < .025) and fall rapidly during calcium loading.[23] Patients with calcium nephrolithiasis and patients with primary hyperparathyroidism often exhibit elevated plasma 1,25(OH)$_2$D levels,[24,25] while plasma levels of the hormone fall progressively with advanced renal failure.[26]

In adult rats, mice, chickens, and dogs, plasma 1,25(OH)$_2$D levels are similar to those in humans. Young male rats tend to have higher plasma 1,25(OH)$_2$D levels than young females.[27] Plasma 1,25(OH)$_2$D levels fall in aged rats.[29] When chickens are fed a low-phosphate diet, plasma 1,25(OH)$_2$D levels rise as has been observed in rats,[2,27] mice,[29] and healthy women.[4]

References

1. Brumbaugh PF, Haussler DH, Bursac KM, Haussler MR: Filter assay for 1α,25-dihydroxyvitamin D$_3$. Utilization of the hormone's target tissue chromatin receptor. Biochemistry 13:4091–4096, 1974.
2. Hughes MR, Haussler MR, Wergedal J, Baylink DJ: Regulation of serum 1α,25-dihydroxyvitamin D$_3$ by calcium and phosphate in the rat. Science 190:578–580, 1975.
3. Eisman JA, Hamstra AJ, Kream BE, DeLuca HF: 1,25-dihydroxyvitamin D in biological fluids: a simplified and sensitive assay. Science 193:1021–1023, 1976.
4. Gray RW, Wilz DR, Caldas AE, Lemann J Jr: The importance of phosphate in regulating plasma 1,25-(OH)$_2$-vitamin D levels in humans: studies in healthy subjects, in calcium-stone formers and in patients with primary hyperparathyroidism. J Clin Endocrinol Metab 45:299–306, 1977.
5. Haussler MR, McCain TA: Basic and clinical concepts related to vitamin D metabolism and action. N Eng J Med 297:1041–1050, 1977.
6. Lambert PW, Syverson BJ, Arnaud CD, Spelsberg TC: Isolation and quantitation of endogenous vitamin D and its physiologically important metabolites in human plasma by high pressure liquid chromatography. J Steroid Biochem 8:929–937, 1977.
7. Lund BJ, Lund BI, Sorenson OH: Measurement of circulating 1,25-dihydroxyvitamin D in man. Changes in serum concentrations during treatment with 1α-hydroxycholecalciferol, Acta Endocrinol (Copenh) 91:338–350, 1979.
8. Mason RS, Lissner D, Grunstein HS, Posen S: A simplified assay for dihydroxylated vitamin D metabolites in human serum: applications to hyper- and hypovitaminosis D. Clin Chem 26:444–450, 1980.
9. Bishop JE, Norman AW, Coburn JW, Roberts PA, Henry HL: Studies on the metabolism of calciferol XVI. Determination of the concentration of 25-hydroxyvitamin D, 24,25-dihydroxyvitamin D and 1,25-dihydroxyvitamin D in a single two-milliliter plasma sample. Min Elect Metab 3:181–189, 1980.
10. Taylor CM, Hann J, St. John J, Wallace JE, Mawer EB: 1,25-dihydroxycholecalciferol in human serum and its relationship with other metabolites of vitamin D$_3$. Clin Chem Acta 96:1–8, 1979.
11. Yamaoka K, Seino Y, Ishida M, Ishii T, Shimotsuji T, Tanaka Y, Kurose H, Matsuda S, Satomura K, Yabuvchi H: Effect of dibutyryl adenosine 3′,5′-mono-

phosphate administration on plasma concentrations of 1,25-dihydroxyvitamin D in pseudohypoparathyroidism type I. J Clin Endocrinol Metab 53:1096–1100, 1981.

12. Caldas AE, Gray RW, Lemann J Jr: The simultaneous measurement of vitamin D metabolites in plasma: studies in healthy adults and in patients with calcium nephrolithiasis. J Lab Clin Med 91:840–849, 1978.

13. Lambert PW, Taft DO, Hodgson SF, Lindmark EA, Witrak BJ, Roos BA: An improved method for the measurement of 1,25-$(OH)_2$-D_3 in human plasma. Endocrin Res Commun 5:293–310, 1978.

14. Holick MF, Deluca HF: A new chromatographic system for vitamin D_3 and its metabolites: resolution of a new vitamin D_3 metabolite. J Lipid Res 12:460–465, 1971.

15. Shepard RM, Horst RL, Hamstra AJ, DeLuca HF: Determination of vitamin D and its metabolites in plasma from normal and anephric man. Biochem J 182:55–69, 1979.

16. Norman AW: Biological and chemical assay of vitamin D, its metabolites, and analogs. In Norman AW (Ed.): Vitamin D: The Calcium Homeostatic Steroid Hormone. Academic Press, New York, p. 85, 1979.

17. Gray RW, Lemann J Jr, Adams ND: The regulation of plasma 1,25$(OH)_2$-D concentrations in healthy adults: In Norman AW (Ed.): Vitamin D: Basic Research and its Clinical Application. Walter de Gruyter, Berlin, pp. 545–551, 1979.

18. Adams ND, Gray RW, Lemann J Jr, Cheung HS: Effects of calcitriol administration on calcium metabolism in healthy men. Kidney Int 21:90–97, 1982.

19. Gray RW, Lemann J Jr: Unpublished findings.

20. Adams ND, Gray RW, Lemann J Jr: The effects of oral $CaCO_3$ loading and of dietary calcium deprivation on plasma 1,25$(OH)_2$-vitamin D concentrations in healthy adults. J Clin Endocrinol Metab 48:1008–1016, 1979.

21. Lemann J Jr, Gray RW, Adams ND: Vitamin D metabolism in patients with nephrolithiasis. In AW Norman (Ed.), Vitamin D: Basic Research and its Clinical Application. Walter de Gruyter, Berlin, pp. 957–961, 1979.

22. Maierhofer WJ, Gray RW, Adams ND, Smith GD, Lemann J Jr: Synthesis and metabolic clearance of 1,25-Dihydroxyvitamin D as determinants of serum concentrations: a comparison of two methods. J Clin Endocrinol Metab 53:472–475, 1981.

23. Slatopolsky E, Gray RW, Adams ND, Lewis J, Hruska K, Martin K, Klahr S, DeLuca WF, Lemann J Jr: Low serum levels of 1,25$(OH)_2$-D_3 are not responsible for the development of secondary hyperparathyroidism in early renal failure. Kidney Int 14:733, 1978.

24. Gray RW: The effects of age and sex on the regulation of plasma 1,25$(OH)_2$-D by phosphorus in the rat. Calcif Tissue Int 33:477–484, 1981.

25. Gray RW, Gambert SR: Effect of age on plasma 1,25$(OH)_2$-vitamin D in the rat. Age 5:54–56, 1982.

26. Meyer RA Jr, Gray RW, Meyer MH: Abnormal vitamin D metabolism in the x-linked hypophosphatemic mouse. Endocrinology 107:1577–1581, 1980.

Chapter 6

Assay for Multiple Vitamin D Metabolites

Phillip W. Lambert, Irene Y. Fu,
David M. Kaetzel, and Bruce W. Hollis

During the past decade there have been major advances in our knowledge of vitamin D metabolism and its role in both physiologic and pathophysiologic situations of calcium (Ca) homeostasis. Critical to these advances has been the development of ligand-binding assays for a wide spectrum of vitamin D_3 metabolites. Separate ligand-binding assays have been reported for vitamin D,[1,2] 25-OHD,[3-11] 24,25-$(OH)_2D$,[12-15] 25,26-$(OH)_2D$,[16-18] and 1,25-$(OH)_2D$.[19-35] More recently, methodology has evolved for the measurement of two or more vitamin D metabolites in a single, relatively small plasma or serum sample.[23,28,34-42]

Recent studies have shown that with specific high-performance liquid chromatography (HPLC) systems certain vitamin D_2 metabolites may elute separately from their counterparts and/or possibly coelute with other vitamin D_3 metabolites.[41,42] In addition, it has been demonstrated that some vitamin D_2 and D_3 metabolites have differential affinity for their specific assay-binding protein.[41,42] Thus, it is possible that plasma vitamin D metabolite levels (i.e., vitamin D_2 and D_3) could be under- or overestimated with our methodology. Resolution of this specific question would require the examination of the retention time of the various vitamin D_2 metabolites on the HPLC solvent gradient system described in this paper. The biosynthesis and purification of these latter metabolites is currently in progress in our laboratory.

In this chapter we outline the methodology for the determination of vitamin D_3, 25-OHD$_3$, 24,25-$(OH)_2D_3$, 25,26-$(OH)_2D_3$, and 1,25-$(OH)_2D_3$ by ligand-binding assay in a single plasma sample. Although many similarities exist among the previously published methods for vitamin D_3 metabolite measurements, critical differences do exist. These specific differences will be discussed in relation to the methodology we have selected in the development of accurate and sensitive ligand-binding assays for vitamin D_3 and its major metabolites. The methodology presented provides a sensitive and accurate means of studying both physiologic and pathophysiologic fluctuations in circulating levels of vitamin D_3 and a broad spectrum of its metabolites in a single small plasma sample.

I. Materials and Methods

A. Reference Sterols

Commercial

Reference sterols obtained commercially in crystalline form are vitamin D_3 (Sigma Chemical Co., St. Louis, Missouri), and 25-OHD$_3$, 24R,25-(OH)$_2$D$_3$, 25,26R-(OH)$_2$D$_3$, and 1,25-(OH)$_2$D$_3$, which are a generous gift from Dr. M. Uskokovic, Hoffmann–LaRoche Inc., Nutley, New Jersey. The following radioactive sterols are obtained commercially: [1,2-^{3}H]vitamin D_3 (12.3 Ci/mmol) and [26,27-^{3}H]24,25-(OH)$_2$D$_3$ (80 Ci/mmol) from Amersham/Searle Corp., Arlington Heights, Illinois; and [26,27-^{3}H]1,25-(OH)$_2$D$_3$ (160 Ci/mmol) from New England Nuclear, Boston, Massachusetts.

Biosynthesis [^{3}H]25,26-(OH)$_2$D$_3$

Because of its unavailability commercially, [^{3}H-26,27]25,26-(OH)$_2$D$_3$ is biosynthesized with chick crude renal homogenates using modifications of previously published methods.[43,44] One-day-old fasting white leghorn cockerels housed in cages maintained at 23–30°C are provided ad libitum a standard rachitogenic diet (Teklad Mills, Madison, Wisconsin) for 4–6 weeks. 500 ng of 1,25-(OH)$_2$D$_3$ in 50 μl of ethanol is then administered subcutaneously four times over a 72-h period. The kidneys from 4 chicks are rapidly excised following arterial irrigation with cold normal saline, minced, and homogenized (20% w/v) in a Tris-acetate buffer (15 mM Tris-acetate, 2.0 mM magnesium chloride, 25 mM sodium succinate, pH 7.4 at 23°C) at 0–5°C with 4 periods (15 s/period at setting No. 10 with a 10 s rest between periods) of a polytron homogenizer (Brinkman Instruments, Westbury, New York). The crude renal homogenate is divided equally between four 25-ml Erlenmeyer flasks. Following the addition of 1 μCi of [^{3}H-26,27]25-OHD$_3$ in <20 μl ethanol to each flash, they are flushed for 30 s with O_2 and incubated at 37°C for 30 min in a Dubnoff Metabolic Shaking Incubator (Scientific Products, McGaw Park, Illinois) at 120 oscillations/min. The homogenates are then subjected to the same methanol–methylene chloride extraction, Lipidex-5000 preparative chromatography, and normal phase HPLC methods described below for the isolation of 25,26-(OH)$_2$D$_3$ in plasma.

Purification of Reference Sterols

The concentration of stock solutions of nonradioactive vitamin D_3 metabolite standards is periodically checked by comparison on HPLC with previously recorded integrated ultraviolet (UV) peak area values (approximately 100 ng of metabolite applied in 25 μl chloroform). The HPLC equipment used for the vitamin D_3 metabolite analyses consists of: U6K injector with a 2-ml sample loop, 6000A solvent pumps, 660 solvent programmer, and a 450 UV detector

(Waters Associates, Inc., Milford, Massachusetts); and a 3380A reporting integrator (Hewlett Packard, Avondale, Pennsylvania). It is not within the scope of this paper to discuss the advantages or disadvantages of these HPLC items compared with other models. A normal-phase HPLC system, consisting of a 0.4 × 30-cm μ-porasil column (Waters Associates, Inc., Milford, Massachusetts) and CO:PELL PAC guard column (Whatman Inc., Clifton, New Jersey) in tandem, is used with the following isocratic n-hexane/isopropranol solvent systems at a flow rate of 2.0 ml/min.: vitamin D_3, 99.5:0.5 (v/v); 25-OHD_3, 96:4; 24,25-$(OH)_2D_3$, 94:6; 25,26-$(OH)_2D_3$, 93:7; and 1,25-$(OH)_2D_3$, 92:8. Identity of all [^3H]vitamin D_3 metabolites is initially confirmed by cochromatography with crystalline standard on the appropriate isocratic normal phase HPLC system noted above. Because of the potential for breakdown of [^3H]vitamin D_3 metabolites (especially the more polar, high-specific-activity radioactive metabolites), determinations of the purity of the labeled compounds are carried out on HPLC every 1–3 months.

The collection of all chromatographic elution fractions is carried out with a Gilson MFK Mini-Escargot fractionator (Gilson Medical Electronics, Inc., Middleton, Wisconsin). The chromatographic fractions used for standard calibrations and radioactive metabolite purity checks are collected and air dried in 7-ml polyethylene scintillation vials (Rochester Scientific Co., Inc., Rochester, New York), 5 ml of toluene based scintillation cocktail (PPO, POPOP) added, and [^3H] determinations made on a Beckman LS-8000 scintillation counter (Beckman Instruments, Inc., Fullerton, California) with a mean counting efficiency of 50%. This routine scintillation cocktail consists of 160 ml of Scintiprep-1 (PPO, 125 gm/l; dimethyl POPOP, 2.5 gm/l) and 3.8 l scintanalyzed toluene (Fischer Scientific Co., Fairlawn, New Jersey).

Storage of Reference Sterols

Until used, all sterols are stored under N_2 in deoxygenated absolute ethanol in borosilicated glass vials (storage < 3 months) or in sealed glass ampules (storage ≥ 3 months) at −76°C. Deoxygenation can be accomplished by bubbling N_2 through the ethanol for approximately 30 min. To seal the sterols in glass under N_2, the ampule is placed in ice and a transfer pipette is positioned just within the opening of the ampule to deliver N_2 at a low flow rate. The ampule is then sealed by flame while the N_2 is flowing (the temperature difference between the neck and base of the ampule drawing N_2 in until the point of sealing).

B. Solvents and Glassware

The spectroanalyzed grade glass-distilled solvents (Burdick and Jackson Laboratories, Inc., Muskegon, Michigan) are filtered through 0.5-μm pore Millipore fluoropore filters (Millipore Corp., Bedford, Massachusetts) and then degassed under vacuum with magnetic stirring prior to preparing the various chromatographic solvent systems each day. The solvent combination stocks, prepared fresh each week, are periodically degassed through the day by 15-min periods of sonication. To reduce inert binding of sterols, all glassware involved in extraction, chromatography, assays, and storage may be silanized.[23] Silanization is accomplished by immersing the glassware in a 1% organosilane solution (Prosil-28, PCR Research Chemicals, Inc. Gainesville, Florida) for 5 sec, rinsing with water, and then heating at 100°C for 10 min.

C. Biological Samples

Plasma or serum samples are ideally obtained in a fasting state to minimize the interference of the ligand-binding assays by lipid contaminants. One should be aware of the significant seasonal variation in circulating levels of 25-OHD$_3$ and 24,25-(OH)$_2$D$_3$ in humans (increased levels July–September).[45] To reduce the potential breakdown of vitamin D$_3$ and its metabolites (especially the more polar ones), the plasma samples are stored in polyethylene vials (Rochester Scientific Co., Rochester, New York) at −76°C. Under these conditions one can expect ≤ 5% metabolite degradation/year. Plasma or serum samples are centrifuged after thawing at 2000 g for 10 min at 0–5°C in a HB-4 swinging bucket rotor to remove any protein precipitates. Using the methodology described below, the minimal volume of plasma required for an accurate determination of a total profile of vitamin D$_3$ and its major metabolites in normal adult humans is 2–3 ml.

II. Extraction

A. Methanol–Methylene Chloride

A schematic diagram of the overall methodology is shown in Figure 1. The samples are extracted with methanol–methylene chloride with separation of phases by centrifugation as previously described.[23] Plasma, 2–5 ml, is placed in a 50-ml polypropylene centrifuge tube (Nalgene, Scientific Products, McGaw Park, Illinois) at 0–5°C. To monitor recovery through extraction and chromatography procedures, 3000 dpm of [³H]vitamin D$_3$ and each of its radioactive metabolite standards are added to each plasma sample and intermittently agitated for 15 min prior to extraction. With the addition of the labeled metabolite markers to the plasma sample, an equivalent volume is added to a scintillation vial for the eventual calculation of metabolite recovery. The scintillation vial is air dried and scintillation cocktail immediately added to prevent the loss of [³H]. Three volumes of cold methanol–methylene chloride (2:1, v/v) are added to the plasma sample and the capped tube vortexed for 1 min. One volume of methylene chloride is then added and vortexed for an additional minute. Vortexing is carried out on individual tubes through the use of a Deluxe Mixer (Scientific Products, McGaw Park, Illinois). The extract is centrifuged at 23,300 g for 15 min at 0–5°C in a Sorvall HB-4 swinging

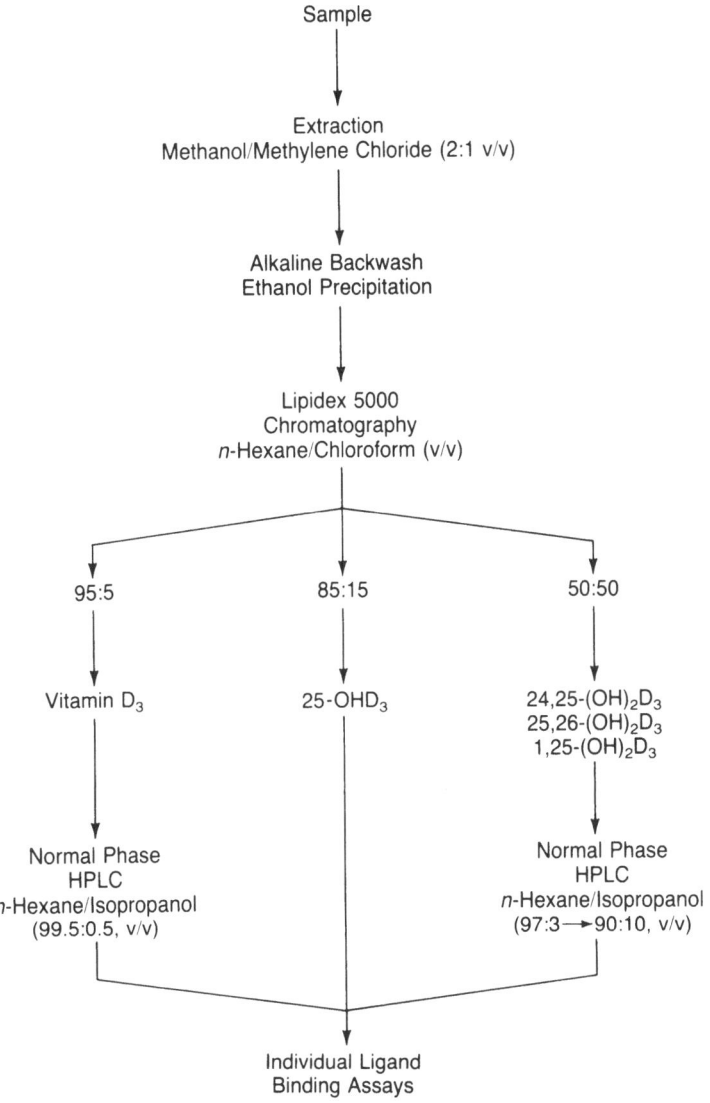

Figure 1. Flow diagram of the methodology for quantitative analysis of vitamin D_3 and its metabolites in plasma.

bucket rotor and RC-5B centrifuge (Dupont Instruments, Newton, Connecticut), resulting in an upper aqueous and lower organic phase with a solid protein interphase. The use of a separatory funnel or fixed angle rotor does not permit as excellent a separation of aqueous and organic phase. The lower organic phase is transferred to a glass vial by either siphoning with polypropylene tubing or using a glass syringe fitted with a 15-cm 16-gauge stainless steel needle and two-way stopcock. A 3–5-ml chloroform rinse of the tubing or syringe is added to the organic phase. The aqueous phase is reextracted with 2 vol methylene chloride and recentrifuged at 23,300 g for 10 min. The organic phases are combined.

B. Alkaline Backwash

To reduce lipid contamination the plasma extract is then subjected to a minor modification of a previously published alkaline backwash method.[30,40] The N_2-dried residue of the organic extract is resolubilized and transferred to a 50-ml polypropylene tube with serial methylene chloride washes (total volume of 5 ml). Ten ml 0.1 M Na_2HPO_4, pH 10.5 is added and the tube vortexed for 5 min using a Deluxe Mixer Vortexer. Following centrifugation at 4000 g for 15 min at 0–5°C in an HB-4 swinging bucket rotor, the lower organic phase is transferred to a glass scintillation vial. The aqueous phase is reextracted with 10 ml methylene chloride, vortexed for 5 min, and centrifuged at 4000 g for 10 min. Depending on the degree of lipid contamination, the organic phase may be subjected to an additional alkaline backwash. The combined organic phases of the extract are N_2-dried on a N-EVAP analytical evaporator (Organomation Associates, Inc., Northborough, Massachusetts) at a temperature of 30–40°C and stored in absolute ethanol under N_2 at −76°C until subjected to chromatography.

C. Ethanol Precipitation

In situations of significant lipid contamination of plasma samples, the organic extract can be subjected to ethanol precipitation following the alkaline backwash. The N_2-dried residue of the organic extract is resolubilized and transferred to a 15-ml corex glass centrifuge tube with serial absolute ethanol washes (total volume of 3 ml). After immersion in acetone–dry ice (freezing the specific lipids but not the ethanol), the tube is centrifuged at 27,138 g in a Sorvall SS-34 fixed-angle rotor at −20°C for 5 min. The tube is reimmersed in acetone–dry ice and the supernatant transferred to a glass scintillation vial. To recover any vitamin D_3 or its metabolites trapped within the precipitate, the precipitate is resolubilized at room temperature in a 2-ml ethanol wash of the original extraction vial and again subjected to centrifugation. The latter procedure is repeated a third time. The combined ethanol supernatant is stored under N_2 at −76°C.

D. Alternative Extraction Methods

The use of the standard Bligh-Dyer methanol–chloroform solvent system for extraction results in excellent recoveries of vitamin D_3 and its metabolites, but is plagued by a substantial contamination with interfering lipophilic substances.[23] With the use of methylene chloride to extract plasma,[22] the amount of extracted lipophilic material is greatly reduced, but there is a significant decrease in the recovery of certain metabolites. The methanol–methylene chloride extraction method offers a compromise, providing both a significant reduction in the lipid contaminants and excellent recovery of vitamin D_3 and its metabolites. Both peroxide-free diethyl ether[40,46] and n-hexane/isopropanol/n-butanol (93:3:4, v/v)[33] have been recently used as extraction solvent systems, permitting the freezing of the aqueous phase and subsequent decanting of the organic phase. The latter systems have proven very successful in the extraction of 1,25-$(OH)_2D_3$, but have been plagued with exceedingly low recoveries for 25-OHD_3. Other investigators have utilized chromatographic systems for extraction, applying plasma samples directly to Clin Elut (diatomaceous earth)[32] for Extrelut (granular celite)[47] with subsequent elution of the vitamin D metabolites with various organic solvent systems. These methods have been efficient in reducing lipid contamination and in isolating 1,25-$(OH)_2D_3$, but have also been hampered by low yields of 25-OHD_3.

III. Lipidex 5000 Chromatography

A. Preparation of Column

The lipophilic, hydrophobic gel Lipidex 5000 (Packard Instrument Co., Inc., Downers Grove, Illinois) is an alkoxy group derivative of Sephadex LH-20 (50% substitution with an average alkoxy group chain length of 15 carbons). Purchased as a methanol slurry, the material is gently stirred, centrifuged at 1000 g for 10 min, and the supernatant with fines decanted. Following initial equilibration with 95:5 n-hexane/chloroform (v/v) (by gentle washing and centrifugation), a 0.9 $\times$ 5.0-cm bed volume of Lipidex 5000 is poured into glass columns equipped with Teflon seals, penton couplings, and Teflon outlet tubing (Bio-Rad Laboratories, Richmond, California). The Lipidex 5000 column is then eluted with 100 ml of 95:5 n-hexane/chloroform for final equilibration. Prior to routine use, the Lipidex 5000 column should be calibrated with a normal plasma extract containing [^{3}H]vitamin D_3 and each of its radioactive metabolite standards.

B. Application and Elution of Organic Extract

The N_2-dried residue of the organic extract is resolubilized in 300 μl of 95:5 n-hexane/chloroform (v/v) (minimizing the vial surface contact of the sample) and applied with a P-200 Pipetman (Ranin Instrument Co., Inc., Woburn, Massachusetts) to the Lipidex 5000 column immediately above the packing bed. After the initial sample extract application is permitted to run into the column bed, two 100-μl washes of the extraction vial are applied in a similar manner (the column flow is turned off during each application). Vitamin D_3 and its metabolites are then eluted with a stepwise solvent gradient system[18] (Figure 2). The majority of the lipid contaminant elutes with the initial 5 ml

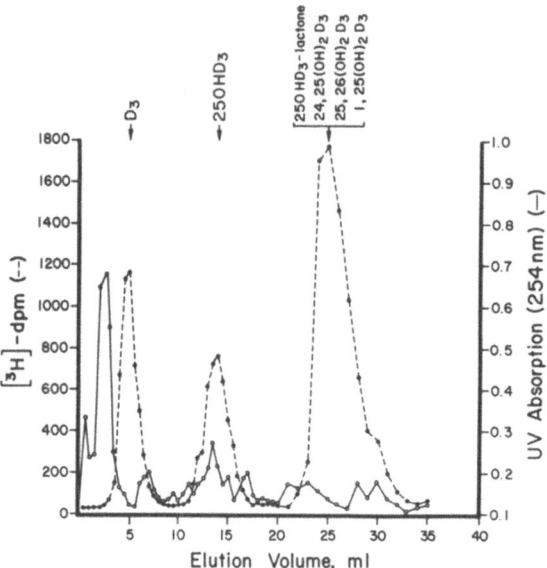

Figure 2. Elution profile of the organic solvent extract of normal human plasma containing labeled vitamin D_3 and metabolite markers on Lipidex 5000 gel chromatography. UV adsorption (———) and [^{3}H]-dpm (---) are plotted as a function of elution volume. The arrow markers indicate the elution position of radioactive vitamin D_3 and metabolite standards. The 0.9 × 5-cm Lipidex column was eluted sequentially with 10 ml of 95:5, 11 ml of 85:15, and 20 ml of 50:50 *n*-hexane/chloroform (v/v) at a flow rate of 2.3 ml/min.

of 95:5 *n*-hexane/chloroform (v/v) at a flow rate of 210 drops/min. ($\simeq$2.3 ml/min.). The vitamin D_3 peak is collected in the subsequent 5 ml of 95:5 *n*-hexane/chloroform (v/v). The column is stopped, the solvent above the column bed changed to 85:15 *n*-hexane/chloroform (v/v), and 25-OHD$_3$ is collected in the subsequent 11-ml eluant at 210 drops/min. The more polar vitamin D_3 metabolites [i.e., 24,25-(OH)$_2$D$_3$, 25,26-(OH)$_2$D$_3$, and 1,25-(OH)$_2$D$_3$] are then batch-eluted with 20 ml of 50:50 *n*-hexane/chloroform (v/v) at the same flow rate. The use of the 50:50 *n*-hexane/chloroform (v/v) will cause a characteristic 30% expansion of the Lipidex 5000 column bed. The column is then sequentially washed with 10 ml of 50:50 and 95:5 *n*-hexane/chloroform (v/v) before application of another plasma extract. Based on the type of plasma sample extracts applied (i.e., degree of lipid contamination) the Lipidex 5000 column is periodically washed ($\simeq$every 5–10 extract applications) with 10 ml of methanol (resulting in a 30% contraction of the bed volume) following the 50:50 *n*-hexane/chloroform elution. With appropriate periodic cleaning with methanol, an individual Lipidex 5000 column should perform in an efficient and accurate manner (regarding metabolite resolution and recovery) for $\geq$20 applications of organic extracts of routine human or experimental animal plasma samples.

C. Plasma Samples with Significant Lipid Contamination

In situations of extremely heavy lipid contamination of biological samples (e.g., patients with marked hyperlipidemia, human milk, and Japanese quail or fish plasma samples) the organic extracts of the plasma samples are applied to silicic acid chromatography prior to their application to Lipidex 5000.[48] The N_2-dried residue of the organic extract is resolubilized in 300 μl of 95:5 n-hexane/ethyl ether (v/v) and applied with a P-200 Pipetman to a 0.9 $\times$ 7-cm silicic acid column (previously equilibrated with 95:5 n-hexane/ethyl ether) with two additional 100-μl washes. Vitamin D_3 and its metabolites are then eluted with the following stepwise solvent gradient: vitamin D_3 (95:5 n-hexane/ethyl ether, v/v); 25-OHD$_3$ (65:35 n-hexane/ethyl ether, v/v); 24,25-$(OH)_2D_3$, 25,26-$(OH)_2D_3$, and 1,25-$(OH)_2D_3$ (95:5 ethyl ether/methanol, v/v). The latter peaks are then individually applied to Lipidex 5000 and eluted as described above.

D. Alternative Preparative Chromatography Methods

Recently rapid, batch-eluting, preparative chromatographic methods using disposable silica cartridges (Silica Sep Pak, Waters Associates, Milford, Massachusetts) and solvent systems of ethyl acetate/n-hexane[49] or diethylether/n-hexane[47] have been reported for the separation of vitamin D_3 and its metabolites from an organic lipid extract. These new preparative chromatographic methods offer a rapid and effective resolution of vitamin D_3 and its metabolites similar to that achieved with Lipidex 5000 chromatography.

IV. High-performance Liquid Chromatography

A. Vitamin D_3

To enhance recovery, the N_2-dried residue of the vitamin D_3 peak eluting from the Lipidex 5000 chromatography is taken up in a 100-μl Precision syringe (Precision Sampling Corp., Baton Rouge, Louisiana) with serial 100-μl chloroform washes in conjunction with partial N_2 evaporation, resulting in a total volume in the syringe of 50–100 μl. The vitamin D_3 peak is then applied to a normal-phase HPLC system with an additional 50-μl wash of the syringe and conical vial. This type of application takes advantage of the increased metabolite solubility in chloroform (a solvent compatible with n-hexane/isopropanol) and the 50-μl wash displaces only 2.5% of the initial sample application from the 2-ml injection loop. The HPLC system consists of a 0.4 $\times$ 30-cm μ-porasil column and CO:PELL PAC guard column in tandem with an isocratic solvent

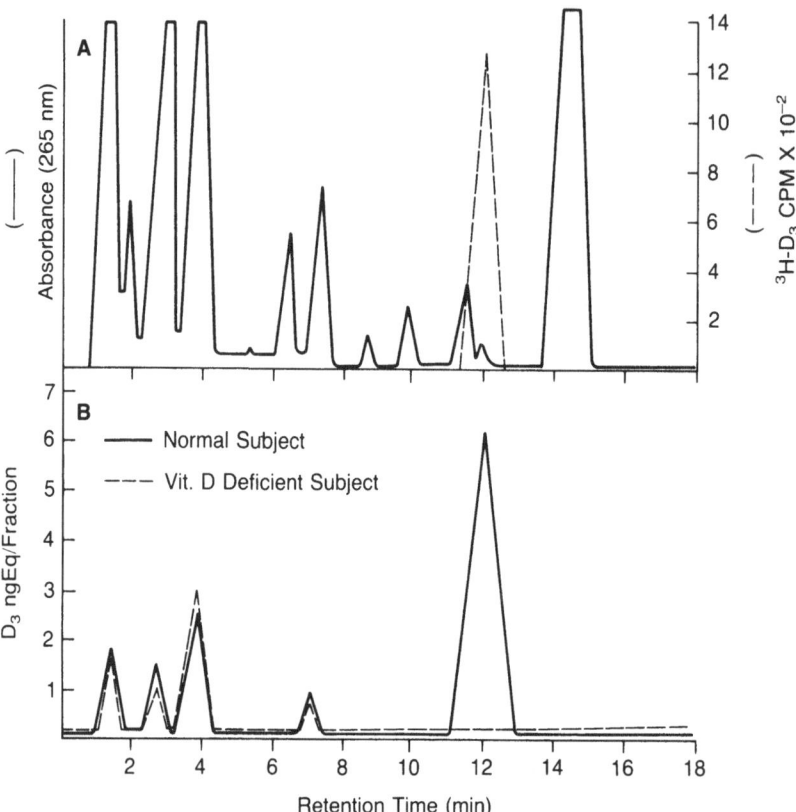

Figure 3. Elution profile of the Lipidex 5000 vitamin D_3 peak on normal phase HPLC. Panel **A** shows the elution profile of the UV-absorbing material (———) and [^{3}H]-dpm (---) after the Lipidex 95:5 n-hexane/chloroform (v/v) vitamin D_3 peak was applied to the HPLC. Labeled vitamin D_3 standard (4500 dpm) was added to the plasma prior to extraction. The ligand-binding assay profiles of the Lipidex 95:5 n-hexane/chloroform vitamin D_3 peak from plasma extracts of a normal human (———) and a subject with vitamin D deficiency (---) are shown in panel **B**. The HPLC system consisted of a 0.4 × 30-cm μ-porasil column in tandem with a CO:PELL PAC guard column. An isocratic solvent system of 99.5:0.5 n-hexane/iso-propanol (v/v) was run at 2.0 ml/min.

system of n-hexane/isopropanol (99.5:0.5 v/v) run at 2.0 ml/min.[1] Refer to the materials section for a description of the module units of the HPLC system. A representative elution profile is illustrated in Figure 3.

Recent studies from our laboratory and others[2] suggest that under certain conditions (most notably in the assay of samples containing high levels of lipids) the vitamin D_3 peak eluting from the above HPLC system can be contaminated with a substance that leads to erroneously low vitamin D_3 levels in the ligand-binding assay. Preliminary data from our laboratory indicate that the application of the Lipidex 5000 vitamin D_3 peak to an isocratic methylene chloride/isopropanol HPLC system (99.75:0.25, v/v at 1.5 ml/min) following the *n*-hexane/isopropanol system described above, eliminates this potential problem.

B. 25-OHD$_3$

Although normally the 25-OHD$_3$ peak from the Lipidex 5000 chromatography is applied directly to a ligand-binding assay, under conditions where the lipid contamination of the sample appears somewhat greater than normal, the 25-OHD peak may be further purified on a normal-phase HPLC system. This isocratic HPLC system consists of a μ-porasil column and CO:PELL PAC guard column in tandem with a solvent system of 98:2 *n*-hexane/isopropanol (v/v) run at 1.5 ml/min. The latter HPLC system resolves the 25-OHD peak from both lipid contaminants and the previously reported earlier eluting "peak X."[16,17] The latter unidentified peak may potentially interfere in the ligand-binding assay for 25-OHD$_3$.[16,17]

C. 24,25-(OH)$_2$D$_3$, 25,26-(OH)$_2$D$_3$, and 1,25-(OH)$_2$D$_3$

The N$_2$-dried residue of the pooled vitamin D$_3$ metabolites in the 50:50 *n*-hexane/chloroform (v/v) elution from the Lipidex 5000 chromatography of the plasma extract is resolubilized and applied to a normal-phase HPLC system in an identical manner as vitamin D$_3$. The HPLC system consists of a μ-porasil column and CO:PELL PAC guard column in tandem. A solvent system of *n*-hexane/isopropanol is run in a concave gradient mode (setting No. 9, Waters 660 programmer) from 97:3 to 90:10 (v/v) over a 15 min period at 1.5 ml/min (pressure $\leq$ 600 psi).[18,35,36] To achieve this gradient elution mode and yet avoid running either the A or B pump (Waters 6000A pumps) below the minimum flow rate (0.25 ml/min) for a prolonged period, the A pump solvent reservoir is filled with 100% *n*-hexane and the B pump with 87:13 *n*-hexane/isopropanol (v/v). The B pump on the 660 solvent programmer is then set to run from 23% to 77%. The mean retention times for the different vitamin D$_3$ metabolites and UV absorption profile from a normal plasma extract on such an HPLC system are illustrated in Figure 4. More recently we have noted an even better resolution of multiple vitamin D$_3$ metabolites with an HPLC system consisting of a Zorbax Sil μ-analytical (Dupont Co., Wilmington, Delaware) and CO:PELL PAC guard column in tandem, and an *n*-hexane/isopropanol gradient elution (setting No. 9, 94:6 to 90:10, v/v) run over a 20 min period at 1.5 ml/min.

Following completion of the sample run, the HPLC is returned to the original solvent conditions over a 5-min period by means of the 660 solvent programmer. At periodic intervals ($\simeq$every five sample applications) the normal phase HPLC column is cleaned with $\simeq$50 ml of 50:50 *n*-hexane/isopropanol. To detect any subtle but significant variations in vitamin D$_3$ metabolite retention times, calibrations of the HPLC system are performed at the beginning, middle, and end of the day with the application of Lipidex 5000 metabolite

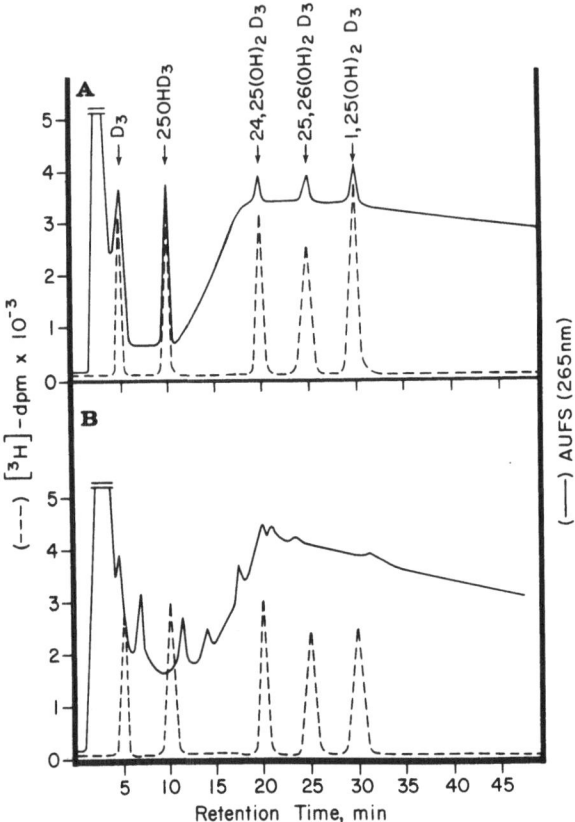

Figure 4. Elution profile of vitamin D_3 metabolites on normal phase HPLC. Panel **A** represents the elution profile of the UV-absorbing material (———) and [^{3}H]-dpm (---) after unlabeled (100 ng) and labeled (4500 dpm) vitamin D metabolite standards were applied to the HPLC. The elution profile of the Lipidex 50:50 n-hexane/chloroform (v/v) metabolite peak from a normal human plasma extract is shown in panel **B**. [^{3}H]-vitamin D_3 metabolite standards (4500 dpm) were added to the plasma sample as elution markers. The arrow markers indicate the elution position of radioactive and nonradioactive vitamin D_3 metabolites. The HPLC system consisted of a 0.4 × 30-cm μ-porasil column in tandem with a CO:PELL PAC guard column. A concave solvent gradient (curve setting No. 9, Waters 660 solvent programmer) of n-hexane/isopropanol (97:3 to 90:10, v/v) was run over a 15-min period at 1.5 ml/min.

peaks from normal human plasma extracts with added labeled metabolite markers.

D. Alternative HPLC Methods

Other types of normal-phase HPLC analytical columns packed with 5-μ silica particles (e.g., Zorbax Sil, Dupont Co., Wilmington, Delaware) may offer somewhat finer resolution of metabolites than achieved with the 10-μ silica particles in the μ-porasil column. Appropriate isocratic solvent systems will always offer a better resolution of two vitamin D_3 metabolites or separation of lipid contaminants from a metabolite on HPLC than solvent systems run in a gradient elution mode. The isocratic systems, however, are not always practical when dealing with the single application of a larger number of metabolites. Reduced peak tailing and improved resolution of vitamin D metabolites on HPLC have been recently achieved with the use of ternary solvent mixtures, with hexane/isopropanol/methanol mixtures (87:10:3, v/v) being recommended in samples with high lipid content.[41]

E. Preparation and Storage of Chromatographic Peaks of Vitamin D_3 and Its Metabolites

The chromatographic peaks of vitamin D_3 and its metabolites are initially evaporated with N_2. In situations of long-term storage of the peaks (i.e., ≥ 1 week before being applied to ligand-binding assays) the N_2-dried residue is transferred with chloroform to borosilicated glass vials (Rochester Scientific Co., Rochester, New York), N_2-evaporated, and stored in 5 ml of absolute ethanol under N_2 at $-76°C$. On occasions where the peaks will be assayed in <1 week the N_2-dried residue is transferred to 1–5-ml glass conical Reactivials (Pierce Chemical Co., Rockford, Illinois) and stored under N_2 at $-76°C$ in various volumes of ethanol: (1) vitamin D_3, 24,25-$(OH)_2D_3$, and 25,26-$(OH)_2D_3$ = 0.5–1.0 ml; (2) 25-OHD_3 = 2.5–5.0 ml; and (3) 1,25-$(OH)_2D_3$ = 0.1–0.2 ml.

V. Ligand-binding Assays

A. Vitamin D_3 Assay

Preparation of the Assay Buffer and Binding Protein

The vitamin D_3 peak eluting from HPLC is applied to a ligand-binding assay previously described.[1] Normal rat plasma, diluted 1:15,000 with assay diluent (0.05 M sodium barbital–acetate, 12 mM thioglycerol, 0.01% gelatin, pH 8.6 at 23°C) is used as the binding protein in the assay. The proper dilution of the rat plasma should result in a maximum specific binding of [³H]25-OHD_3 (i.e., the percentage of [³H]25-OHD_3 bound by the rat plasma vitamin-D-binding protein minus nonspecific binding) of $\simeq 40$–60%. The rat plasma (obtained from normal 3–5-month-old male rats) is stored at $-76°C$ in 20-μl aliquots, with each aliquot being subjected to thawing–freezing a maximum of 3 times. The barbital–acetate buffer system for the vitamin D_3 ligand-binding assay is freshly prepared each week. One volume of stock solution A (58.5 mg/ml Na barbital and 38.85 mg/ml Na acetate in double-distilled water) and 0.4 volumes of stock solution B (0.83% v/v concentrated HCL in double-distilled water) is mixed with monothioglycerol (Sigma Chemical Co.), final concentration 12 mM, and 4.6 vol double-distilled water, with a final pH of 8.6 at 23°C. Both the stock solutions and final buffer system are stored at 0–5°C. The gelatin is added to the buffer with gentle heating and magnetic stirring the day before the assay, and the buffer system then stored at 0–5°C. The addition of 0.1% gelatin has been reported to increase the sensitivity of the assay.[2]

Assay

The appropriate amount of vitamin D$_3$ standard (from 50 pg to 1.0 ng in 100 μl ethanol/assay tube) and the vitamin D$_3$ peak eluting from HPLC is added with a P-200 Pipetman to 12 × 75-mm disposable glass culture tubes (Scientific Products, McGaw Park, Illinois). Details of the assay format are shown in Table 1. A 20% aliquot of the chromatographic peak of vitamin D$_3$ is taken just prior to the assay for final recovery determinations (see calculations). The vitamin D$_3$ peak (in 0.5–1.0 ml ethanol in a conical vial) is assayed in duplicate at one or more dilutions (e.g., 100 μl × 2, 50 μl × 2, and 100 μl control). Determination of the appropriate dilutions in the assay of vitamin D$_3$ (as well as other metabolite peaks from plasma samples) is made on the basis of the clinical or experimental setting of the sample, the volume of sample extracted, the metabolite recovery, and the detection limits of the assay. Following the addition with a P-1000 Pipetman of 900 μl of assay diluent to all control tubes (nonspecific binding) and 900 μl of rat plasma-assay diluent to all remaining tubes in conjunction with a brief vortexing, the assay tubes are tightly covered with parafilm and incubated for 18 h in a Dubnoff metabolic shaking incubator at 0–5°C and 180 oscillations/min. To date almost all vitamin D metabolite assays have been carried out with equilibrium ligand-binding conditions (ie.., simultaneous addition of tracer and binding protein). Nonequilibrium ligand-binding assays (i.e., delayed addition of tracer) have been effectively employed in the measurements of various peptide hormones.[50] The delay in the addition of tracer enhances the sensitivity of the assay. At the end of the initial incubation period 5000 dpm of [^{3}H]25-OHD$_3$ is added to each assay tube, vortexed briefly, and incubated an additional 2 h at 0–5°C. Because of the greater affinity of [^{3}H]25-OHD$_3$ for the vitamin-D-binding protein than [^{3}H]D$_3$, the former radioactive metabolite is employed as the tracer to enhance the sensitivity of the ligand-binding assay.

Separation of Bound and Free Vitamin D$_3$

Standard dextran-charcoal methods are used for the separation of bound from free [^{3}H]vitamin D$_3$ in the ligand-binding assay.[3,9,18,35] A 0.2% w/v dextran (Dextran T-70, Pharmacia Fine Chemicals, Piscataway, New Jersey) and 2.0% charcoal (Norit Carbon Decolorizing Neutral, Fischer Scientific Co., Fairlawn, New Jersey) mixture is freshly prepared in barbital–acetate buffer system for each assay. The dextran-charcoal mixture is magnetically stirred overnight at 0–5°C. Following centrifugation in polypropylene tubes at 3079 g for 15 min at 0–5°C in a Sorvall HB-4 swinging bucket rotor (Dupont

Table 1. Ligand-binding Assay Formats

MSB[a] No. 1	Standards: Vitamin D₃, 25-OHD₃, 24,25-(OH)₂D₃ or 25,26-(OH)₂D₃ (ng/tube)								MSB No. 2
MB[b]	0.05	0.075	0.1	0.15	0.2	0.3	0.6	1.0	MB
MB	0.05	0.075	0.1	0.15	0.2	0.3	0.6	1.0	MB
C[c]	C	C	C	C	C	C	C	C	C

Samples: Vitamin D₃, 25-OHD₃, 24,25-(OH)₂D₃, or 25,26-(OH)₂D₃ Chromatographic Peak (volume/tube)[d]

Sample Series	MSB No. 3	Sample Series	MSB No. 4
100 μl	MB	100 μl	MB
100 μl	MB	100 μl	MB
50 μl	C	50 μl	C
50 μl		50 μl	
C		C	

C[e]	MSB No. 1	Standard: 1,25-(OH)₂D₃ (pg/tube)								MSB No. 4
5 ng	MB	5	10	15	20	25	30	50	100	MB
5 ng	MB	5	10	15	20	25	30	50	100	MB

Sample: 1,25-(OH)₂D₃ Chromatographic Peak (volume/tube)

Sample Series	MSB No. 3	Sample Series	MSB No. 4
50 μl	MB	50 μl	MB
50 μl	MB	50 μl	MB
25 μl		25 μl	
25 μl		25 μl	

[a]Maximum specific binding of tracer; derived from maximum binding (MB) minus nonspecific binding (C).
[b]Maximum binding of tracer; assay tube contains tracer and binding protein without metabolite standard or chromatographic peak.
[c]Control or nonspecific binding; assay tube contains tracer with or without metabolite standard (standard control) or chromatographic peak (sample control).
[d]Metabolite peaks assayed in duplicate at one or more dilutions; see section on total volume and aliquots for specific metabolite peaks and criteria for selection of dilutions.
[e]Control or nonspecific binding; assay tube contains tracer and excess (5 ng) 1,25-(OH)₂D₃ without binding protein.

Instruments, Newton, Connecticut), the supernatant is decanted and the dextran-charcoal pellet resuspended in an equivalent volume of buffer–0.01 percent gelatin. The final dextran–charcoal mixture is magnetically stirred for 30 min before the addition of 400 μl of the mixture to each assay tube. The assay tubes are vortexed briefly and then centrifuged at 3079 g for 15 min at 0–5°C in an HS-4 rotor. Charcoal-contaminated items can inertly bind vitamin D₃ and its metabolites. To eliminate this type of nonspecific binding it is important that all polypropylene tubes, glassware, and magnetic stirrers employed in dextran–charcoal preparation by only used for that purpose, and not for other aspects of the analyses.

[³H] Determination

The supernatant (bound radioactive and nonradioactive vitamin D_3) is carefully decanted into a 20-ml polyethylene scintillation vial (Rochester Scientific Co., Inc., Rochester, New York) and 10 ml of toluene-based scintillation cocktail (PPO, POPOP, Bio-Solv-BBS-III) is added. This aqueous scintillation cocktail consists of: 5 g scintanalyzed PPO and 13 g dimethyl POPOP (Fischer Scientific Co., Fairlawn, New Jersey); 160 ml Bio-Solv-BBS-III (Beckman Instruments, Inc., Fullerton, California); and 840 ml scintanalyzed toluene (Fischer Scientific Co., Fairlawn, New Jersey). [³H] determinations are made in a Beckman LS-8000 scintillation counter (Beckman Instruments, Inc., Southfield, Michigan) with a mean counting efficiency of 50%.

Assay Characterization and Results

The nonspecific binding for the vitamin D_3 ligand-binding assay is routinely < 10%. The detection limits, determined as >2 SD of the mean maximum specific binding, are 0.05 ng/assay tube. A representative standard curve for vitamin D_3 is illustrated in Figure 5. The intra- and interassay variation, based on repetitive determinations in pooled normal human blood band plasma, is < 15%. The final recovery (mean ± SD) for vitamin D_3 following extraction and recovery is 74.0 ± 8.1%. The mean normal vitamin D_3 level in human plasma is 1.6 ± 0.4 ng/ml.

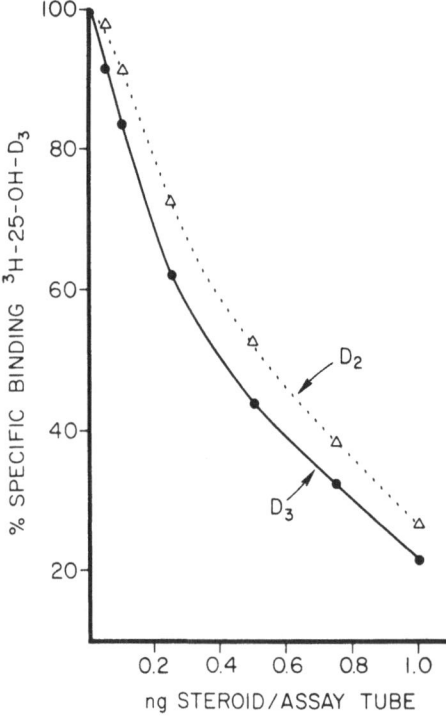

Figure 5. Competitive displacement of [³H]25-OHD₃ from the rat plasma vitamin-D-binding protein by vitamin D_2 (-----) and vitamin D_3 (———). These standard curves were achieved using a nonequilibrium incubation system.

B. 25-OHD₃, 24,25-(OH)₂D₃, and 25,26-(OH)₂D₃ Assay

Preparation of the Assay Buffer and Binding Protein

The Lipidex 5000 chromatography peak for 25-OHD$_3$ and the individual HPLC peaks for 24,25-(OH)$_2$D$_3$ and 25,26-(OH)$_2$D$_3$ from the plasma extracts are applied to ligand-binding assays modified[18,35,36] from previously published methods.[3,9] Normal rat plasma, diluted 1:15,000 with assay diluent (0.05 M sodium barbital-acetate, 12 mM thioglycerol, 0.1% bovine serum albumin fraction V [Pentex, Inc., Kankakee, Illinois] pH 8.6 at 23°C) is used as the binding protein in the assays. The barbital–acetate buffer system is prepared in the same manner as described for the vitamin D$_3$ assay. The monothioglycerol (Sigma Chemical Co., St. Louis, Missouri) is added at the time the buffer system is prepared. The bovine serum albumin is added to the buffer 1 day prior to the assay. In addition to bovine serum albumin other investigators have used ethanol[47] and β-lipoprotein[4] for the solubilization of the vitamin D metabolites in the aqueous incubation media.

Assay

The appropriate amount of metabolite standards (50 pg to 1.0 ng in 100-μl ethanol/assay tube) and the plasma metabolite peaks eluting from chromatography are added to 12 × 75-mm disposable glass culture tubes with a P-200 Pipetman. Details of the format of these assays are shown in Table 1. Determination of the appropriate dilutions in the assays of these metabolite peaks is made on the same basis as for vitamin D$_3$. Because 25-OHD$_3$, 24,25-(OH)$_2$D$_3$, and 25,26-(OH)$_2$D$_3$ are equipotent in displacing [³H]25-OHD$_3$ from the rat plasma vitamin-D-binding protein (Figure 6), [³H]25-OHD$_3$ is used as the tracer in the ligand-binding assays for all three metabolites. 5000 dpm of [³H]25-OHD$_3$ is added to each 25-OHD$_3$ assay tube. To compensate for the addition of [³H]24,25-(OH)$_2$D$_3$ and [³H]25,26-(OH)$_2$D$_3$ to the plasma prior

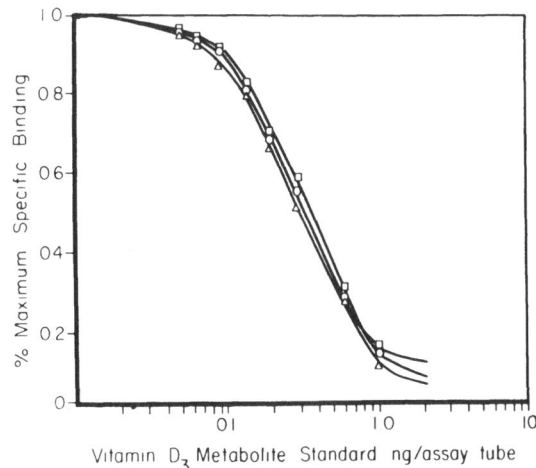

Figure 6. Representative standard curves of the ligand binding assays for 25-OHD$_3$($\triangle$), 24R,25-(OH)$_2$D$_3$ (O), and 25R,26-(OH)$_2$D$_3$ ($\square$). Increasing amounts of each vitamin D$_3$ metabolite were added to normal rat plasma and 5000 dpm of [³H]25-OHD$_3$. The percent of maximum specific binding is plotted as a function of the amount of metabolite per assay tube on a semilogarithic scale.

to extraction for monitoring recovery, an appropriate reduction in the amount of $[^3H]$25-OHD$_3$ added to each of these metabolites assay tubes is made to equalize the total radioactivity. Following the addition of $[^3H]$25-OHD$_3$ to each assay tube, 900 μl of assay diluent is added with a P-1000 Pipetman to all control tubes (nonspecific binding) and 900 μl of rat plasma-assay diluent to all remaining tubes. The assay tubes are briefly vortexed and then incubated for 2 h at 0–5°C.

Separation of Bound and Free Metabolite

The standard dextran–charcoal methods described for the vitamin D$_3$ assay are used to separate protein bound from free metabolite, except that following centrifugation the dextran–charcoal pellet is resuspended in an equivalent volume of buffer–0.1% bovine serum albumin. The final dextran–charcoal mixture is magnetically stirred for 30 min before the addition of 400 μl of the mixture to each assay tube. The assay tubes are vortexed briefly and then centrifuged at 3079 g for 15 min at 0–5°C in an HS-4 swinging bucket rotor. The supernatant (bound radioactive and nonradioactive metabolite) is carefully decanted into a 20-ml polyethylene scintillation vial, 10 ml of toluene-based scintillation cocktail (PPO, POPOP, Bio-Solv-BBS-III) added, and $[^3H]$ determinations made in a Beckman LS-8000 scintillation counter (see the vitamin D$_3$ assay for details of the preparation of the aqueous scintillation cocktail).

Assay Characterization and Results

The nonspecific binding for these ligand-binding assays is routinely <10%. The detection limits, determined as >2 SD of the mean maximum specific binding, are 0.05 ng/assay tube for 25-OHD$_3$, 24,25-(OH)$_2$D$_3$, and 25,26-(OH)$_2$D$_3$. Representative standard curves for the latter three metabolites are illustrated in Figure 6. The intra- and interassay variation, based on repetitive determinations in pooled normal human blood bank plasma, is <12% for all three metabolites. The final recoveries (mean ± SD) for these metabolites following extraction and chromatography of plasma are 25-OHD$_3$, 86.8 ± 7.3%; 24,25-(OH)$_2$D$_3$, 75.8 ± 10.2%; and 25,26-(OH)$_2$D$_3$, 72.6 ± 9.9%. Mean normal metabolite levels in human plasma are 25-OHD$_3$, 26.5 ± 5.3 ng/ml; 24,25-(OH)$_2$D$_3$, 1.3 ± 0.4 ng/ml; and 25,26-(OH)$_2$D$_3$, 0.5 ± 0.1 ng/ml.

C. 1,25-(OH)$_2$D$_3$ Assay

Preparation of the Cytosol Receptor

Plasma levels of 1,25-(OH)$_2$D$_3$ are determined by a highly sensitive ligand-binding assay described previously.[18,23,35] This method involves the application of the 1,25-(OH)$_2$D$_3$ peak eluting from HPLC to an assay employing cytosol receptor from the intestinal mucosa of vitamin-D-deficient chicks. One-day-old fasting white leghorn cockerels, housed in cages maintained at 23–30°C without UV light, are provided ad libitum a 0.6% Ca, 0.4% PO$_4$, vitamin-D-deficient, purified soy-protein diet (Teklad Mills, Madison, Wisconsin) and deionized water until sacrifice by decapitation at 4–6 weeks. With the methods described below, receptor preparations from normal chicks have characteristically shown approximately 10–20% less binding activity than the vitamin-D-deficient preparations in the 1,25-(OH)$_2$D$_3$ ligand-binding assay. It is important to decapitate the chick rapidly to avoid intestinal hemorrhagic necrosis. The duodenal loop is quickly removed, the pancreas lying immediately between the duodenal loop carefully trimmed away, and the bowel irrigated with 10–20 cc cold normal saline. The opened duodenum is placed on a stainless steel plate resting on ice, and the mucosa harvested by gentle scraping with a glass slide. To minimize serum protein contamination and to maintain stability of the receptor preparation, the mucosa is gently washed 4 times with 8 vol (ml/g) of cold normal saline, stirred with a glass rod, and centrifuged each time at 1000 g for 5 min at 0–5°C. The mucosal pellet is then resuspended in 3 volumes of buffer (0.05 M Tris–HCl, 0.15 M KCl, 12 mM thioglycerol, 10% glycerol, pH 7.4 at 23°C) and homogenized in an ice bath with 4 periods (15-s bursts each at setting No. 10 with 10-s rests between periods) of a polytron homogenizer (Brinkman). Significant foaming of the homogenate should be avoided to minimize receptor inactivation.

The homogenate is ultracentrifuged in either 2.0 × 0.5-in polyallomer tubes (Beckman Instruments, Inc., Palo Alto, California) at 233,000 g for 1 hr. at 0–5°C in an SW 50.1 ultracentrifuge rotor (Beckman Instruments, Inc., Palo Alto, CA) or in 3.5 × 0.5-in. cellulose nitrate tubes (Beckman Instruments, Inc.) at 200,000 g for 1.5 h at 0.5°C in an SW 40.1 ultracentrifuge rotor (Beckman Instruments, Inc.). After the surface lipid layer is carefully removed with a transfer pipette, the cytosol supernatant is rapidly frozen with acetone-dry ice and stored in 0.5–5-ml aliquots in borosilicated glass scintillation vials under N$_2$ at −76°C. In addition to easier handling, we have noted slightly greater maximum specific binding when the cytosol receptor preparation is stored frozen rather than lyophilized.[35] Immediately before the assay the cytosol receptor preparation is thawed and diluted at 0–5°C with buffer (0.05 M Tris–HCl, 0.15 M KCl, 12 mM thioglycerol, 10% glycerol, pH 7.4 at 23°C). The final protein concentration should yield a maximal specific binding of [^{3}H]1,25-(OH)$_2$D$_3 \simeq$ 40–60% (usually 0.5–1.0 mg/ml protein). Because glycerol interferes with the standard Lowry assay, protein concentrations are determined by the Bradford assay method.[56] The addition of a reducing agent, such as thioglycerol[32,35] or dithiothreitol,[33,40,47] to the buffer system enhances the storage stability of the receptor preparations. The receptor preparations show ≤5% decrease in maximum specific binding after 12 months of storage in a frozen state at −76°C. However, the use of receptor preparations previously thawed and then refrozen results in ≥25% decrease in binding activity.

Alternative Methods for the Preparation of the Cytosol Receptor

The Tris should be ultrapure grade (Schwarz/Mann, Orangeburg, New York). As previously noted by other investigators,[19,21] the addition of the appropriate amount of KCl to the homogenate and assay buffer is critical for the specific binding of [³H]1,25-(OH)$_2$D$_3$ to the cytosol receptor. The major buffer systems employed in the published 1,25-(OH)$_2$D$_3$ cytosol receptor assays have been either Tris-HCl[19,20,32,35,40,47] or KHPO$_4$.[21,22,24,30,33] In addition to the addition of KCl and a reducing agent (e.g., thioglycerol, dithiothreitol, mercaptoethanol), several laboratories have included EDTA in the buffer system to enhance the stability and activity of the receptor preparation.[28,32,40,47] More recently, one group of investigators has used an ethanol extract of vitamin-D-deficient chick serum to solubilize the 1,25-(OH)$_2$D$_3$ and lyophilized receptor, create a similar lipid environment among all assay tubes, and increase the sensitivity of the assay.[32] The efficiency of mucosal cell disruption, as assessed by phase contrast microscopy and yield of receptor judged by total specific binding activity recovered, is greater with the polytron homogenizer than with the Potter-Elvehjen teflon-glass homogenizer (90% versus 75% cell disruption, respectively).[35]

Assay

Each assay mixture in 12 × 75-mm disposable glass culture tubes consists of, in order of addition: (1) nonradioactive 1,25-(OH)$_2$D$_3$ standard or the HPLC peak of the 1,25-(OH)$_2$D$_3$ plasma extract in 20 μl ethanol; (2) 5000 dpm [³H]1,25-(OH)$_2$D$_3$ in ≤ 20 μl ethanol; and (3) the appropriate amount of cytosol protein in 1 ml buffer. To minimize receptor inactivation, the total ethanol concentration should be ≤5% of the assay mixture. 1,25-(OH)$_2$D$_3$ standard curves are carried out in duplicate over a range of 5–100 pg and the 1,25-(OH)$_2$D$_3$ HPLC peak from human plasma samples is performed in duplicate at one or more dilutions where feasible. Determination of the dilutions is based on the same criteria as noted for the vitamin D$_3$ assay. Details of the 1,25-(OH)$_2$D$_3$ assay format are shown in Table 1. Nonspecific binding is assessed by the inclusion of an excess of 1,25-(OH)$_2$D$_3$ (5 ng) in the standard curve. To compensate for the addition of [³H]1,25-(OH)$_2$D$_3$ to the plasma prior to extraction for monitoring recovery, an appropriate reduction in the radioactivity added to each assay tube is made to equalize the total radioactivity. Following a rapid vortexing the assay tubes are incubated at 25°C for 1 h at 120 oscillations/min. The tubes are then placed in an ice bath for 10 min prior to separation of bound from free 1,25-(OH)$_2$D$_3$ by a dextran–charcoal–human plasma method. Final recovery for 1,25-(OH)$_2$D$_3$ is calculated from the [³H]–dpm of the metabolite added to the plasma sample prior to extraction and the [³H]-dpm present in a 20% aliquot of the 1,25-(OH)$_2$D$_3$ HPLC peak taken at the time of the assay.

Separation of Bound and Free 1,25-(OH)$_2$D$_3$

A 0.5% (w/v) charcoal (Norit Carbon Decolorizing Neutral, Fischer Scientific Co., Fairlawn, New Jersey) and 0.05% dextran (Dextran T-70, Pharmacia Fine Chemicals, Piscataway, New Jersey) mixture is made up in double-distilled water, mixed, and allowed to stand for 5 min, being stirred occasionally. The mixture is centifigued at 250 g for 10 min at 0–5°C, and the precipitate is resuspended in buffer (0.05 M Tris–HCl, 0.15 M KCl, 12 mM thioglycerol, 10% glycerol, pH 7.4 at 23°C) with the addition of normal human plasma to yield a 1.0% charcoal, 0.1% dextran, and 5% plasma concentration. The mixture is magnetically stirred for 12–16 h at 0–5°C and centrifuged at 250 g for 10 min at 0–5°C. The precipitate is then resuspended in Tris Buffer with magnetic stirring at 0–5°C to yield a final 5% charcoal and 0.5% dextran concentration. The dextran–charcoal–human plasma mixture, 250 μl, is added to each assay tube. The tubes are briefly vortexed, allowed to stand in an ice bath 30 min, mixed at 10-min intervals and then centrifuged at 3079 g for 15 min at 0–4°C in an HS-4 swinging bucket rotor. The supernatant (bound [³H]-1,25-(OH)$_2$D$_3$) is carefully decanted into a 20-ml polyethylene scintillation vial, 10 ml of toluene-based scintillation solution (PPO, POPOP, Bio-Solv-BBS-III) added, and [³H] determinations made in a Beckman LS-8000 Scintillation counter (see the vitamin D$_3$ assay for the details of the preparation of the aqueous scintillation cocktail). The dextran–charcoal–human plasma technique described for the separation of receptor-bound from free [³H]1,25-(OH)$_2$D$_3$ is less tedious and equally or more effective than separation by either a chromatin binding,[19] DEAE-filtration,[20] or polyethylene glycol technique.[21] With this separation method, enhanced assay sensitivity and a low and reproducible background (i.e., nonspecific binding) of <5% are achievable. Equally low values for nonspecific binding have also been noted with methods using hydroxylapatite[40,47] and lyophilized equivalent-sized charcoal with dextran.[32]

Assay Characterization and Results

The nonspecific binding for this cytosol-receptor assay, as noted above, is characteristically <5%. The detection limit, determined as >2 SD of the mean maximum specific binding, is 2 pg 1,25-(OH)$_2$D$_3$ per assay tube. A representative standard curve for 1,25-(OH)$_2$D$_3$ is shown in Figure 7. The intra- and

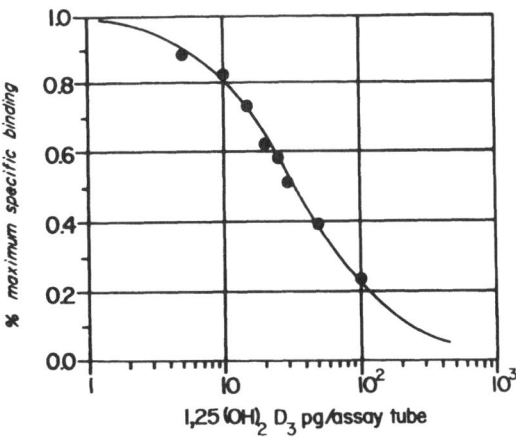

Figure 7. 1,25-(OH)$_2$D$_3$ receptor assay standard curve. Specific binding (i.e., total binding minus nonspecific binding) is expressed as a percent of maximum specific binding and is plotted against the amount of unlabeled 1,25-(OH)$_2$D$_3$ standard added per assay tube on a semilogarithmic scale.

interassay variation, based on repetitive determinations of $1,25\text{-}(OH)_2D_3$ in pooled normal human blood bank plasma, is <5%. The final recovery (mean ± SD) for $1,25\text{-}(OH)_2D_3$ following extraction and chromatography of plasma is 76.3 ± 9.0%. Mean normal $1,25\text{-}(OH)_2D_3$ levels in human plasma are 34.1 ± 9.8 pg/ml.

Recently, two unique and sensitive cytoreceptor assays for $1,25\text{-}(OH)_2D_3$ have been reported employing either cultured osteogenic sarcoma cells[31] or ammonium sulfate fractionated thymic cytosol.[57] Both techniques apparently do not require HPLC for accurate determinations of $1,25\text{-}(OH)_2D_3$. In addition to ligand-binding assay techniques employing cytosol receptor, methodology has been reported for the measurement of specific vitamin D_3 metabolites utilizing radioimmunoassays,[25-27,53] a sensitive bioassay system,[54] and mass fragmentography.[55]

VI. Calculations

All calculations described below are performed by computer program (Digital PDP 1145 hardware, Tektronix 4006-1 software, and modified Fortran language), utilizing [³H] data on Teletype (Beckman Instruments, Inc., Fullerton, California) paper tape generated by the Beckman LS-8000 scintillation counter. The [³H] counts/min are converted to disintegrations per minute (dpm) through the use of a quench standard curve ([³H] quench standards, Beckman Instruments, Inc., Fullerton, California). Specific dpm are derived from total dpm minus nonspecific binding dpm (control tubes). The specific dpm for the standards and invidual metabolite peaks is then expressed as a percentage of maximum specific binding (i.e., specific dpm/mean maximum specific binding). The percentage of maximum specific binding is plotted in both a semilogarithmic and logit (log–log) fashion as a function of the quantity of standard vitamin D_3 or metabolite. The semilogarithmic plot of the standard curve provides a visual check of the general quality of the assay. Specific factors in the qualitative assessment of the assay include: (1) the standard metabolite value for 50% displacement from maximum specific binding; (2) the mean maximum specific binding value (i.e., the percentage of tracer specifically bound without competition from vitamin D_3 or metabolite); and (3) the nonspecific binding value. In addition to these latter qualitative checks, one can periodically assess the assay values derived from pooled normal human or vitamin-D-deficient rat plasma samples containing known amounts of added vitamin D_3 or metabolite standard.

The percentage of maximum specific binding values for individual metabolite peaks of plasma samples is applied to the logit plot of the standard curve. The nanogram or picogram quantity of vitamin D_3 or metabolite derived from these latter calculations is converted to a final plasma concentration by consideration of: (1) the dilution of the metabolite peak in the assay; (2) percentage recovery; and (3) plasma sample volume. Final recoveries for individual metabolites are calculated from the [^{3}H]–dpm of the metabolite marker added to the plasma sample prior to extraction and the [^{3}H]–dpm present in a 20% aliquot of the metabolite chromatographic peak taken at the time of the assay.

Acknowledgments   This research was supported by the Veterans' Administration, the National Institutes of Health, the Kidney Foundation of Ohio, and the Euclid Clinic Research Foundation. The authors gratefully acknowledge the secretarial help of Geraldine Galloway and Loretta Montgomery.

References

1. Hollis BW, Roos BA, Lambert PW: Vitamin D in plasma: quantitation by a non-equilibrium ligand binding assay. Steroids 37:609–619, 1981.
2. Horst RL, Reinhardt TA, Beitz DC, Littledike ET: A sensitive competitive protein binding assay for vitamin D in plasma. Steroids 37:581–591, 1981.
3. Haddad JG, Chyu KJ: Competitive protein binding radioassay for 25-hydroxycholecalciferol. J Clin Endocrinol Metab 33:992–995, 1971.
4. Belsey R, DeLuca HF, Potts JT: Competitive binding assay for vitamin D and 25-OH vitamin D. J Clin Endocrinol Metab 33:554–557, 1971.
5. Bayard F, Bec P, Louvet JP: Measurement of plasma 25-hydroxycholecalciferol in man. Eur J Clin Invest 2:195–198, 1972.
6. Edelstein S, Charman M, Lawson DEM, Kodicek E: Competitive protein-binding assay for 25-hydroxycholecalciferol. Clin Sci Mol Med 45:231–240, 1974.
7. Preece MA, O'Riordan JLH, Lawson DEM, Kodecek E: A competitive protein-binding assay for 25-hydroxycholecalciferol and 25-hydroxyergocalciferol in serum. Clin Chim Acta 54:235–242, 1974.
8. Mason RS, Posen S: Some problems associated with assay of 25-hydroxycalciferol in human serum. Clin Chem 23:806–810, 1977.
9. Dorantes LM, Arnaud SB, Arnaud CD: Importance of the isolation of 25-hydroxyvitamin D before assay. J Lab Clin Med 91:791–796, 1978.
10. Delvin EE, Glorieux FH, Dussault M, Bourbonnais R, Walters G: Simultaneous measurement of serum 25-hydroxycholecalciferol and 25-hydroxyergocalciferol. Med Biol 57:165–170, 1979.
11. Hollis BW, Burton JH, Draper HH: A binding assay for 25-hydroxycalciferols and 24R, 25-dihydroxycalciferols using bovine plasma globulin. Steroids 30:285–293, 1977.
12. Haddad JG, Min C, Walgate J, Hahn T: Competition by 24, 25-dihydroxycholecalciferol in the competitive protein binding radioassay of 25-hydroxycalciferol. J Clin Endocrinol Metab 43:712–715, 1976.
13. Taylor CM, Hughes SE, de Silva P: Competitive protein binding assay for 24, 25-dihydroxycholecalciferol. Biochem Biophys Res Commun 70:1243–1249, 1976.
14. Haddad JG, Min C, Mendelsohn M, Slatopolsky E, Hahn TJ: Competitive protein-binding radioassay of 24, 25-dihydroxyvitamin D in sera from normal and anephric subjects. Arch Biochem Biophys 182:390–395, 1977.
15. Kremer R, Guillemont S: A simple and specific competitive protein binding assay for 24, 25-dihydroxyvitamin D in human serum. Clin Chim Acta 86:187–194, 1978.

16. Horst RL, Shepard RM, Jorgensen NA, DeLuca HF: The determination of 24, 25-dihydroxy-vitamin D and 25-26-dihydroxyvitamin D in plasma from normal and nephrectomized man. J Lab Clin Med 93:277–285, 1979.

17. Shepard RM, Horst RL, Hamstra AJ, DeLuca HF: Determination of vitamin D and its metabolites in plasma from normal and anephric man. Biochem J 182:55–69, 1979.

18. Lambert PW, DeOreo PB, Hollis BW, Fu IY, Ginsberg BJ, Roos BA: Concurrent measurement of plasma levels of vitamin D₃ and five of its metabolites in normal humans, chronic renal failure patients, and anephric subjects. J Lab Clin Med 98:536–548, 1981.

19. Brumbaugh PF, Haussler DH, Bressler R, Haussler MR: Radioreceptor assay for 1α, 25-dihydroxyvitamin D₃. Science 183:1089–1091, 1974.

20. Brumbaugh PF, Haussler DH, Bursac DM, Haussler MR: Filter assay for 1α, 25-dihydroxyvitamin D₃. Utilization of the hormone's target tissue chromatin receptor. Biochemistry 13:4091–4097, 1974.

21. Eisman JA, Hamstra AJ, Kream BE, DeLuca HF: 1, 25-dihidroxyvitamin D in biological fluids: A simplified and sensitive assay. Science 193:1021–1023, 1976.

22. Eisman JA, Hamstra AJ, Kream BE, DeLuca HF: A sensitive, precise and convenient method for determination of 1,25-dihydroxyvitamin D in human plasma. Arch Biochem Biophys 176:235–243, 1976.

23. Lambert PW, Syverson BF, Arnaud CD, Spelsburg TC: Isolation and quantitation of endogenous vitamin D and its physiologically important metabolites in human plasma by high pressure liquid chromatography. J Steroid Biochem 8:929–937, 1977.

24. Dokoh S, Morita R, Fukunaga M, Yamamoto I, Torizuka K: Competitive protein binding assay for 1, 25-dihydroxyvitamin D in human plasma. Endocrinol Jpn 25:431–436, 1978.

25. Clemens TL, Hendy GN, Papapoulos SE, Fraher LJ, Care AD, O'Riordan JLH: Measurement of 1,25-dihydroxycholecalciferol in man by radioimmunoassay. Clin Endocrinol 11:225–234, 1979.

26. Peacock M, Taylor GA, Brown WB: Measurement of plasma 1, 25(OH)₂ vitamin D by protein binding assay and radioimmunoassay. In Norman AW, Schaefer K, Herrath D, Grigoleit HG, Coburn JW, Deluca HF, Mawer EB, and Suda T (Eds), Vitamin D: Basic Research and its Clinical Applications. Walter de Gruyter, Berlin, pp. 179–182, 1979.

27. Schaefer PC, Goldsmith RS: Radioimmunoassay of vitamin D metabolites. In Norman AW, Schaefer K, Herrath D, Grigoliet HG, Coburn JW, Deluca HF, Mawer EB, and Suda T (Eds), In Vitamin D: Basic Research and its Clinical Applications. Walter de Gruyter, Berlin, pp. 205–212, 1979.

28. Taylor CM, Hann J, St. John J, Wallace JE, Mawer EB: 1,25-dihydroxycholecalciferol in human serum and its relationship with other metabolites of vitamin D₃. Clin Chim Acta 96:1–8, 1979.

29. Bouillon R, DeMoor P, Baggiolini EG, Uskokovic MR: A radioimmunoassay for 1, 25-dihydroxycholeciferol. Clin Chem 26:562–567, 1980.

30. Mallon JP, Hamilton JG, Nauss-Karol C, Kavol RJ, Ashley CJ, Matuszewski DS, Tratnyek CA, Bryce GF, Miller ON: An improved competitive protein binding assay for 1, 25-dihydroxyvitamin D. Arch Biochem Biophys 201:277–285, 1980.

31. Manolagas SC, Howard J, Abare J, Culler FL, Brickman AS, Deftos LJ: Cyto-

receptor assay for 1,25-dihydroxy vitamin D_3: A convenient method and its application to clinical studies, In Norman AW, Schaefer K, Herrath D, Grigoliet HG, Coburn JW, Deluca HF, Mawer EB, and Suda T (Eds), Vitamin D: Chemical, Biochemical, and Clinical Endocrinology of Calcium Metabolism. Proceedings of the Fifth Workshop on Vitamin D, Williamsburg, Walter de Gruyter, Berlin, pp. 769–771, 1982.

32. Dokoh S, Pike JW, Chandler JS, Peters JM, Haussler MR: An improved radioreceptor assay for 1, 25-dihydroxyvitamin D in human plasma. Anal Biochem 116:211–222, 1981.

33. Jongen MJM, Van Der Vijgh WJF, Willems HJJ, Netelenbos JC: Analysis for 1, 25-dihydroxyvitamin D in human plasma, after a liquid-chromatographic purification procedure, with a modified competitive protein-binding assay. Clin Chem 27:444–450, 1981.

34. Hughes MR, Baylink DJ, Jones PG, Haussler MR: Radioligand receptor assay for 25-hydroxyvitamin D_2/D_3 and 1α, 25-dihydroxyvitamin D_2/D_3. Application to hypervitaminoses D. J Clin Invest 58:61–70, 1976.

35. Lambert PW, Toft DO, Hodgson SF, Lindmark EA, Witrak BJ, Roos BA: An improved method for the measurement of 1, 25-$(OH)_2D_3$ in human plasma, Endocrinol Res Commun 5:293–310, 1979.

36. Lambert PW, Hollis BW, Bell NH, Epstein S: Demonstration of a lack of change in serum 1, 25-dihydroxyvitamin in response to parathyroid extract in pseudohypoparathyroidism. J Clin Invest 66:782–791, 1980.

37. Lambert PW, Stern PH, Avioli RC, Brackett NC, Turner RT, Greene A, Fu IY, Bell NH: Evidence for extrarenal production of 1α, 25-dihydroxyvitamin D in man. J Clin Invest 69:722–725, 1982.

38. Lambert PW, DeOreo PB, Fu IY, Hollis BW, Kaetzel DM, Roos BA: Urinary and plasma vitamin D metabolism in normal humans and the nephrotic syndrome. Metab Bone Dis Relat Res 4:7–15, 1982.

39. Caldas AE, Gray RW, Lemann J: The simultaneous measurement of vitamin D metabolites in plasma studies in healthy adults and in patients with calcium nephrohthiasis. J Lab Clin Med 91:840–849, 1978.

40. Bishop JE, Norman AW, Coburn JW, Roberts PA, Henry HL: Studies on the metabolism of calciferol. XVI Determination of the concentration of 25-hydroxyvitamin D, 24,25-dihydroxyvitamin D and 1,25-dihydroxyvitamin D in a single two-milliliter plasma sample. Min Elect Metabol 3:181–189, 1980.

41. Jones G: Ternary solvent mixtures for improved resolution of hydroxylated metabolites of vitamin D_2 and vitamin D_3 during high-performance liquid chromatography, J Chromatogr Biomed Appl 221:27–37, 1980.

42. Horst RL, Littledike ET, Gray RW, Napoli J: Impaired 24,25-dihydroxyvitamin D production in anephric human and pig. J Clin Invest 67:274–280, 1981.

43. Suda T, DeLuca HF, Schnoes HK, Tanaka Y, Holick MF: 25, 26-dihydroxycholecalciferol, a metabolite of vitamin D_3 with intestinal calcium transport activity. Biochemistry 9:4776–4780, 1979.

44. Hollis BW, Roos BA, Lambert PW: 25, 26-dihydroxycholecalciferol: a precursor in the renal synthesis of 25-hydroxycholecalciferol-26, 23-lactone. Biochem Biophys Res Commun 95:520–528, 1980.

45. Kano K, Yoshida H, Yata J, Suda T: Age and seasonal variations in the serum levels of 25-hydroxyvitamin D and 24, 25-dihydroxyvitamin D in normal humans. Endocrinol Jpn 27:215–221, 1980.

46. Horst RL, Littledike ET, Riley JL, Napoli JL: Quantitation of vitamin D and its metabolites, and their plasma concentrations in five species of animals. Anal Biochem 116:189–203, 1981.

47. Mason RS, Lissner D, Grunstein HS, Posen S: A simplified assay for dihydroxylated vitamin D metabolites in human serum: Applications to hyper- and hypovitaminosis D. Clin Chem 26:444–450, 1980.

48. Hollis BW, Roos BA, Draper HH, Lambert PW: Vitamin D metabolites in human and bovine milk. J Nutr 111:1240–1248, 1981.

49. Adams, JS, Clemens TL, Holick MF: Sep-pak preparative chromatography for vitamin D and its metabolites. J Chromatogr Biomed Appl 226:198–201, 1981.

50. Arnaud CD, Tsao KS, Littledike T: Radioimmunoassay of human parathyroid hormone in serum. J Clin Invest 50:21–34, 1971.

51. O'Riordan JLH, Graham RF, Dolev E: An assay for 24, 25- and 24, 26-dihydroxycholecalciferols in sera. In Norman AW, Schaefer K, Coburn JW, DeLuca HF, Fraser D, Grigoliet HG, and von Herrath D (Eds), Vitamin D: Biochemical, Chemical and Clinical Aspects Related to Calcium Metabolism. Walter de Gruyter, Berlin, pp. 519–521, 1977.

52. Chesney RW, Rosen JF, Hamstra AJ, Smith C, Mahaffey K, DeLuca HF: Absence of seasonal variation in serum concentrations of 1, 25-dihydroxyvitamin D despite a rise in 25-hydroxyvitamin D in summer. J Clin Endocrinol Metab 53:139–142, 1981.

53. Gray TK, McAdoo T, Pool D, Lester GE, Williams ME, Jones G: A modified radioimmunoassay for 1, 25-dihydroxycholecalciferol. Clin Chem 27:458–463, 1981.

54. Stern PH, Phillips TE, Mavreas TO: Bioassay of 1, 25-dihydroxyvitamin D in human plasma purified by partition, alkaline extraction, and high pressure chromatography. Anal Biochem 102:22–30, 1980.

55. Seamark DA, Trafford DJH, Makin HLJ: The estimation of vitamin D and some metabolites in human plasma by mass fragmentography, Clin Chim Acta 106:51–62, 1980.

56. Bradford MM: A rapid and sensitive method for the quantitation of microgram quantities of protein utilizing the principle of protein-dye binding. Anal Biochem 72:248–254, 1976.

57. Reinhardt TA, Horst RL, Orf J, Hollis BW: A microassay of 1,25-dihydroxy vitamin D which does not require high performance liquid chromatography. Calc Tiss Int, in press, 1983.

Chapter 7

Radioimmunoassay of the Vitamin D Metabolites

Bernard P. Halloran

The preferred method for measurement of most hormones, and in particular the steroid hormones, is the radioimmunoassay (RIA).[1,2] Radioimmunoassays have been developed and are in wide use for virtually all of the major steroid hormones. There are, however, other methods for measuring hormone concentrations; one of the most popular is the competitive protein-binding (CPB) assay. This type of assay makes use of naturally occurring binding proteins found in blood and various tissues. Although the CPB assay is very similar to the RIA, the latter offers certain advantages. For example, the proteins used in CPB assays are often less stable than antibodies, necessitating frequent preparation of fresh binding protein. In addition, normally occurring binding proteins do not always offer the same high degree of specificity as antibodies.

In the study of vitamin D, quantitation of the vitamin D metabolites is typically done using CPB assays and not RIAs. The reason for this has been the inability to develop RIAs superior to existing CPB assays. Recently, however, interest has grown in producing highly specific antibodies for various vitamin D metabolites and using these antibodies in RIAs.[3-16] This chapter examines how these antibodies are made, how they are incorporated into RIAs, and what advantages they have over existing CPB assays.

I. Production of Antibodies

A. Techniques for Creating Immunogens

In general the vitamin D metabolites are not antigenic. In order to produce an antigen they must be used as haptens and covalently linked to some protein molecule. The most widely used protein for this purpose has been bovine serum albumin (BSA), although other proteins and even polypeptides such as poly-L-lysine could theoretically be used. The objective is to link the protein to some form of vitamin D in such a way that the resulting complex will induce the formation of antibodies specific for a given metabolite.

The procedure for linking a protein such as BSA to vitamin D first involves producing a vitamin D derivative that terminates in a reactive group such as carboxylic acid. Ideally this group should be located at a site distant from the important functional groups of the molecule (such as the C-1 position). Exam-

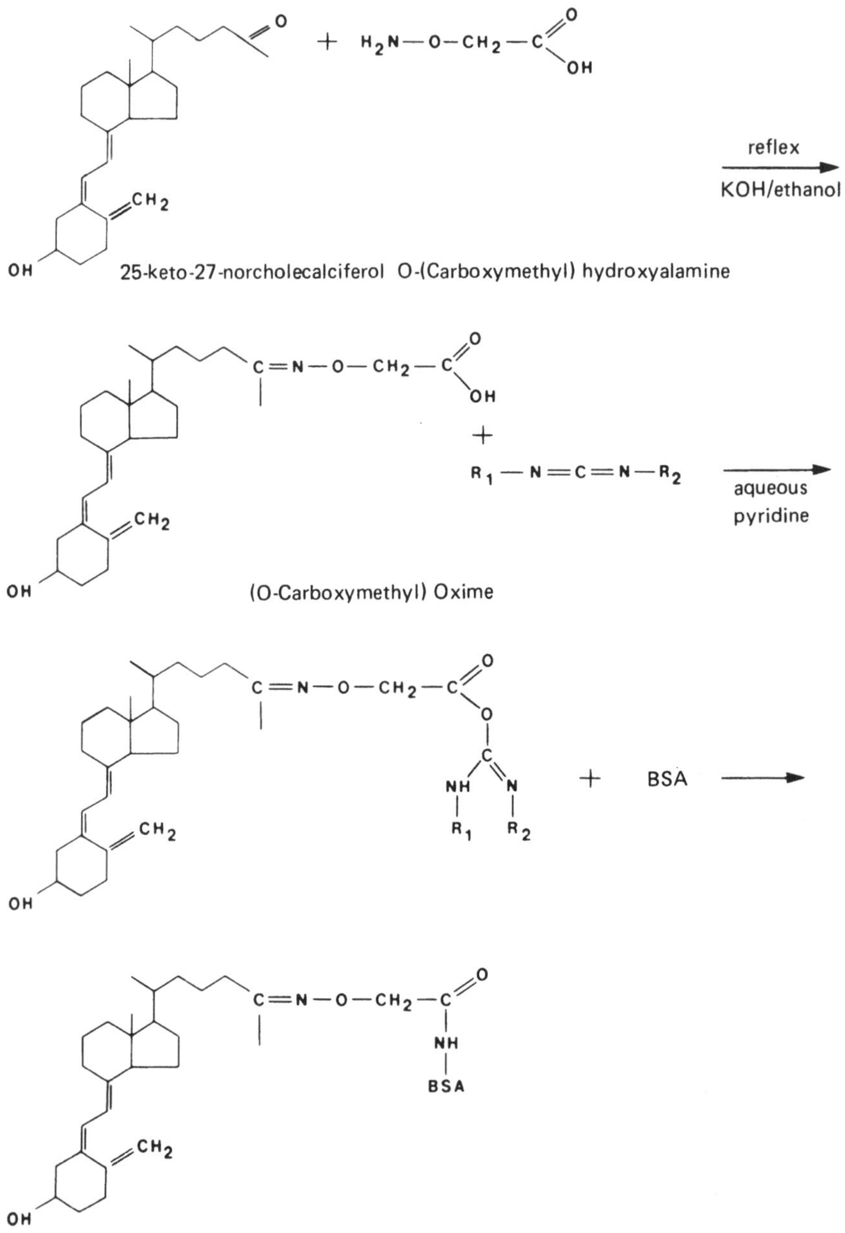

Figure 1. Synthesis of the 25-carboxymethyloxime of 25-keto-27-norcholecalciferol and peptide bond formation between free carboxyl group and the ε-amino groups of lysine residues on BSA.

ples of some of the vitamin D derivatives used include 1,25-$(OH)_2D_3$-25-hemisuccinate,[3,5,6,8] 1,25-$(OH)_2D_3$-3-hemisuccinate,[9,12,15] 25-$(OH)D_3$-25-hemisuccinate,[7,10] 25,26,27-trinor-C24-carboxylic acid,[13] and the 25-carboxymethyloxime of 25-keto-27-norcholecalciferol.[4] These derivatives are linked to BSA using either the mixed anhydride reaction or carbodiimide condensation.[17] The synthesis of the 25-carboxymethyloxime of 25-keto-27-norcholecalciferol and carbodiimide condensation is illustrated in Figure 1. 25-keto-27-norcholecalciferol was obtained from Philip Bell, Tenovus Institute for Cancer Research, Welsh National School of Medicine, Cardiff, United Kingdom.

B. Immunization and Harvesting of Antibody

The vitamin-D-derivative–BSA complex constitutes the antigen and at this point is ready to be injected into an animal. A number of species, immunization schedules, dosages, and adjuvants have been used to induce antibody formation. The most popular species in the vitamin D field has been the rabbit. The antigen is emulsified in 1 ml complete Freund's adjuvant and injected intradermally or subcutaneously at multiple sites along the back of the rabbit. Reports of the amount of antigen injected range from 100 to 450 μg. Typically, 4–6 weeks after the initial injections, a booster injection is given. Antibody can be harvested as soon as 10 days after the last injection, or monthly booster injections can be continued for several months before harvesting. Blood is obtained from the ear vein using a special vacuum flask; serum is stored at $-20°C$. Fifty milliliters of blood can be collected from a rabbit at weekly intervals.

It is important at this point to isolate partially the γ-globulin (IgG) fraction from the antiserum. This is done primarily to remove the serum vitamin-D-binding protein which may interfere in the RIA. With low-titer antisera this is particularly important. Two techniques have been used to effect partial purification of the IgG. In one, Rivanol(6,9 diamino-2 ethoxyacridine lactate), at a concentration of 0.4%, is added to the antiserum to precipitate the α-globulins. In the second, sodium sulfate (14%) is used to precipitate the IgG fraction. The precipitate is then redissolved in phosphate-buffered saline (0.05 M phosphate, pH 7.5) and dialyzed to remove the sodium sulfate.[18] Further purification can be accomplished by DEAE Sephadex chromatography if necessary.

C. Characterization of Antibody

The three most important properties in characterizing an antibody are titer, affinity, and specificity. Titer, as reflected by the degree to which the antiserum can be diluted, can vary enormously. The best dilution of an antiserum is generally that which binds 30–50% of the radioactive hormone under assay conditions. Using this as a criterion, reported titers for vitamin D antibodies range from 1:50 to 1:50,000 dilution. The antiserum used in our laboratory, which was produced and characterized by Drs. P. Schaefer and R. Goldsmith,[4] is usually diluted 1:200.

The binding affinity can be determined by Scatchard analysis. Typical equilibrium dissociation constants are on the order of 10^{-10} M or similar to that observed for the intestinal cytosolic binding protein for 1,25-$(OH)_2D_3$. The smaller the dissociation constant, the more sensitive the assay. Most antibody assays are as sensitive as, or perhaps a little more sensitive than, receptor assays.

The specificity of binding can be determined by comparing the relative ability of various vitamin D metabolites, as well as other compounds, to displace radiolabeled hormone. The specificity of antibodies varies considerably. No two are alike. On the one hand it would be convenient to have an antibody with high affinity for all the vitamin D metabolites. This would ensure that it could be used for measurement of any given metabolite. On the other hand, it would be ideal if an antibody could be produced with high enough specificity to be used to assay serum concentrations of the metabolite of interest without purification. In the case of 25-OHD, this seems entirely feasible (see Section II.B). With 1,25-$(OH)_2D$, however, it is less likely. The reasons for this are the relatively low concentrations of 1,25-$(OH)_2D$ in serum compared to 25-OHD, requiring an extraordinary degree of specificity to permit accurate measurements. Since the concentration of 25-OHD in most serum samples is roughly a thousand times higher than the concentration of 1,25-$(OH)_2D$, it would be necessary for the specificity of the antibody to be four or five orders of magnitude greater for 1,25-$(OH)_2D$ than 25-OHD. To date, this has not been the case. As a result, all RIAs for the vitamin D metabolites, with the exception of 25-OHD, must be preceded by some form of chromatography.

The antibody used in our laboratory will not bind progesterone, testosterone, estradiol, or a number of other steroids even at concentrations of 1.0 mg/ml. The relative specificity for the vitamin D metabolites using our antibody is 1,25-$(OH)_2D_3$ > 25-OHD_3 ≈ 24,25-$(OH)_2D_3$ > 1α-OHD_3 ≫ D_3.

II. Radioimmunoassay

A. General Assay Techniques

The methods by which antibodies are used to measure the vitamin D metabolites are, for the most part, very similar to those existing for the naturally occurring binding proteins. In many cases they are identical. The antiserum, or partially purified IgG fraction, simply takes the place of the vitamin-D-binding protein or cytosol receptor.

The inadequate specificity of existing antibodies and the presence of interfering substances in serum make it necessary, in most cases, first to isolate the vitamin D metabolites of interest before assay. The typical procedures for isolation are virtually identical to those used prior to CPB assay. As an illustration, the following methods are used in our laboratory for the isolation and assay of 25-OHD and 1,25-(OH)$_2$D from serum.

Extraction and Chromatography

1. Typically, 20 serum samples are run simultaneously. Approximately 4000 dpm each of 1,25(OH)$_2$[23,24(n)-^{3}H]D$_3$ and 25-OH[23,24(n)-^{3}H]D$_3$ (70–110 Ci/mmole, Amersham, Arlington Heights, New Jersey) are added to 0.5–2.0 ml serum in a screw top test tube. The sample is vortexed and allowed to sit for 10 min.

2. Three volumes of peroxide-free diethyl ether (product No. 9248, J. T. Baker Co., Phillipsburg, New Jersey) are added, and the sample tubes are shaken vigorously for 5 min.

3. After the tubes are allowed to sit for several minutes, the bottom or aqueous phase is frozen by immersing the tubes in a dry ice-acetone bath.

4. The ether phase is decanted, the aqueous phase thawed, and the extraction procedure repeated. Completion of steps 1–4 takes approximately 1 h. The extraction efficiencies for 25-OHD$_3$ and 1,25-(OH)$_2$D$_3$ are 77 and 97%, respectively.

5. The ether extracts are pooled, dried under nitrogen, and redissolved in 0.5 ml of hexane/chloroform/methanol (9:1:1) in preparation for Sephadex LH-20 chromatography. All solvents are redistilled or high-pressure liquid chromatography (HPLC) grade.

6. Samples are applied to a 0.7 × 12-cm Sephadex LH-20 column and eluted with hexane/chloroform/methanol (9:1:1). 25-OHD elutes between 4 and 9.5 ml, and 1,25-(OH)$_2$D elutes between 9.5 and 25 ml. Each fraction is collected separately, dried, and redissolved in the appropriate solvent in preparation for high-pressure liquid chromatography. Note: Ether extraction alone does not result in high extraction efficiency for 25-OHD or 25-OHD$_3$-26,23 lactone. At the same time, however, substantially less lipid is extracted. This results in a cleaner sample and eliminates the need for further low-pressure chromatography in preparing the 25-OHD fraction for HPLC (see Chapter 2).

7. The 25-OHD fraction from LH-20 chromatography (which includes both 25-OHD$_2$ and 25-OHD$_3$) is applied to a Waters Radial Compression Separation System (RCSS) using an 8-mm Si radial compression column (RCC) with 10-μm particles. The solvent system is isopropanol/hexane (4:96). The design of the RCSS allows higher flow rates than conventional steel columns with little loss of resolution. Using a flow rate of 4.0 ml/min, the 25-OHD$_3$ elutes in approximately 3 min, 15 s (13 ml) (Figure 2). The time between application of samples is on the order of 8 min. Overall recovery, including extraction and chromatography, is approximately 60%. Quantitation of the 25-OHD is accomplished in one of two ways. If enough 25-OHD is present, it is measured directly using ultraviolet (UV) absorbance (see Chapter 2). If adequate amounts of 25-OHD are not present, the

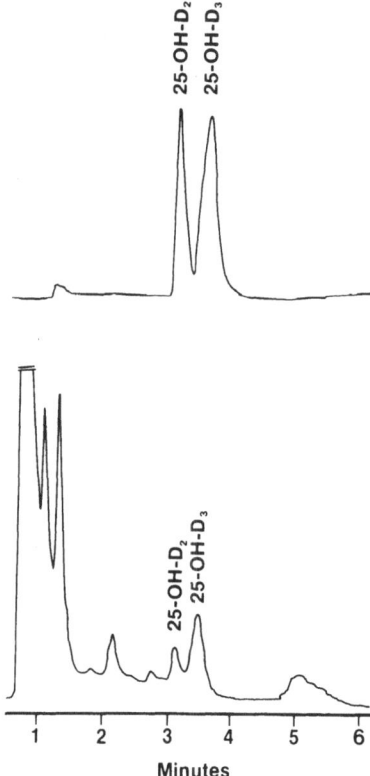

Figure 2. Elution profile of 25-OHD₂ and 25-OHD₃ using a Waters Radial Compression Separation System, a solvent system of isopropanol/hexane (4:96), and a flow rate of 4 ml/min. **Top:** Standards of 25-OHD₂ and 25-OHD₃. **Bottom:** Typical sample profile. The peak eluting at the same time as authentic 25-OHD₃ is equivalent to approximately 12 ng of hormone.

25-OHD₂ and 25-OHD₃ fraction is collected, dried under nitrogen, and 100% ethanol added in preparation for RIA. Note: Caution must be taken to clean all glassware with ethanol to ensure that contaminants do not interfere with UV quantitation of the 25-OHD. This is a common problem and can be recognized by the appearance of unusually large amounts of UV-absorbing material in the chromatogram.

8. The 1,25-(OH)₂D fraction from LH-20 chromatography is also applied to a Waters RCSS. The column used is a 8-mm Si RCC with 5-μm particles. At a flow rate of 4.5 ml/min and a solvent system of isopropanol/hexane (8:92), 1,25-(OH)₂D₃ elutes in approximately 8 min, 30 s. Time between application of samples is on the order of 12 min.

Assay

The basic procedures presented here for RIA of 25-OHD and 1,25-(OH)₂D were developed by Drs. P. Schaefer and R. Goldsmith.[4] They are essentially identical; therefore, only the assay of 1,25-(OH)₂D will be presented.

1. The 1,25-(OH)₂D fraction from HPLC is transferred to a polypropylene assay tube (product No. 223-6430, Biorad, Richmond, California) and

brought up in 30 μl of 100% ethanol. The assay tubes have an attached cap which can be used to seal the tubes during mixing and incubation. No additional 1,25-(OH)$_2$D$_3$ is added to the sample tubes. The labeled 1,25-(OH)$_2$D$_3$ initially added to the serum samples before extraction is used as tracer.

2. Standards are run in triplicate. Each standard tube receives 3000 dpm of [^{3}H]1,25-(OH)$_2$D$_3$ in 10 μl of 100% ethanol plus an additional 20 μl of ethanol containing 0, 6, 10, 20, 30, 40, 60, 80, 100, or 200 pg of unlabeled 1,25-(OH)$_2$D$_3$.

3. Diluted antibody, 200 μl (1:200), is added to each tube. The tubes are capped, vortexed, and incubated at room temperature for 2 h without being shaken. The diluted antibody is prepared by combining 1 part of specific rabbit IgG (antibody) with 200 parts of a phosphate buffer solution (0.05 M sodium phosphate, pH 7.5) containing 0.18 mg/ml nonspecific rabbit IgG (product No. 15006, Sigma, St. Louis, Missouri). Addition of nonspecific IgG helps solubilize the 1,25-(OH)$_2$D.

4. After 2 h incubation, a 50-μl aliquot is removed from each tube, combined with scintillation fluor, and counted. The counts from this aliquot, corrected for volume, allow calculation of the amount of 1,25-(OH)$_2$D (in standard or samples) assayed. These counts are also used in calculating the ratio of bound to total (B/T) [^{3}H]1,25-(OH)$_2$D$_3$.

5. Following removal of the 50-μl aliquot for determination of total counts, 50 μl of dextran-coated charcoal (0.5% charcoal, 0.05% dextran in 0.05% phosphate buffer, pH 7.5) is added. The tubes are capped, vortexed, and allowed to incubate for 15 min at room temperature. Note: The charcoal solution is prepared in the following manner. 20 mg of dextran (dextran T-40, product No. 17-0270-01, Pharmacia, Piscataway, New Jersey) is dissolved in 40 ml of 0.05 M sodium phosphate buffer, pH 7.5. This solution is added to 200 mg of charcoal (carbon, neutral, norit, Fisher Scientific, Fair Lawn, New Jersey), and the charcoal-dextran suspension is mixed for 2 h at 4°C using a magnetic stirrer. After mixing, the suspension is centrifuged in a Sorvall GLC-1 centrifuge at 2000 rpm for 5 min, the supernatant decanted, and the charcoal-dextran pellet resuspended in 40 ml of fresh phosphate buffer before use.

6. After charcoal adsorption of free ligand, the assay tubes are centrifuged in a Sorvall GLC-1 centrifuge at 2000 rpm for 5 min at room temperature.

7. Following centrifugation, 160 μl of supernatant is removed from each tube, combined with scintillation fluor, and counted. The counts from this aliquot are used to calculate the bound fraction.

Calculations

All raw count data are corrected to disintegrations per minute (dpm). From the number of dpm in the initial 50-μl aliquot (dpm$_{50}$) removed before addition of charcoal and the 160-μl aliquot (dpm_{160}) removed after charcoal adsorption of free ligand (both corrected for volume), the ratio of bound to total [^{3}H]1,25-$(OH)_2D_3$ is calculated:

$$T = \text{total dpm in assay tube} = \frac{230}{50} dpm_{50}$$

$$B = \text{total bound dpm in assay tube} = \frac{230}{160}\frac{230}{180} dpm_{160}$$

$$B/T = 0.399 \frac{dpm_{160}}{dpm_{50}}$$

This ratio, expressed as percent bound, is plotted as a function of total 1,25-$(OH)_2D_3$ assayed for each of the standards. Note that the total 1,25-$(OH)_2D_3$ assayed is not equal to the amount of 1,25-$(OH)_2D_3$ added to each standard tube. Two corrections must be made. One involves a correction for solubilization of the 1,25-$(OH)_2D_3$. Not all of the 1,25-$(OH)_2D_3$ added to the standard tubes is solubilized: a certain fraction adheres to the walls of the assay tubes. This is the reason for the addition of the nonspecific rabbit IgG to the assay: it helps solubilize the 1,25-$(OH)_2D_3$. The soluble fraction can be determined from the 50-μl aliquot removed before addition of the charcoal, and knowledge of how many dpm were initially added to each tube. For example, if 3000 dpm were added to each standard tube but the number of dpm in the tubes as determined by counting the 50-μl aliquot (corrected for volume) is 2400 dpm, it would imply that only 80% of the total 1,25-$(OH)_2D_3$ in the tube is soluble. Typically, the percent soluble, or percent assayed ($A_\%$), for the standards ranges between 65 and 85%:

$$A_\% = \frac{230}{50}\frac{dpm_{50}}{dpm_{added}}$$

The second correction takes into account the amount of 1,25-$(OH)_2D_3$ added to each tube in the form of tracer. Adding 3000 dpm of [^{3}H]1,25-$(OH)_2D_3$ at 110 Ci/mmole is equivalent to adding 5.1 pg of 1,25-$(OH)_2D_3$. If 3000 dpm are added to each standard tube in a given assay, the total 1,25-$(OH)_2D_3$ in the "0" standard tube would be 5.1 pg, and in the "6" pg standard tube it would be 11.1 pg, and so on. The total 1,25-$(OH)_2D_3$ assayed for each standard (T_a) is calculated from the following formula.

$$T_a = A_\%[1,25\text{-}(OH)_2D_3 \text{ standard added}$$
$$+ 1,25\text{-}(OH)_2D_3 \text{ added as tracer}]$$

To illustrate these corrections, a typical standard curve is shown in Figure 3. Note that the first point on the standard curve is 9.8 pg. This results from the addition of approximately 6750 dpm (110 Ci/mmole) of 1,25-$(OH)_2D_3$, the equivalent of 11.5 pg, and an 85% solubilization of the hormone (11.5 $\times$ 85% = 9.8 pg).

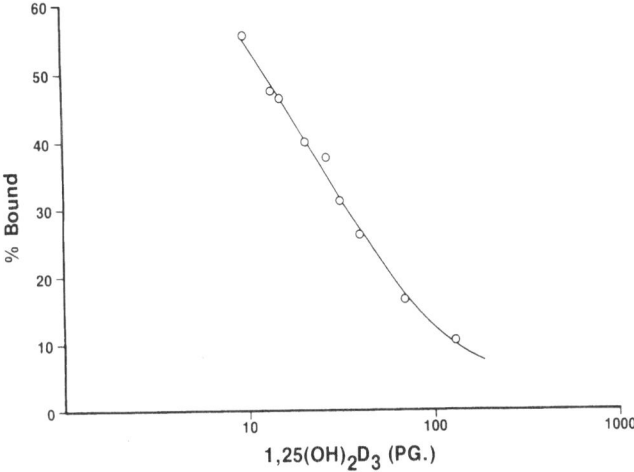

Figure 3. Standard curve for 1,25-(OH)$_2$D$_3$.

To determine the amount of 1,25-(OH)$_2$D$_3$ in each sample, the ratio of bound to total or percent bound is calculated. From this value, the number of picograms of 1,25-(OH)$_2$D$_3$ in the assay tube can be read from the standard curve (pg$_{curve}$). This number is then corrected for the contribution of the tracer (usually on the order of 3–10 pg, depending on the number of dpm initially added to the serum sample and the specific activity). At this point the concentration of 1,25-(OH)$_2$D$_3$ in the serum sample can be calculated by dividing the corrected amount of 1,25-(OH)$_2$D$_3$ in the assay tube by the product of the percent of 1,25-(OH)$_2$D$_3$ assayed of the sample and sample volume. The percent recovered (usually on the order of 65–70%) is calculated from the number of dpm in the 50-μl aliquot removed before charcoal addition and knowledge of the number of dpm added to the samples initially.

$$\text{Serum concentration } 1,25\text{-(OH)}_2\text{D} = \frac{\text{pg}_{curve} - \text{tracer assayed}}{\text{percent assayed} \times \text{ml}}$$

Intraassay and interassay coefficients of variation are on the order of 10%. The minimum detectable amount of 1,25-(OH)$_2$D is approximately 10 pg.

B. New Applications

Taking advantage of the relatively high specificity and affinity of antibodies, our laboratory has recently demonstrated the feasibility of measuring 25-OHD directly on ethanol extracts of serum. The success of previous attempts to measure 25-OHD in this manner using the vitamin-D-binding protein have proved controversial. Some investigators report that 25-OHD can be accurately measured and some report that it cannot. The general consensus is that it can not.

The reason for this appears to be the presence of interfering substances in plasma and serum. It has been suggested that if small enough serum equivalents could be used, the interfering substances would be diluted out and the assay would work. The sensitivities of existing vitamin-D-binding protein assays appear to preclude this possibility. On the other hand, using our antibody, it appears quite possible to measure 25-OHD directly. This is accomplished in the following manner.

Extraction

1. 50 μl of serum is combined with 300 μl of 100% ethanol in a 500-μl polypropylene tube (Section II.A), the contents vortexed and the tubes allowed to sit for 15 min.
2. The tubes are then centrifuged in a Sorvall GLC-1 centrifuge at 2000 rpm for 5 min.
3. Following centrifugation, 10- and 20-μl aliquots of supernatant are removed and placed in 500-μl polypropylene assay tubes (see Section II.A). The extraction efficiency for 25-OHD under these conditions is nearly 100%. No attempt is made to calculate recoveries for each individual sample. Instead, the amount of 25-OHD in the 10- and 20-μl aliquots, expressed as a percentage of the total amount of 25-OHD in the initial 50 μl of serum, was measured repeatedly in separate experiments. Approximately 2.80% and 5.70% of the total 25-OHD in the 50 μl of serum was contained in the 10- and 20-μl aliquots, respectively indicating an extraction efficiency of >98%.

Assay

1. Approximately 3000 dpm of [^{3}H]25-OHD in 10 μl of 100% ethanol are added to each tube (including standards). Those tubes containing the 10-μl sample aliquot receive an additional 10 μl of ethanol. Note that, unlike the 1,25-$(OH)_2$D assay, the samples contain no radiolabeled 25-OHD prior to this step, and that each tube, samples and standards, contains the same number of counts.
2. 200 μl of diluted antibody(1:200) is added to each tube.
3. After 2 h incubation at room temperature, 70 μl of 0.5% charcoal, 0.05% dextran in phosphate buffer [same preparation as used in the 1,25-$(OH)_2$D assay] is added. The tubes are vortexed and allowed to incubate at room temperature for 15 min.
4. The assay tubes are then centrifuged in a Sorvall GLC-1 centrifuge at 2000 rpm for 5 min; 200 μl of supernatant is then removed and counted.

Calculations

Corrections for labeled hormone added and solubilization of hormone have not been found to improve assay results. For this reason, these corrections are not made. Instead the number of dpm in each standard tube is plotted as a function of 25-OHD in the standard. The number of dpm in the samples is then used to calculate the number of pg in each sample tube (S_{pg}). After volume correction, the concentration of 25-OHD is expressed in ng/ml.

$$\text{Serum concentration of 25-OHD in ng/ml} = \frac{(S_{pg})(20)}{0.028 \text{ or } 0.057}$$

Comparing measured concentrations of 25-OHD using the direct assay and a chromatographic-HPLC-UV absorbance assay in 14 patients (range = 9–45 ng/ml), we have observed a difference of only 5% (r = 0.95), with the direct assay results being consistently higher. This small discrepancy can be explained by the fact that the antibody also recognizes $24,25\text{-}(OH)_2D$, $1,25\text{-}(OH)_2D$, and perhaps other vitamin D metabolites. Therefore, what is being measured is not just 25-OHD. However the contribution of these other metabolites is relatively small (approximately 5%), and for most clinical purposes (e.g., assessment of nutritional status) can be ignored.

C. Future Developments

RIAs for the vitamin D metabolites are relatively new and, until recently, have not been shown to offer major advantages over CPB assays. However, recent work is beginning to demonstrate the potential utility of antibodies in the measurement of vitamin D metabolites. Once the antibody has been obtained, the RIA offers the convenience of not having to routinely prepare fresh binding protein. In addition, some RIAs offer increased sensitivity compared to CPB assays. New assay techniques are also beginning to emerge which may improve assay efficiency,[11] and there is at least one report of the production of monoclonal antibodies.[16] Antibodies with relatively high specificity for certain of the vitamin D metabolites have recently been obtained,[13,15] and hint at the possibility of producing antisera with sufficiently high specificity to eliminate the need for chromatography. Already, direct RIA of 25-OHD has been demonstrated and, as antibodies with increasing specificities are developed, it is likely that routine measurement of other vitamin D metabolites will be accomplished using the RIA.

References

1. Antoniades HN, Ed.: Hormones in Human Blood. Harvard University Press, Cambridge, Mass., 1976.
2. Jaffe BM, Behrman HR, (Eds.): Methods of Hormone Radioimmunoassay. Academic Press, New York, 1979.
3. Clemens TL, Hendy GN, Graham RF, Baggiolini EG, Uskokovic MR, O'Riordan JLH: A radioimmunoassay for 1,25-dihydroxycholcalciferol. Clin Sci Mol Med 54:329–332, 1978.

4. Schafer PC, Goldsmith RS: Radioimmunoassay of vitamin D metabolites. In Norman AW, et al. (Eds.): Vitamin D: Basic Research and its Clinical Application. Walter de Gruyter, Berlin, p. 205, 1979.

5. Peacock M, Taylor GA, Brown WB: Measurement of plasma 1,25(OH)$_2$D by protein binding assay and radioimmunoassay. In Norman AW et al (Eds.): Vitamin D: Basic Research and its Clinical Application. Walter de Gruyter, Berlin, p. 179, 1979.

6. O'Riordan JLH, Clemens TL, Hendy GN, Fraher LJ, Sandler LM, Papapoulos SE: Radioimmunoassay for circutating 1,25- and 24,26-dihydroxycholecalciferols in man. In Norman AW et al. (Eds.): Vitamin D: Basic Research and its Clinical Application. Walter de Gruyter, Berlin, p. 221, 1979.

7. Schmidt-Goyk H, Mayer E, Schuer R, Lichtwald K, Bouillon R, Clemens TL: Studies on antisera against vitamin D metabolites. In Norman AW et al. (Eds.): Vitamin D: Basic Research and its Clinical Applications. Walter de Gruyter, Berlin, p. 233, 1979.

8. Peacock M, Taylor GA, Brown W: Plasma 1,25(OH)$_2$D measured by radioimmunoassay and cytosol radioreceptor assay in normal subjects and patients with primary hyperparathyroidism and renal failure. Clin Chim Acta 101:93–101, 1980.

9. Bouillon R, DeMoor P, Baggiolini EG, Uskokovic MR: A radioimmunoassay for 1,25-dihydroxycholecalciferol. Clin Chem 26:562–567, 1980.

10. Bouillon R, VanBaelen H, Van Herck E: Simple, direct, nonchromatographic assay for 25-hydroxyvitamin D. Presented at the Fifth Workshop on Vitamin D, Williamsburg, Va., 1982.

11. Fraher LJ, Baker TS, O'Riordan JLH: Development of a solid phase radioimmunoassay for 1,25-dihydroxycholecalciferol. In Norman AW, Schaefer K, Herrath Dv, Grigoleit H-G (Eds.): Vitamin D: Chemical, Biochemical and Clinical Endocrinology of Calcium Metabolism. Walter de Gruyter, Berlin, p. 793, 1982.

12. Gray TK, McAdoo T, Pool D, Williams ME, Lester GE, Thierry-Palmer M: Development and application of a radioimmunoassay for 1,25-dihydroxycholecalciferol. In Norman AW, Schaefer K, Herrath Dv, Grigoleit H-G (Eds.): Vitamin D: Chemical, Biochemical, and Clinical Endocrinology of Calcium Metabolism. Walter de Gruyter, Berlin, p. 763, 1982.

13. Henderson SL, Clemens TL, Hannafin N, Holick MF, Baggiolini E, Uskokovic M: Development of a sensitive and specific radioimmunoassay for 1,25-dihydroxyvitamin D$_3$ using antibodies raised against 1-hydroxycholecalcioic acid-bovine serum albumin. Presented at the Fifth Workshop on Vitamin D, Williamsburg, Va., 1982.

14. O'Riordan JLH, Adami S, Sandler LM, Fraher LJ: Clinical application of vitamin D radioimmunoassays. In Norman AW, Schaefer K, Herrath Dv, Grigoleit H-G (Eds.): Vitamin D: Chemical, Biochemical, and Clinical Endocrinology of Calcium Metabolism. Walter de Gruyter, Berlin, p. 751, 1982.

15. Peacock M, Brown WB: Characteristics of antisera to 1,25(OH)$_2$D$_3$ raised against two antigens. Presented at the Fifth Workshop on Vitamin D, Williamsburg, Va., 1982.

16. Perry HM, Chappel JC, Clevinger BL, Haddad JG, Teitelbaum SL: An assay for 1,25-dihydroxyvitamin D using a monoclonal antibody. In Norman AW, Schaefer K, Herrath Dv, Grigoleit H-G (Eds.): Vitamin D: Chemical, Biochemical, and

Clinical Endocrinology of Calcium Metabolism. Walter de Gruyter, Berlin, p. 181, 1982.

17. Abraham GE, Grover PK: Covalent linkage of steroid hormones to protein carriers for use in radioimmunoassay. In Odell WD, Daughaday WH (Eds.): Principles of Competitive Protein Binding Assays. J. B. Lippincott, Philadelphia, p. 140, 1971.

18. Strauss AJL, Kemp PG, Vannier WE, Goodman HC: Purification of human serum γ-globulin for immunologic studies. J Immunol 93:24–35, 1964.

Chapter 8

The Cytoreceptor Assay for 1,25-Dihydroxyvitamin D

Stavros C. Manolagas

It is established that 1,25-dihydroxyvitamin D_3 [1,25-(OH)$_2D_3$] acts in a manner similar to that proposed for other steroid hormones. Target cells for 1,25-(OH)$_2D_3$ contain in their cytoplasm specific "receptor" proteins which bind 1,25-(OH)$_2D_3$ with high affinity. Once the hormone-receptor complex is formed, it is transferred to the cell nucleus, where it initiates the synthesis of new messenger RNA. The messenger RNA in turn is translated in the ribosomes of the cells to new proteins responsible for the specific effects of the hormone.

In studies performed in cytosol preparations, although 1,25-(OH)$_2D_3$ exhibits higher affinity, other analogs of vitamin D such as 25- and 24,25-(OH)$_2D_3$ have also been shown to interact with the receptor but with an affinity of two to three orders of magnitude less than the affinity of 1,25-(OH)$_2D_3$. Nevertheless, considering that the circulating levels of 25-(OH)D$_3$ ($10^{-7} M$) and 24,25-(OH)$_2D_3$ ($10^{-8} M$) are, respectively, 1000 and 100 times higher than the level of 1,25-(OH)$_2D_3$ ($10^{-10} M$), these metabolites could substitute in vivo for 1,25-(OH)$_2D_3$ at the receptor level. However, such a possibility is not consistent with the carefully and tightly controlled plasma levels of 1,25-(OH)$_2D_3$. In our laboratory we have recently examined this puzzling issue and have demonstrated that metabolites of vitamin D_3 other than 1,25-(OH)$_2D_3$ at their normal circulating levels have no access to the receptor because of their higher affinity to the plasma-binding protein and relatively lower affinity for the intracellular receptor.[1] By contrast, 1,25-(OH)$_2D_3$, because of its higher affinity for the receptor and lower affinity for the plasma protein, diffuses more readily across the cell membranes and is retained inside its target cells. Hence, the differential affinities between plasma vitamin D binding protein and intracellular 1,25-(OH)$_2D$ receptor provide a biological separation allowing 1,25-(OH)$_2D_3$ to be delivered to its target cells without interference, in spite of the relatively low circulating levels of the hormone. This biological concept has been exploited to develop a simple and convenient method for measuring 1,25-(OH)$_2D_3$ in the blood.[2]

The measurement of 1,25-(OH)$_2D_3$ has been particularly difficult in the past. The most widely used traditional methods are competitive binding radioassays that utilize as binding proteins either 1,25-(OH)$_2D_3$ receptors isolated from tissues of vitamin-D-deficient animals or antibodies raised against the hormone.[3-6] Unfortunately, none of these binding proteins is selective enough to distinguish directly the hormone at its low concentration from other circulating vitamin D compounds. Thus, 1,25-(OH)$_2$D has to be isolated and purified from blood by a variety of steps including high-pressure liquid chromatography (HPLC) before its binding can be assessed with the binding protein. These methods, although effective in determining accurately 1,25-(OH)$_2$D concentrations, are time-consuming, cumbersome, and not readily adapted to wide clinical application.

The cytoreceptor assay obviates several technical problems of the previous methods, including the need for the chromatographic isolation of 1,25-(OH)$_2$D$_3$. In this novel method, extracts of blood samples are incubated with intact cells which possess the specific 1,25-(OH)$_2$D$_3$ receptor in medium that contains human α-globulin, a protein rich in plasma D-binding protein. The intact cell membrane in combination with the D-binding protein on the outside of the cells and the receptor system inside the cells perform, in a biological manner, the separation of 1,25-(OH)$_2$D$_3$ from the other metabolites of vitamin D present in the blood extract. 1,25-(OH)$_2$D$_3$ is retained inside the cells bound to the receptor; the other metabolites remain outside the cells bound to the D-binding protein and are subsequently removed by washing the cells free of the medium. The quantitation of 1,25-(OH)$_2$D$_3$ in the sample is accomplished by comparing the displacement of radioactive 1,25-(OH)$_2$D$_3$ bound inside the cells produced by the sample extract to the displacement produced by known amounts of unlabeled 1,25-(OH)$_2$D$_3$.

The principle of the cytoreceptor assay is illustrated diagrammatically in Figure 1.

I. Methodology

A. Cells

The cells we have been using are rat osteogenic sarcoma lines (ROS).[7] These cells are cultured on the surface of plastic flasks T175 (Falcon) at 37°C in 5% CO_2 in air with Coon's F12 medium, 5% fetal calf serum (GIBCO), penicillin-streptomycin 10 μg/ml, and mycostatin 20 μg/ml. Every 3 days confluent cells are split 1:2 to 1:3 and are subcultured in new flasks. For the final subculture before harvesting the cells to be used in the assay, the culture medium is enriched with 10^{-7} M triamcinolone acetonide (Sigma). After 3 days of growth in the presence of triamcinolone the medium is aspirated and the flasks are rinsed with 10 ml phosphate buffered saline (PBS) buffer. The cells are then freed by 5-min treatment with 2 ml Trypsin-EDTA solution in PBS buffer; this solution is prepared by 1:8 dilution of a concentrated preparation of Trypsin-EDTA (10×, Irvine Scientific). After the 5-min treatment, fresh medium is added and the cells are transferred into plastic conical tubes where they are spun at 1000 rpm for 5 min. The supernatant is then aspirated, and the pelleted

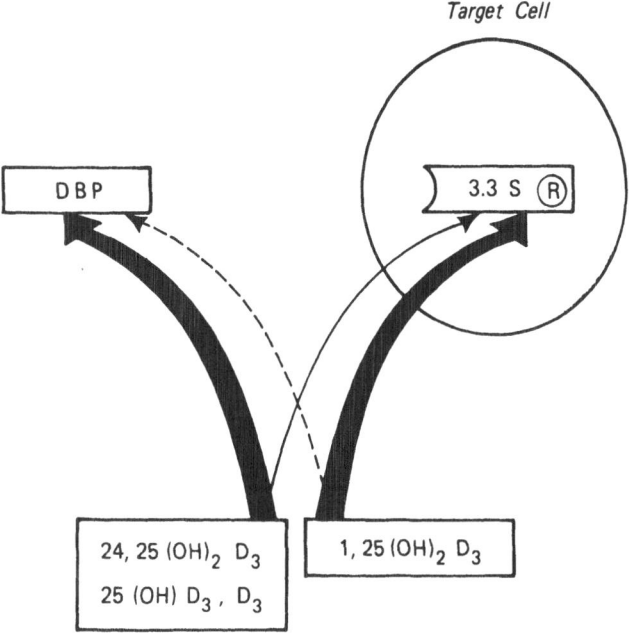

Figure 1. Schematic illustration of the cytoreceptor assay principle.

cells are resuspended in regular medium containing 10% DMSO. While cells are being harvested from consecutive flasks the freed cells are kept at $-4°C$; the total yield is finally pooled and mixed gently to obtain a homogeneous suspension. An aliquot of this suspension is counted with hemocytometer to determine the total cell number and the suspension is adjusted to $\sim 12 \times 10^6$ cells/ml; aliquots of this are then frozen quickly at $-60°C$ in plastic tubes in a Revco ultralow freezer. Alternatively, cells can be frozen in liquid nitrogen.

B. Chemicals

1,25-$(OH)_2$D [26,27-^{3}H]D_3 (163 Ci/mmol) is purchased from the Radiochemical Centre (Amersham, Arlington Hts., Illinois). 1,25-$(OH)_2D_3$ is provided by Dr. Milan Uskoković of Hoffmann-LaRoche (Nutley, New Jersey). Human α-globulin (Fraction IV) is purchased from Miles Lab., Inc. Benzene (spectra-grade) is purchased from Mallinckrodt. Disposable minicolumns prepacked with diatomaceous earth (4 cm height in 8 × 1.5-cm plastic syringes, Clin-Elute 1003) are purchased from Analytichem International, Inc. (Harbor City, California).

Isotonic buffer (washing solution): 0.25 M sucrose; 0.025 M KCl; 0.001 M MgCl$_2$; 0.001 M EDTA; 0.012 M thioglycerol; 2 mg BSA/ml; and 0.05 M Tris-HCl, pH 7.4.

C. Methods

The step-by-step procedure of the cytoreceptor assay employed in our laboratory is as follows.

Extraction of Samples

1. Wash extraction columns (Clin-Elute 1003) with 10 ml benzene. Discard. Let columns dry.
2. Spin samples (serum or plasma) at 1,500 rpm $\times$ 10 min to remove debris (cryoproteins, etc.) and aliquot 1–2 ml in glass counting vials.
3. Add $\simeq$1500 cpm of [^{3}H]1,25-(OH)$_2$D$_3$ (in 20 μl ETOH) for recovery monitoring. (Save 3 aliquots for Recovery Total Activity.) Vortex samples. Leave 30 min at room temperature. Add 45 μl of 1 N NaOH solution/ml of sample and make up volume to 3 ml with distilled H$_2$O.
4. Place sample on the column. Elute with 20 ml benzene (5 ml $\times$ 4). Collect eluent in glass counting vials. Dry under N$_2$. Resuspend in 1 ml ETOH.

Assay

1. Aliquot 0.24 ml (in triplicate) of the ETOH resuspended sample and standard amounts of unlabeled 1,25-(OH)$_2$D$_3$ (0. 2, 4, 8, 16, 32, 64, and 128 pg/100 μl ETOH) into 12 $\times$ 75 glass tubes. Concentrated standard solution of 1,25-(OH)$_2$D$_3$ (1280 pg/ml ETOH) is kept at -20°C in sealed ampules in the presence of N$_2$ and appropriate dilutions are done with ETOH prior to their use. Pipette another 0.24 ml aliquot of sample into glass vials and count for recoveries.
2. Evaporate to *complete dryness* samples and standards under N$_2$ stream.
3. While samples are drying prepare the incubation medium: MEM with Hank's salts + 50 mM HEPES + 0.5 mg α-globulin/ml, pH 7.4. *You need 0.2 ml $\times$ number of tubes.*
4. Thaw out freeze-stored cells in a water bath at 37°C. Add 6 ml of regular medium. Mix gently. Spin at 1000 rpm $\times$ 5 min. Discard medium and resuspend cell pellet in the incubation medium of step 3. This suspension should contain $\sim$0.5 $\times$ 10^6 cells/150 μl medium.
5. Pipette 0.15 ml cell suspension in each tube and start incubations at 37°C. Maintain gentle agitation of tubes during incubation using either a shaking water bath or a shaking platform; this will prevent cells from settling in the bottom of the tube. Continue incubation for 45 min.
6. At the end of the first 45-min incubation, add [^{3}H]1,25-(OH)$_2$D$_3$ to each tube ($\simeq$5000 cpm/20 μl of a mixture of 75% incubation medium and 25% ETOH). Continue incubation for another 30 minutes.
7. Place tubes on iced H$_2$O and leave for 15 min. Add 3 ml washing solution

and wait for another 15 min. Spin tubes at 1000 rpm for 10 min (4°C) and decant supernatant. Add another 3 ml washing solution to the cell pellet and repeat as above.

8. Add 0.4 ml of cell solubilizer (Omnisol, West Chem, San Diego, CA) and incubate for 30 min at 37°C.
9. Decant in counting vials and count with 8 ml scintillation mixture (beta-max, West Chem, San Diego, CA).

Calculation of 1,25-(OH)$_2$D concentration per 1 ml sample is done using the following formula:

$$\frac{pg}{ml} = \frac{pg \text{ per tube} \times \text{recovery factor}}{\text{initial sample volume}}$$

where

$$\text{Recovery factor} = \frac{\text{total radioactivity added for recoveries}}{\text{recovered radioactivity}}$$

II. Discussion

The extraction of 1,25-(OH)$_2$D from blood samples (1–2 ml serum or plasma) in this assay is accomplished with benzene using prepacked diatomaceous earth minicolumns. The aqueous sample is retained by the matrix while benzene passes freely through with a rate of 5 ml per 3–5 min. The eluent contains no pigments or particulate matter. In our laboratory we have determined that the efficiency of this column size for extracting 1,25-(OH)$_2$D is directly related to the volume of the aqueous sample within a range of 1–3 ml. With 20 ml of organic solvent added to the column, $\simeq$80% of 1,25-(OH)$_2$D$_3$ is recovered in the eluent (only 17 ml of solvent eluted because there is a dead space of 6 ml, 3 ml sample + 3 ml solvent). With the same volume of solvent, however, the recovery of 24,25-(OH)$_2$D$_3$ is 50% and the recovery of 25-(OH)D$_3$ is only 30%. Although additional small amounts of 1,25-(OH)$_2$D$_3$ can be recovered by using more than 20 ml solvent, for practical purposes we routinely use only 20 ml.

This mini-column-extraction technique is simple, fast, and highly reproducible. Furthermore, this system performs simultaneously the purification of the extract from lipids ($>$98%) which are retained by the diatomaceous earth matrix. The removal of lipids from blood samples has been a prerequisite of the earlier 1,25-(OH)$_2$D assay methods, and it is equally important for the cytoreceptor assay procedure; in the latter, lipids produce interference probably by affecting the permeability or integrity of the cell membrane. The retention of lipids by the diatomaceous earth column increases by increasing the alkalinity of the sample (optimum retention at pH = 10.5) with NaOH (Figure 2).

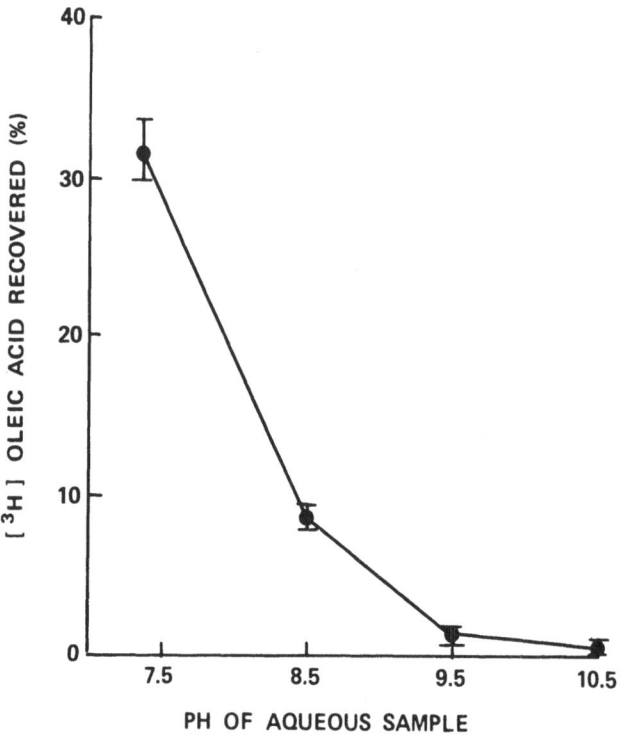

Figure 2. Effect of sample pH on the retention of FFA by diatomaceous earth columns.

    The sensitivity of the cytoreceptor assay standard curve, defined as two standard deviations of the zero tube, is 2 pg/tube. Fifty percent displacement of the bound tracer is obtained at a tube dose of $\simeq 10$ pg, and the useful range is 2–100 pg/tube. Only a small percent of the bound [^{3}H]1,25-(OH)$_2$D$_3$ is non-specific; that is, not displaceable by 200-fold molar excess of unlabeled 1,25-(OH)$_2$D (Figure 3). Cross-reactivity of other analogs of vitamin D with the 1,25-(OH)$_2$D$_3$ binding in the cytoreceptor assay has been found to be: 75% for 1,25-(OH)$_2$D$_2$; 0.06% for 25-(OH)D$_3$; and 0.003% for 24,25-(OH)$_2$D$_3$. Vitamin D$_3$ itself even at 1×10^6 pg is not capable of displacing [^{3}H]1,25-(OH)$_2$D$_3$. It should be appreciated that potential interference by these metabolites in the cytoreceptor assay is lower than that indicated by the cross-reactivities because of their lower recovery compared to 1,25-(OH)$_2$D$_3$ during the extraction of the sample. Considering cross-reactivities, recoveries, and the final volume of sample assayed per tube, we have determined that other major analogs of vitamin D$_3$ at their circulating levels do not interfere with the 1,25-(OH)$_2$D measurement. It is, nevertheless, possible that the extremely high levels of 25-(OH)D that occur in vitamin D intoxication could lead to an over-estimation of the 1,25-(OH)$_2$D measurement. However, the capacity of the cytoreceptor assay to resist interference from very high levels of 25-(OH)D, if this is suspected, can be increased by increasing the concentration of human α-globulin in the incubation medium. Because the incidence of vitamin D intoxication, and hence very high levels of 25-(OH)D, is unusual, and routine use of higher amounts of α-globulin would reduce the sensitivity of the assay, we have chosen for the purpose of routine clinical measurements to use the conditions described in Section I. Human α-globulin is a protein preparation rich in vitamin D-binding protein (DBP). Its function in the cytoreceptor system is to retain outside the cells vitamin D metabolites with high affinity for DBP preventing their interference with the intracellular binding of 1,25-(OH)$_2$D$_3$. Nevertheless, human α-globulin also binds 1,25-(OH)$_2$D$_3$ although with relatively lower affinity and thus lowers the free fraction of the hormone capable of intracellular binding. Therefore, at lower human α-globulin concentrations the 1,25-(OH)$_2$D$_3$ displacement curve becomes more sensitive, and vice versa.

The cross-reaction of 1,25-(OH)$_2$D$_2$ with 1,25-(OH)$_2$D$_3$ binding is 75%.

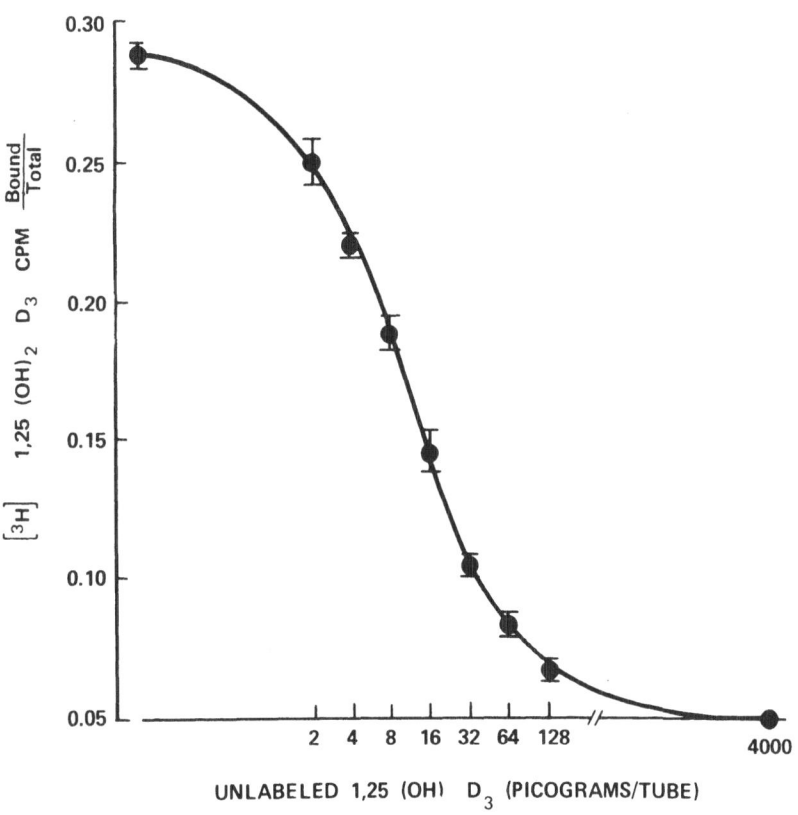

Figure 3. Representative standard curve of the cytoreceptor assay.

This value is similar to that observed in the traditional radioreceptor assay.[8] In view of this evidence it is apparent that the values read by the cytoreceptor assay represent both 1,25-(OH)$_2$D$_3$ and 1,25-(OH)$_2$D$_2$. However, it has been shown that greater than 95% of all vitamin D metabolites in humans occur in the D$_3$ form with the exception of extremely high vitamin D$_2$ intake. The high specificity of the cytoreceptor assay for 1,25-(OH)$_2$D does not depend only on the differential affinities of various metabolites for the receptor versus that for DBP; it is also most likely enhanced further by events related to the translocation of the sterol–receptor complex to the nucleus. There is evidence that under the conditions in which binding is established in intact cells, 95% of the hormone is located in the cell nucleus.[9] We have recently demonstrated that, in contrast to the 1,25-(OH)$_2$D$_3$–receptor complex, which can interact with nuclear components, 25-(OH)D$_3$ and 24,25-(OH)$_2$D$_3$ do not have such interactions.[10]

In the procedure described here, we also exploit the effect of glucocorticosteroids on the up-regulation of the 1,25-(OH)$_2$D$_3$ receptor in cultured osteogenic sarcoma cells by cultivating cells used for the assay in the presence of the synthetic glucocorticosteroid triamcinolone–acetonide.[11] This treatment increases the binding of 1,25-(OH)$_2$D$_3$ to the cells as much as 100%. An additional factor of practical importance in this assay is the use of freeze-stored cells. We have determined that cells can be stored frozen in DMSO without deterioration of their binding capacity, and when thawed rapidly can perform in the assay as well as freshly isolated cells. Although we have until now used rat osteogenic sarcoma cells, other 1,25-(OH)$_2$D$_3$ receptor-possessing cell lines can also be used. Haussler and his colleagues have recently found that the embryonic mouse skin fibroblast cell line 3T6 (American Type Culture Collection, Rockville, Maryland) contains 1,25-(OH)$_2$D$_3$ receptors in similar concentrations to the rat osteogenic sarcoma cells ROS 17/2 (personal communication). In preliminary studies using the 3T6 cells in the cytoreceptor assay, we have obtained results identical to those obtained using the ROS cells.

In validating the cytoreceptor assay method we have determined that the analytical recovery of 1,25-(OH)$_2$D$_3$ added to serum samples is 94.5%. 1,25-(OH)$_2$D concentrations determined in samples using 2 ml undiluted serum and 2 ml serum diluted with distilled H$_2$O (1:1) show that the 1:1 diluted sample value is 93% of the expected half value of the undiluted sample. The intraassay coefficient of variation has been estimated at 10% and the interassay coefficient of variation at 15%. In experiments in which 25-(OH)D$_3$ was added to serum samples (up to 100 ng/ml), we have found no significant difference in 1,25-(OH)$_2$D levels between plain and spiked samples.

III. Applications

The 1,25-(OH)$_2$D measurement can be performed using either serum or plasma samples. Figure 4 shows values obtained in 14 individuals from whom

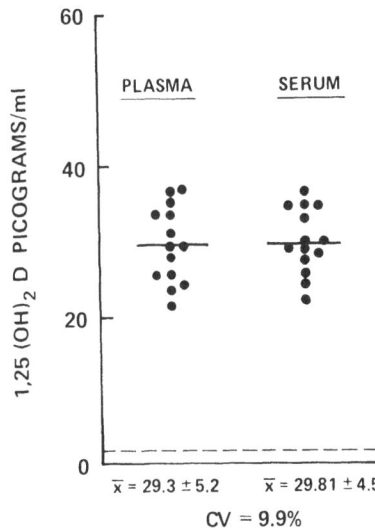

Figure 4. 1,25-(OH)₂D levels measured in plasma and serum.

we had obtained both serum and plasma samples. The mean coefficient of variation between the serum and plasma measurements was 9.9%.

Summary of the results obtained by the application of the cytoreceptor assay method in clinical studies[12] is shown in Table 1.

The 1,25-(OH)₂D levels determined by this assay in normal and diseased children are illustrated in Figure 5.

In the patients with hyperparathyroidism, 1,25-(OH)₂D levels correlated positively with intestinal calcium absorption ($r = 0.81$, $P < .001$) confirming in a biological manner the validity of the assay. Furthermore, in the entire normal group ($n = 120$), 1,25-(OH)₂D levels correlated negatively with age ($r = -0.4$, $P < .001$); this latter finding supports previous suggestions for an age-related decrease in 1,25-(OH)₂D synthesis and might be relevant to the pathogenesis of age-related loss of bone mass.

Table 1. 1,25-(OH)D Levels (pg/ml) in Normal Subjects and Patients Determined by the Cytoreceptor Assay

	Normal Adults[a] (n = 80)	Normal Children[b] (n = 40)	Pregnancy (3d trimester) (n = 6)	Anephrics (n = 3)	Renal Failure: Hemo-dialysis (n = 11)	Primary Hyper-parathy-roidism (n = 25)
X̄	33.8	43.3	96.5	undetectable	11.5	50.2
SE	±10.6	±13.9	±17.1	—	±6.7	±15.3
Range	14–58	22–79	70–123	—	0–21	27–85

[a]Mean age = 46 years.
[b]Mean age = 9.3 years.

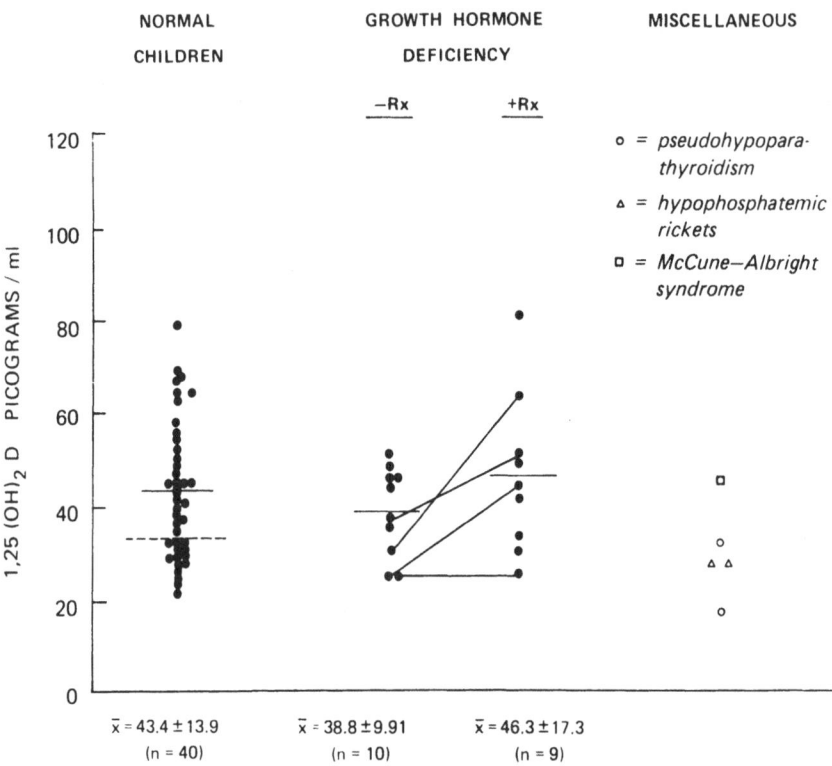

Figure 5. 1,25-(OH)$_2$D levels in normal children, growth hormone deficient children untreated ($-$Rx) or treated with growth hormone ($+$Rx), and children with miscellaneous metabolic bone diseases. Solid horizontal line indicates the mean of the group; dashed line indicates the mean 1,25-(OH)$_2$D levels in adults. Dots connected with a line indicate the same individual before and after treatment.

The 1,25-(OH)$_2$D levels obtained in these studies using the cytoreceptor assay are in close agreement with those determined by more cumbersome techniques and establish the validity of this assay as a clinically useful and reliable test. We have also applied the cytoreceptor assay to animal studies. Results in normal rats, rats in whom 1⅚ of their kidneys have been removed, and rats fed a low-PO$_4$ diet are shown in Figure 6.

In conclusion, the cytoreceptor assay method described here simplifies the quantitation of 1,25-(OH)$_2$D in blood. The samples are extracted and purified in a single fast and efficient step, and the measurement is performed directly on the extracts. In addition, the separation of bound from free ligand is accomplished by a simple washing step. These technical improvements over previous techniques permit the measurement of 1,25-(OH)$_2$D in a large number of samples by one person within a working day.

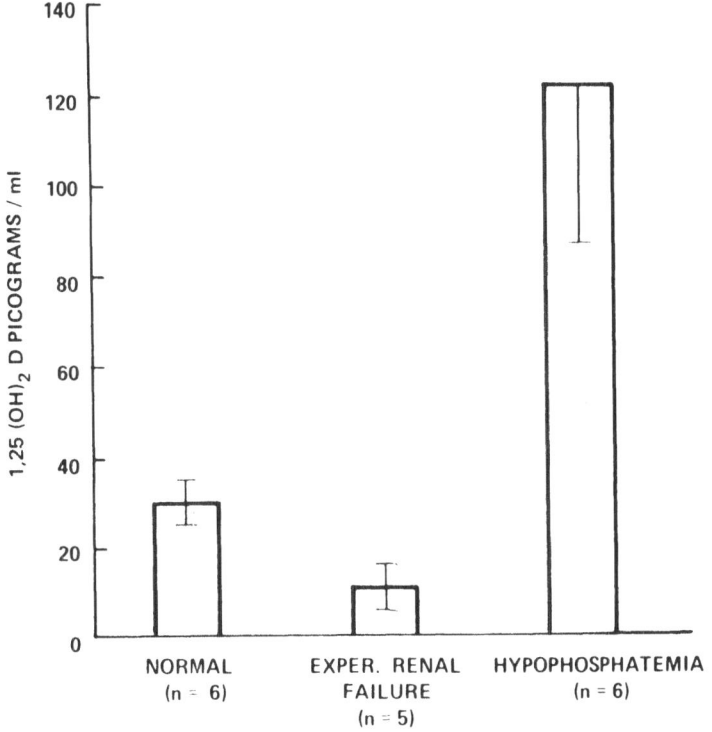

Figure 6. Serum levels of 1,25-(OH)$_2$D in rats determined by the cytoreceptor assay.

References

1. Manolagas SC, Deftos LJ: Studies of the internalization of vitamin D$_3$ metabolites by cultured osteogenic sarcoma cells and their application to a non-chromatographic cytoreceptor assay for 1,25-dihydroxyvitamin D$_3$. Biochem Biophys Res Commun 95:596, 1980.

2. Manolagas SC, Deftos LJ: Cytoreceptor assay for 1,25-dihydroxyvitamin D$_3$: a novel radiometric method based on binding of the hormone to intracellular receptors in vitro. Lancet 2:401, 1980.

3. Brumbaugh PF, Haussler DH, Bursac KM, Haussler MR: Filter assay for 1α-dihydroxyvitamin D$_3$. Utilization of the hormone's target tissue chromatin receptor. Biochemistry 13:4091, 1974.

4. Eisman JA, Hamstra AJ, Kream BE, DeLuca HF: A sensitive, precise and convenient method for determination of 1,25-dihydroxyvitamin D in human plasma. Arch Biochem Biophys 176:235, 1976.

5. Clemens TL, Hendy GN, Papapoulos SE, Fraher LJ, Care AD, O'Riordan JLH: Measurement of 1,25-dihydroxycholecalciferol in man by radioimmunoassay. Clin Endocrinol 11:225, 1979.

6. Bishop JE, Norman AW, Coburn JW, Roberts PA, Henry HL: Studies on the metabolism of calciferol. XVI: Determination of the concentration of 25-hydroxy-

vitamin D in a single two-milliliter plasma sample. Min Elect Metabol 3:181, 1980.

7. Manolagas SC, Haussler MR, Deftos LJ: 1,25-Dihydroxyvitamin D_3 receptor-like macromolecule in rat osteogenic sarcoma cell lines. J Biol Chem 255:4414, 1980.

8. Hughes MR, Baylink DJ, Jones PG, Haussler MR: Radioligand receptor assay for 25-hydroxyvitamin D_2/D_3 and 1,25-dihydroxyvitamin D_2/D_3. J Clin Invest 58:61, 1976.

9. Haussler MR, Pike JW, Dokoh S, Chandler JS, Chandler SK, Donaldson CA, Marion SL: 1,25-dihydroxyvitamin D receptor in cultured cell lines: Occurence, subcellular distribution and relationship to bioresponses. In: Norman AW, Schaefer K, Herrath Dv, Gigoleit H-G (Eds.): Vitamin D: Chemical, Biochemical and Clinical Endocrinology of Calcium Metabolism. Walter de Gruyter, Berlin, 1982.

10. Manolagas SC, Deftos LJ: Comparison of 1,25-, 25-, and 24,25-hydroxylated vitamin D_3 binding in fetal rat calvariae and osteogenic sarcoma cells. Calcif Tissue Int 33:655, 1981.

11. Manolagas SC, Abare J, Howard J and Deftos LJ: Regulation of 1,25-$(OH)_2D_3$ receptor in osteoblast-like cells during prolonged culture in the absence or presence of glucocorticoids. Calcif Tiss Intern Supp. 1 to 34:S55.

12. Manolagas SC, Culler FL, Howard JE, Brickman AS, Deftos LJ: Cytoreceptor assay for 1,25-dihydroxyvitamin D and its application to clinical studies. J Clin Endocrinol Metab 56:751, 1983.

Chapter 9

General Principles, Problems, and Interpretation in the Radioimmunoassay of Calcitonin and Parathyroid Hormone

Bayard D. Catherwood and
Leonard J. Deftos

Although radioimmunoassay (RIA) methods for parathyroid hormone (PTH) and calcitonin (CT) have been available for more than a decade, there remain problems in assay performance due to plasma and other protein artifacts, circulating peptide heterogeneity, and tumor products having calcitonin- or PTH-like immunoreactivity. The development of sensitive radioimmunoassays remains challenging. We will review some of the approaches to these problems.

I. Radioimmunoassay of Calcitonin

A. Biosynthesis and Metabolism of Calcitonin

Several features of the biosynthesis and metabolism of calcitonin are pertinent to the design and interpretation of calcitonin radioimmunoassays. The existence of a prohormone for calcitonin was suggested by pulse-chase studies[1] and cell-free translation of calcitonin mRNA.[2] Sequencing of cDNA complementary to rat calcitonin mRNA shows a pair of dibasic residues[3] separating the calcitonin sequence from an amino-terminal extension with propeptide and signal regions.[4] Three dibasic residues also separate calcitonin on the carboxyl side from structurally unrelated 16-amino-acid peptide (calcitonin-related peptide, CRP)* which is contained within the biosynthetic precursor for calcitonin and presumably separated posttranslationally by proteolytic cleavage.[3,4] Although little is currently known about the fate of this peptide, radioimmunoassay for CRP may have future diagnostic utility in diseases such as medullary thyroid carcinoma. A number of studies have demonstrated multiple immunoreactive forms of calcitonin in plasma.[5-7] At least four different-sized peptides with calcitonin immunoreactivity are present in medium from med-

---

*Recent evidence suggests the existence of a similar 21-amino-acid peptide in humans.[81]

ullary thyroid carcinoma cells in culture.[8] The molecular size profile of peptides extracted from the tumor cells is similar to that in the medium,[8] suggesting that some of the larger-molecular-weight forms may be secreted rather than formed by peripheral metabolism. Although dimers of human calcitonin were found during its original purification from medullary thyroid carcinoma, only the largest form of calcitonin found in tissue culture medium can be converted to the size of calcitonin monomer by reduction and carboxymethylation.[8] The relationship of the other large-molecular-weight forms of calcitonin to its biosynthetic precursors is unclear. Of these multiple immunoreactive peptides, only calcitonin monomer appears to be biologically active by adenylate cyclase assay.[8] Although calcitonin monomer forms a substantial portion of the circulating immunoreactive calcitonin in some conditions such as medullary thyroid carcinoma,[5,6] in other conditions such as end-stage renal failure the proportion of calcitonin monomer is low in spite of the elevated levels of total immunoreactive calcitonin.[9] Calcitonin monomer is the principal immunoreactive form responding to stimulation with pentagastrin and suppression by somatostatin.[10] The metabolic clearance of injected calcitonin is quite short,[11] with a half-time of 5–15 min. The metabolism of the larger-molecular-weight forms of immunoreactive calcitonin is not as well understood.

B. Antibody Development

High-affinity antisera can be developed against calcitonin using unconjugated peptide injected on a monthly basis in Freund's adjuvant.[11–13] The antisera resulting from this immunization method have been directed against the linear portion of the molecule to the C-terminal side of the disulfide ring formed by the cystines at positions 1 and 7 in all of the known calcitonins. According to Morita et al. (personal communication), [ASU[1,7]]eel calcitonin used to treat patients with Paget's disease has not been observed to lead to antibody formation, unlike native eel or salmon calcitonin. Thus, the strong tendency to middle and carboxy-terminal antisera formation may be the result of disulfide interchange in vivo with structural alteration or steric hindrance of the amino-terminus. Although such antisera have been useful for physiological and diagnostic measurements, they generally recognize both calcitonin monomer and the larger, apparently inactive, molecular forms. Since both procalcitonin and products of posttranslational disulfide interchange involve structural alteration at the amino-terminus, an antibody with specificity for the 1–12 region of calcitonin might have an enhanced capability for the selective recognition of only calcitonin monomer. A number of methods are available for the conjugation of small peptides as haptens to larger, more immunogenic proteins. Such methods include bifunctional reagents including glutaraldehyde and bisdiazo-benzidine[14–16] and water-soluble carbodiimides such as ethyl, dimethylamino-propyl carbodiimide.[17] We have attempted to produce human calcitonin antisera with different antigenic specificities by covalently linking human calcitonin to bovine thyroglobulin with glutaraldehyde. However, this has resulted only in high-capacity, low-affinity antisera. Immunization with conjugated hCT 1–12 has also thus far been unsuccessful.

C. Preparation of Radioligand

Human calcitonin, salmon calcitonin, and the artiodactyl group of calcitonins are readily radioiodinated to high specific activity (200–400 μCi/μg) by the method of Hunter and Greenwood[18] as modified for these peptides.[11,13] Sufficient purification can be achieved by adsorption to microfine silica and removal of residual ^{125}I with an anion exchange resin.[11,13] Electrolytically iodinated calcitonin retains biologic activity, but it does not differ from chloramine-T labeled ^{125}I-CT in several radioimmunoassay systems.[19] Although gel filtration chromatography may achieve some purification of labeled calcitonin and improve binding, a greater number of molecular forms of iodinated calcitonin can be separated by gradient high-performance liquid chromatography. This method may prove useful for tracer purification in the future.[20]

D. Calcitonin-free Plasma

High concentrations of plasma proteins are known to alter the binding of radioligand, standards, or both compared with the same radioimmunoassay performed in dilute low-protein buffers,[21,22] In the calcitonin assays, this is reflected in a decrease in maximum binding and/or an increase in the standard dose required for 50% inhibition of binding. This effect has been observed in our radioimmunoassay for human calcitonin and rat calcitonin with the effect being greater with rat plasma. Our human calcitonin radioimmunoassay was originally developed by adding to the standards plasma from patients who had undergone total thyroidectomy and who had no plasma calcitonin response to a calcium challenge. To avoid this limitation we have developed (B. D. Catherwood and H. G. Bone III, unpublished observations) a method of preparing plasma proteins with a low concentration of calcitonin and PTH. Our procedure uses rat or banked human plasma which is treated with heparin and dialyzed three times against 10 vol of normal saline, then three times against 10 vol distilled water. The plasma is lyophilized, redissolved in normal saline at a concentration of 8 gm/dl, centrifuged, and stored in frozen aliquots. Heparinized rat plasma processed in this manner showed no difference in apparent calcitonin content compared with thyroidectomized rat plasma in an assay containing no plasma substitute in the standard curve. When added with the assay standards the heparinized plasma gave a standard curve superimposable on that obtained by adding thyroidectomized rat plasma with the standards (Figure 1). Similar results were observed with human plasma. Such controls do not, however, correct for patient-specific protein artifacts, a phenomenon which seems to be prominent in patients with breast carcinoma and which may account for the difference between assay systems in the prevalence of ectopic calcitonin production by this tumor.[23]

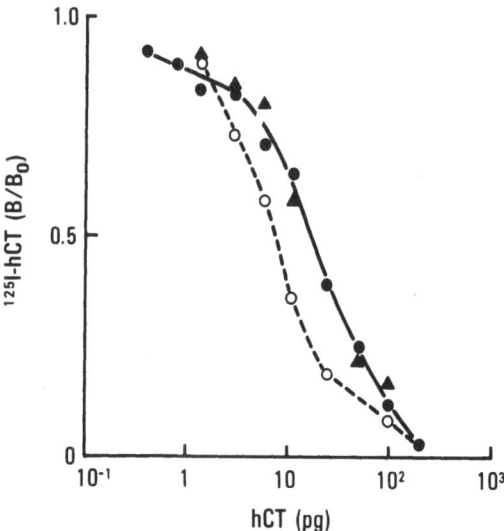

Figure 1. Effect of athyreotic rat plasma (▲) and exhaustively dialyzed rat plasma
(●) on the calcitonin radioimmunoassay (from B. D. Catherwood and H. G. Bone III,
unpublished observations). Pooled calcitonin-free rat plasma was obtained from thyroidectomized rats after demonstrating no calcitonin response to a calcium infusion.
Dialyzed rat plasma was prepared as described in Section I.D. Assays were incubated
for 3 days, then [125I]hCT (10,000 cpm/tube) was added and goat antirabbit IgG
second antibody added after 1 more day. (○ = no plasma, 0.25% bovine serum albumin in buffer; ▲ = 0.1 ml/tube athyreotic rat plasma; ● = 0.1 ml/tube dialyzed rat
plasma.)

E. Incubation Conditions

The incubation buffer should be chosen to maintain the optimum pH for the
antigen–antibody reaction and to minimize adsorption and degradation of the
peptide. Studies of incubation of calcitonin with normal plasma in vitro have
demonstrated a time- and temperature-dependent loss of peptide immunoreactivity, with the artiodactyl calcitonins being more sensitive than human and
salmon calcitonin.[24] Several agents are effective in inhibiting this process,[25]
notably EDTA. A 20 mM, pH 7.4 phosphate buffer containing 1 mM EDTA
and bovine serum albumin (2.5 g/l) with incubation at 4°C is a useful and
effective system in our laboratories for radioimmunoassay of all species of calcitonin. Increasing the ionic strength with either phosphate or sodium chloride
diminishes radioimmunoassay binding.

   An important part of the development of a radioimmunoassay for the measurement of calcitonin levels in normal subjects is the determination of conditions that optimize the sensitivity of the assay while maintaining precision. Previous theoretical studies based on mathematical modeling of simple
equilibrium radioimmunoassay systems have stated that optimum sensitivity is
obtained using a low dose of radioligand to minimize the mass of hormone and

adjusting the antibody concentration[26] to achieve a B/F of 0.5. It has since been observed that preincubation of samples and standards with the antibody ("nonequilibrium" or "sequential saturation" assays) results in a marked improvement in sensitivity. Many such assays have been set up using incubation conditions similar to those applicable to equilibrium radioimmunoassays sometimes requiring 3–4 days of incubation before and 3–4 days after addition of labeled peptide.[22] Given the constraints of the conditions proposed by Yalow and Berson above, we have previously found increasing sensitivity with time of incubation up to 6–10 days before and 4 days after addition of labeled peptide. More recently, however, Ekins[27] computed a mathematical model of a nonequilibrium radioimmunoassay and described the general principles of the optimization of sensitivity in this type of assay. An important conclusion of this study was that higher doses of radioligand should be used in many nonequilibrium assays relative to the optimal amount in equilibrium assays in order to achieve binding of label in a shorter time and minimize exchange of the radioligand with previously bound hormone. We have applied these principles to the calcitonin radioimmunoassay (B. D. Catherwood, unpublished observations). We first determined that the time required for effective maximal tracer binding could be shortened from 4 days to 1 day by increasing the dose of radioligand in the assay from 2000 cpm to 10,000 cpm. Reasonable assay precision and counting accuracy then required that the antibody concentration be adjusted to achieve a specific binding (zero binding minus nonspecific binding) of 0.10–0.15, which was achieved with an increase in antibody concentration of 33% over our previous assay conditions. Finally, we determined that with this increased antibody concentration, binding of hormone in standards and samples was achieved in only 3 days. The use of a higher dose of radioligand thus permitted us to cut the assay incubation time by 60% and simultaneously improve the sensitivity of the assay as assessed by a 40% reduction in the dose required for 50% inhibition of binding (Figure 2).

F. Separation of Bound from Free

Many methods of radioiodination, including the chloramine T method, result in generation of radioligand species which are "damaged" or weakly bound to plasma proteins.[28] This results in an increase in nonspecific binding in both calcitonin and PTH radioimmunoassays when separation of bound and free is performed either by dextran-coated charcoal or dioxane.[22] The use of double-antibody precipitation has largely eliminated these problems in our human calcitonin assay. We add 20 μg rabbit IgG per tube with the radioligand. After the addition of goat antirabbit γ-globulin, binding can be determined in as little as 2 h. The immunoprecipitate is dense after 18 h, permitting centrifugation and decanting of supernatants. Antigen–antibody complexes are insoluble in aqueous polyethyleneglycol,[28] and this agent has been used in conjunction with a double-antibody immunoprecipitation to decrease the time required prior to

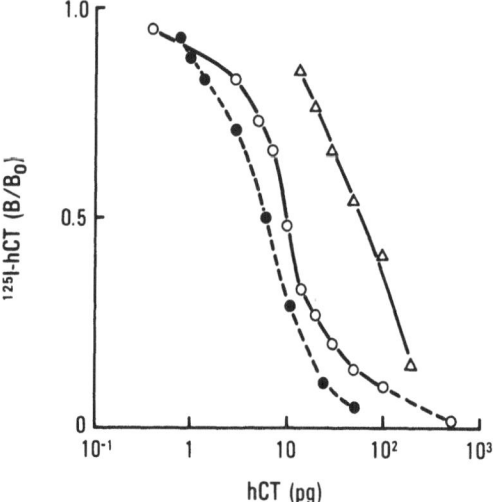

Figure 2. Effect of radioligand dose on incubation time and sensitivity in human calcitonin radioimmunoassay (antiserum LJ-1). Assay conditions (double-antibody separation):

Symbol	Antiserum Dilution	$[^{125}I]$hCT cpm/tube	Before Tracer	After Tracer	NSB (cpm)	SB (cpm)	50% Point
△	1:400,000	3,200	0	3 days	197	703	63 pg
○	1:400,000	2,200	6 days	4 days	106	764	10 pg
●	1:300,000	9,600	3 days	1 day	559	1037	6 pg

separation. However, in our hands this method has required the extra steps of washing the immunoprecipitate to free it of excessive amounts of trapped supernatant, and therefore saves little time.

G. Sensitivity and Specificity

The antigenic determinants of calcitonin antisera can be tested by gel filtration chromatography of serum from a patient with medullary thyroid carcinoma and, more precisely, by cross-reaction of synthetic fragments. As mentioned above, most assays have their primary antigenic determinants in the 11–32 region. Our human calcitonin assay, using antiserum LJ-1, requires the sequence hCT(17–32) to reproduce the effect of hCT(1–32) in inhibition of binding.[19] Given the growing list of peptides localized in "ectopic" tissues,[29–31] antisera should be carefully tested for cross-reactivity. Our calcitonin assay has been extensively tested for chance cross-reactivities with known hormones. ACTH and β-endorphin have been shown to cause inhibition of binding but only at doses greater than 1 μg/tube compared to 1 pg/tube for calcitonin.[32] In 41 routine 4-day plasma calcitonin assays, the sensitivity for determination in a single tube was 1.3 pg, and the dose required for 50% inhibition of binding was 6–10 pg/tube. The intraassay coefficient of variation in a pool of normal human plasma was 15% and the interassay coefficient of variation was 31%. The radioimmunoassay detects multiple peaks of immunoreactivity by gel filtration chromatography of blood plasma from patients with medullary thyroid carcinoma[6] and chronic renal failure.[9] The development of region-specific radioimmunoassays for calcitonin in the future may permit the selective measurement of monomeric calcitonin without chromatography.

H. Clinical Validation and Interpretation

Secretion of calcitonin in humans is stimulated by infusion of calcium (3 mg/ kg) over a 10-min period[12] and by a rapid intravenous injection of pentagastrin,[33,34] with the response being greater in males than females. The determination of the normal range in the basal state and after one or both of these provocative stimuli provides a straightforward initial clinical validation of any new assay. Although the regulation of plasma calcitonin in hypocalcemia has been studied,[35] hypocalcemia is not a useful state for assay validation. The more widespread availability of somatostatin may make this agent helpful in the future for testing suppression of plasma calcitonin in assays which are sensitive enough to detect such changes.[36] A number of clinical conditions have been reported to result in abnormalities of calcitonin secretion or blood levels and must be considered in the interpretation of plasma calcitonin determinations (Table 1). Notable among these are hyperparathyroidism[37–39] and malignancy.[40,41] Our calcitonin assay has recently been used to characterize the calcitonin response to a calcium challenge[42] in males and females aged 10–80. The evidence indicates that the plasma calcitonin response to a standard cal-

Table 1. Conditions Reported or Postulated to be Associated with Abnormalities of Calcitonin Secretion or Metabolism

Medullary thyroid carcinoma
  C-cell hyperplasia
Malignancy
  Calcitonin-secreting tumors
  Hypercalcemia
  Other mechanisms?
Hypocalcemia
Hypercalcemia
Hyperparathyroidism
Hypercalciuria
Pancreatitis
Pregnancy
Renal disease
Metabolic bone disease
  Hyperostotic states
  Hyperresorptive states

cium stimulus declines with age and is less for females than males; it may, therefore, be necessary to take an age-adjusted normal range into account in interpretation of results.

II. Radioimmunoassay of Parathyroid Hormone

A. Biosynthesis and Metabolism

Parathyroid hormone was one of the first peptide hormones for which the existence of a prohormone was established.[43] PTH is synthesized in the form of a 115-amino-acid precursor designated preproparathyroid hormone.[43] PreproPTH is processed during or immediately following translation by the cleavage[44] and hydrolysis[45] of a hydrophobic leader sequence. The resulting 90-amino-acid proparathyroid hormone is packaged in secretory granules and is subsequently processed into PTH.[46] Evidence shows that parathyroid hormone is metabolized intracellularly, especially when secretion is inhibited by high levels of extracellular calcium[47] and biologically inactive fragments of PTH may be secreted by the gland.[48-50] PTH secreted into blood is metabolized by the liver and kidneys in dogs[51] and humans.[52] This peripheral metabolism involves cleavage of the peptide between amino acids 33–34 and 36–37.[53] These proteolytic cleavages result in a biologically inactive midmolecule to carboxyterminal (C-terminal) fragment and could potentially produce an amino-terminal fragment with full biological activity.[54] Proteolytic cleavage of bovine PTH has been demonstrated by isolated rat liver,[54] Kupfer cells,[55] and dog kidney.[56] Intact bovine and human parathyroid hormone are rapidly cleared[57,63] from the circulation with a half-time of 2–10 minutes. A small amino-terminal fragment of PTH has been observed in the circulation of calves infused with exogenous bovine PTH,[57] and its metabolic clearance is also very rapid. The metabolism of the biologically inactive C-terminal fragment of bovine PTH has been studied in the dog and appears to occur by glomerular filtration, resorption, and degradation.[58] Clearance of C-terminal fragments is relatively rapid in the calf[57] and much longer, with a half-time of hours, in humans.[59] Because clearance of C-terminal fragments is much slower than either intact hormone or amino-terminal fragment, they are the predominant peptide in normal blood and can accumulate to a marked degree in renal failure.

B. Production of Antisera

The goal in the development of new antisera for radioimmunoassay of PTH is to achieve a high-affinity antiserum with specificity for a predicted region of the PTH molecule of the species of interest. Preoperative confirmation of a diagnosis of primary hyperparathyroidism has been more easily achieved using antisera that recognize the biologically inactive C-terminal fragment formed from the peripheral metabolism of PTH. Such antisera have been obtained by immunizing various animal species with bovine PTH(1–84), but extracts of human parathyroid glands have proven more useful in obtaining antisera with a high affinity for the human sequence.[49,60-63] Assays developed from the direct immunization with synthetic human PTH(44–68) and human PTH(53–84) have not yet been reported. However, we have used [Tyr[43]] human PTH(43–68) to promote selectively the continued synthesis of antibodies with middle molecule specificity in a goat immunized with human parathyroid extract having weak, interfering C-terminal antibodies (Figure 3). For some applications, including venous catheterization studies for localization of parathyroid tissue, hypercalcemic secondary hyperparathyroidism and chronic renal failure, and certain physiological studies, antisera that recognize the biologically active amino-terminal region or require the entire PTH molecule for recognition may be most useful. Such antisera have been generated by immunization with bovine PTH,[64] human PTH,[62] and the human PTH(1–34) sequence.[65,66] Immunization with bovine PTH has also resulted in antisera that require the entire PTH molecule for recognition.[62] Many of these antisera have a recognition site in the PTH(28–48) region.[67] In general, immunization with preparations other than pure peptide fragments has the potential for evoking multiple antibodies with differing antigenic determinants. In this case, techniques such as absorption with competing peptides,[68] immunoaffinity purification,[62] or peptide radioligands[49] are required to obtain region-specific radioimmunoas-

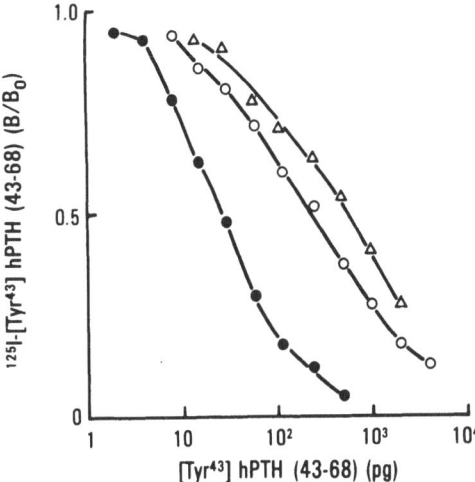

Figure 3. Effect of minor antibody populations on PTH radioimmunoassay. Goat 510 was immunized with human parathyroid extract. Serum 2/20/79 showed high-affinity binding for [^{125}I]-[Tyr[43]]hPTH(43–68) and low-affinity binding for [^{125}I]-[Tyr[52]]hPTH(52–84). Serum 11/3/81 was obtained after boosting with [Tyr[43]]hPTH(43–68). All assays received 10,000 cpm/tube [^{125}I]-[Tyr[43]]hPTH(43–68) and were incubated for 3 days before and 1 day after radioligand (double-antibody separation). △ = antiserum 2/20/79, 1:4,500, 50% point 533 pg; ○ = antiserum 2/20/79, 1:8,000 plus hPTH(53–84) (100 pg/tube), 50% point 232 pg; ● = antiserum 11/3/81, 1:4,000, 50% point 30 pg [no effect of absorption with hPTH(53–84)].

says. Affinity-purified amino-terminal and C-terminal antibodies have been used for a two-site immunoradiometric assay that detects only intact hormone.[69] Hybridoma technology may provide another approach by giving the ability to select in culture a clonal cell line producing a single predominant immunoglobulin with a desired recognition site.

C. Preparation of Radioligand

The chloramine-T method has been widely used for the radioiodination of bovine PTH.[70] PTH(1–84) labeled by the chloramine-T method requires purification prior to use. Generation of peroxide with glucose oxidase followed by enzymatic iodination with lactoperoxidase[71] may give less oxidative damage but has been less commonly used. The availability of a synthetic radioiodination substrate, [Tyr[43]]PTH(46–68) has diminished the need for repurification prior to use in assays with this recognition site.[49] For the region-specific radioimmunoassay of human PTH(1–34), preparation of labeled peptide is complicated by absence of a tyrosine in the peptide, although three histidines can be iodinated. A [Tyr]-substituted peptide might provide a more effective substrate for radioiodination. An alternative approach has been demonstrated by the radioiodination of affinity-purified antibodies to the amino-terminus of human PTH (radioimmunometric assay).[62,63]

D. PTH-free Plasma

In most PTH assays plasma proteins produce a nonspecific inhibition of binding when tested with plasma samples from patients with surgical hypoparathyroidism or samples stripped of hormone by immunoextraction or physical methods such as activated charcoal or microfine silica. The immunoreactive PTH content in aged banked plasma is relatively low and such plasma has frequently been added in low concentration to assay buffers. However, this does not compensate for the nonspecific effects of the additional plasma protein added with an unextracted plasma sample. Several approaches have been used to control for this interference. The first is to test the nonspecific plasma effect in each sample by assaying the sample before and after extraction of PTH with immobilized anti-PTH antibody[72] or charcoal. The data is then analyzed by subtracting the apparent amount of PTH due to nonspecific effects from the total apparent PTH to determine specific immunoreactive PTH. A second approach is to include in every assay one or more plasma samples from patients with complete hypoparathyroidism and perform a similar analysis.[73] The third approach is to add an equal amount of PTH-free plasma to the standard tubes. The volume of plasma required in this last approach makes the use of hypoparathyroid human plasma impractical. In our hands the residual content of immunoreactive PTH in aged banked plasma is too high for this use, and charcoal or silica-stripped plasma do not give reproducible results. Exhaustively dialyzed blood bank plasma as described in Section I.D. contains no detectable

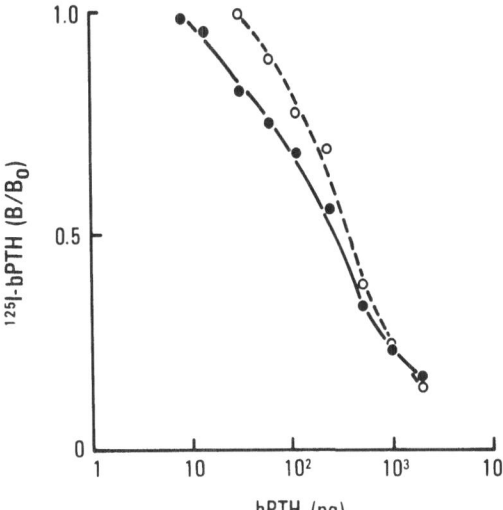

Figure 4. Effect of dialyzed blood bank plasma (Section I.D) in a PTH(53–84) radioimmunoassay. Dialyzed plasma, 100 μl (●) contained no detectable PTH (Bo/total = 0.191) compared with control assay (○: Bo/total = 0.178) but gave nonparallel bovine PTH standard curves.

immunoreactive PTH in our C-terminal (53–84 region) PTH assay as assessed by inhibition of binding but it compensates for the plasma interference (Figure 4).

E. Incubation Conditions and Separation of Bound and Free

PTH is susceptible to degradation in vitro, and antioxidants are protective.[74] The practical effect is dependent on the assay involved. A sodium barbital buffer has been widely adopted for PTH radioimmunoassays,[70] and destruction is limited by incubation at 4°C. Absorption of antisera with synthetic peptides has been used for development of region-specific assays. We have found this technique useful also for the enhancement of assay sensitivity with antisera raised against human parathyroid extract (Figure 3). Although adsorption to dextran-treated charcoal has been widely used for separation of bound from free,[70] this method is being replaced by double antibody precipitation in many laboratories. Due to the wide variety of species used for the production of PTH antibodies, including guinea pigs, chickens, and goats, the second antibody precipitation system must be individualized to the assay.

F. Standards

Several options are available in the choice of assay standard. The heterogeneity of immunoreactive PTH in plasma provides an especially difficult problem for assays using labeled bovine PTH as radioligand and an antiserum which may express multiple antigenic determinants. In this situation the use of multiple dilutions of a large pool of human hyperparathyroid serum has been used to provide the mix of peptides present in actual samples to be assayed and expressed[73] in microliter equivalents of this plasma standard per milliliter.

Such assays have been very effective in confirming the diagnosis of hyperpara-thyroidism.[61] In region-specific C-terminal radioimmunoassays which use a synthetic tyrosine-substituted mid- or C-terminal fragment, it may be conve-nient to use the synthetic peptide for the standard and express results in molar equivalents although there may still be multiple fragments in the circulation that will react in these assays. Some attention must be paid to preservation of standards prepared for long-term use. Synthetic human mid- and C-terminal fragments appear to be more stable than amino-terminal fragments or bovine PTH. If purified peptides are used as standards, precautions must be taken against adsorption in dilute solution. We have found that egg lysozyme (1 g/ l) is effective and provides better stability of standards than buffers containing low concentrations of plasma.

G. Sensitivity and Specificity

Many mid- and C-terminal PTH assays have been reported which are capable of detecting basal levels of hormone in most normal subjects. Sensitivity is more critical for assays specific for the amino-terminus or intact molecule since 25% or less of immunoreactive PTH in plasma by a C-terminal assay will be recognized in an amino-terminal assay.[75] Blind comparison data on precision and reproducibility of established PTH assays have recently been reported.[70] The average within-assay coefficient of variation between two determinations of international reference preparation of bovine PTH was 16%. The within assay variance for a recently produced human PTH reference preparation was 0.88 of the variance for the bovine standard. Interassay and between-labora-tory variances were only slightly larger.

H. Clinical Validation and Interpretation

Provocative stimuli are generally not required for the clinical validation of PTH assays. Dose–dilution parallelism should be assessed with samples con-taining high levels of PTH from patients with primary hyperparathyroidism, secondary hyperparathyroidism, and possibly parathyroid venous effluent, if available. The latter contain higher ratios of intact hormone to biologically inactive fragments if taken from a vein draining a hyperfunctioning gland. The assay should then be tested for discrimination of hypoparathyroidism, primary hyperparathyroidism, secondary hyperparathyroidism, and hypercalcemia of malignancy. Patients with malignant hypercalcemia without evident bony metastases form at least two groups of patients, one having suppression of both urine cyclic AMP and PTH in blood and another having elevated urinary cyclic AMPs and usually normal or only slightly elevated levels of PTH.[64] A significant number of cancer patients also have coexistent primary hyperpara-thyroidism.[76] The ability to discriminate malignant hypercalcemia from pri-mary hyperparathyroidism is a stringent test of the power of the PTH assay and significant differences in discriminative ability exist among PTH assays offered by reference laboratories.[77] Clinical correlation with the serum calcium may help to make this distinction[78] but formal discriminant analysis would require that the discriminant function be determined empirically for each new PTH assay.

Amino-terminal PTH assays have thus far found the most use in analysis of venous catheterization samples in cases of persistent postoperative hyperparathyroidism because they will not recognize the background levels of peripherally derived inactive fragments.[79] In spite of the very short half-life of amino-terminal immunoreactive PTH, several amino-terminal assays have shown some ability to discriminate between normal subjects and those with primary hyperparathyroidism.[65,66] Interpretation of such assays may be complicated by a greater degree of short-term fluctuation in hormone concentrations in blood. Increasing understanding of the neurohumoral control of PTH secretion[80] may lead to the development of simple suppression tests to enhance further the power of these assays. Application of amino-terminal PTH radioimmunoassays to the evaluation of patients with chronic renal failure and very high levels of C-terminal fragments has been less well explored. Whether a sensitive amino-terminal PTH assay will recognize a circulating "PTH-like" substance in humoral hypercalcemia of malignancy also remains to be determined.

III. Conclusion

Understanding of the biosynthesis, post-translational processing, secretion, peptide heterogeneity, and metabolism of calcitonin and parathyroid hormone are important in the design and proper interpretation of radioimmunoassays for these hormones. In addition, performance and interpretation of radioimmunoassays requires awareness of the many potential nonspecific interferences in sample materials. Newer techniques of radioimmunoassay design are beginning to yield assays with defined recognition regions which may ameliorate some of the technical pitfalls in present methods and facilitate clinical interpretation. Despite their limitations, presently established assays have nevertheless been very useful in diagnosis of important clinical disorders and in the understanding of normal and abnormal calcium metabolism.

Acknowledgments This work was supported in part by the National Institutes of Health (AM15888, AM25604, AM07048), the American Cancer Society, and the Veterans Administration.

References

1. Roos BA, Okano K, Deftos LJ: Evidence for a pro-calcitonin. Biochem Biophys Res Commun 60:1134–1140, 1974.

2. Jacobs JW, Potts JT Jr, Bell NH, Habener JF: Calcitonin precursor identified by cell-free translation of mRNA. J Biol Chem 254:10600–10603, 1979.

3. Amara SG, David DN, Rosenfeld MG, Roos BA, Evans RM: Characterization of rat calcitonin messenger RNA. Proc Natl Acad Sci USA 77:4444–4448, 1980.

4. Jacobs JW, Goodman RH, Chin WW, Dee PC, Habener JF, Bell NH, Potts JT Jr: Calcitonin messenger RNA encodes multiple polypeptides in a single precursor. Science 213:457–495, 1981.

5. Singer FR, Habener JF: Multiple immunoreactive forms of calcitonin in human plasma. Biochem Biophys Res Commun 61:660–666, 1974.

6. Deftos LJ, Roos BA, Bronzert D, Parthemore JG: Immunochemical heterogeneity of calcitonin in plasma. J Clin Endocrinol Metab 40:407–410, 1975.

7. Sizemore GW, Heath H, Larson JM: Immunochemical heterogeneity of calcitonin in plasma of patients with medullary thyroid carcinoma. J Clin Invest 55:111–118, 1975.

8. Goltzman D, Tischler AS: Characterization of the immunochemical forms of calcitonin released by a medullary thyroid carcinoma in tissue culture. J Clin Invest 61:449–458, 1978.

9. Lee J, Parthemore JG, Deftos LJ: Calcitonin secretion in renal disease. Calcif Tissue Res 22S:154–157, 1977.

10. Bertagna XY, Bloomgarden ZT, Rabin D, Roberts LJ, Orth DN: Molecular weight forms of immunoreactive calcitonin in a patient with medullary carcinoma of the thyroid: dynamic studies with calcium, pentagastrin, and somatostatin. Clin Endocrinol 13:115–124, 1980.

11. Deftos LJ, Lee MR, Potts JT Jr: A radioimmunoassay for thyrocalcitonin. Proc Natl Acad Sci USA 60:293-299, 1968.

12. Parthemore, J.G., Deftos, L.J.: Calcitonin secretion in normal human subjects. J Clin Endocrinol Metab 47:184-188, 1978.

13. Deftos, L.J.: An immunoassay for human calcitonin. I. The method. Metabolism 20:1120–1128, 1971.

14. Wold F: Bifunctional reagents. In Hirs, C.H.W. (Ed.): Methods in Enzymology, vol. XI, Acadernic Press, New York, pp 617–640, 1967.

15. Avrameas S: Coupling of enzymes to proteins with glutaraldehyde. In Darrington, K (Ed.): Immunochemistry, Volume 6. Pergamon Press, Oxford, pp 43–52, 1969.

16. Gregory DW, Williams MA: The preparation of ferritin-labelled antibodies and other protein-protein conjugates with bis-diazotized benzidine. Biochim Biophys Acta 133:319-332, 1967.

17. Kurzer, F., Douraghi-Zadeh K: Advances in the chemistry of carbodiimides. Chem Rev 67:107–152, 1967.

18. Hunter WM, Greenwood FC: Preparation of iodine-131 labelled human growth hormone of high-specific activity. Nature 194:495–496, 1962.

19. Scarpace PJ, Deftos LJ: Preparation and immunological characteristics of biologically active radioiodinated human CT. Endocrinology 101:1398–1405, 1977.

20. Lambert PW, Roos BA: Paired-ion reversed-phase high-performance liquid chromatography of human and rat calcitonin. J Chromatogr 198:293–300, 1980.

21. Deftos LJ, Bury AE, Habener JF, Singer FR, Potts JT Jr: Immunoassay for human calcitonin. II. Clinical studies. Metabolism 20:1129–1137, 1971.

22. Habener JF, Mayer GP, Powell D, Murray TM, Singer FR, Potts JT Jr: Dextran-charcoal and dioxane separation methods in the radioimmunoassays for parathyroid hormone and calcitonin. Clin Chim Acta 45:225–233, 1973.

23. Roos BA, Lambert PW, Lindall AW, O'Neil J, Frelinger AL, Birnbaum R, Bay-

lin SB: Diagnostic importance of high MW calcitonin in lung cancer. Clin Res 27:376A, 1979.

24. Habener JF, Singer FR, Deftos LJ, Potts JT Jr: Immunological stability of calcitonin in plasma. Endocrinology 90:952–960, 1972.

25. Pento JT: The influence of various enzyme inhibitors on ^{125}I-porcine calcitonin adsorption and incubation damage. Horm Metab Res 7:444–445, 1975.

26. Yalow, RS, Berson SA: General principles of radioimmunoassay. In Hayes, RL, Goswitz FA, Pearson Murphy BE: Radioisotopes in Medicine: In Vitro Studies. US Atomic Energy Commission Oakridge, Tennessee, pp 7–41, 1967.

27. Ekins, R.P.: Automation of radioimmunoassay and other saturation assay procedures. In Symposium on Radioimmunoassay and Related Procedures in Clinical Medicine and Research. Intl Atomic Energy Agency, Vienna, pp 91–109, 1974.

28. Desbuquois, B., Aurbach, G.D.: Effects of iodination on the distribution of peptide hormones in aqueous two-phase polymer systems. Biochem J 143:83-91, 1974.

29. Baylin, S.B., Mendelsohn, G.: Ectopic (inappropriate) hormone production by tumors: mechanisms involved and the biological and clinical implications. Endocr Rev 1:45-77, 1980.

30. Deftos LJ, Burton D, Catherwood BD, Bone HG, Parthemore JG, Guillemin R, Watkins W, Moore RY: Demonstration by immunoperoxidase histochemistry of calcitonin in the anterior lobe of the rat pituitary. J Clin Endocrinol Metab 47:457–460, 1978.

31. Catherwood BD, Deftos LJ: Presence of radioimmunoassay of a calcitonin-like substance in porcine pituitary glands. Endocrinology 106:1886–1891, 1980.

32. Catherwood BD, Deftos LJ: Reactivity of ACTH and synthetic ACTH peptides with antisera to human calcitonin. J Immunol Methods 32:315–322, 1979.

33. Hennessey JF, Gray TK, Cooper CW, Ontjes DA: Stimulation of thyrocalcitonin secretion by pentagastrin and calcium in two patients with medullary carcinoma of the thyroid. J Clin Endocrinol Metab 36:200–203, 1973.

34. Owyang C, Heath H III, Sizemore GW, Go VLW: Comparison of the effects of pentagastrin and meal-stimulated gastrin on plasma calcitonin in normal man. Am J Dig Dis 23:1084–1088, 1978.

35. Deftos LJ, Powell D, Parthemore JG, Potts JT Jr: Secretion of calcitonin in hypocalcemic states in man. J Clin Invest 52:3109–3114, 1973.

36. Gordin A, Lamberg BA, Pelkonen R, Almqvist S: Somatostatin inhibits the pentagastrin-induced release of serum calcitonin in medullary carcinoma of the thyroid. Clin Endocrinol 8:289–294, 1978.

37. Lambert PW, Heath HH III, Sizemore GW: Pre- and postoperative studies of plasma calcitonin in primary hyperparathyroidism. J Clin Invest 63:602–608, 1979.

38. Parthemore JG, Deftos LJ: Secretion of calcitonin in primary hyperparathyroidism. J Clin Endocrinol Metab 49:223–226, 1979.

39. Becker KL, Silva OL, Wisneski LA, Cyrus J, Snider RH, Moore CF, Higgins GA: Limited calcitonin reserve in hyperparathyroidism. Am J Med Sci 280:11–15, 1980.

40. Wallach SR, Royston, I, Taetle, R, Wohl, H, Deftos, LJ: Plasma calcitonin as a marker of disease activity in patients with small cell carcinoma of the lung. J Clin Endocrinol Metab 53:602–606, 1981.

41. Schwartz KE, Wolfsen AR, Forster B, Odell WD: Calcitonin in nonthyroidal cancer. J Clin Endocrinol Metab 49:438–444, 1979.

42. Deftos LJ, Weisman MH, Williams GH, Karpf DB, Frumar AM, Davidson BH,

Parthemore JG, Judd HL: Influence of age and sex on plasma calcitonin in human beings. N Engl J Med 302:1351–1353, 1980.

43. Habener JF, Potts JT Jr: Biosynthesis of parathyroid hormone. Parts 1 and 2. N Engl J Med. 299:580–585; 635–644, 1978.

44. Habener JF, Amherdt M, Ravazzola, M, Orci L: Parathyroid hormone biosynthesis. Correlation of conversion of biosynthetic precursors with intracellular protein migration as determined by electron microscope autoradiography. J Cell Biol 80:715–731, 1979.

45. Habener JF, Rosenblatt M, Dee PC, Potts JT, Jr: Cellular processing of preproparathyroid hormone involves rapid hydrolysis of the leader sequence. J Biol Chem 254:10596-10599, 1979.

46. Habener JF, Kemper BW, Potts JT Jr, Rich A: Calcium-independent intracellular conversion of proparathyroid hormone to parathyroid hormone. Endocr Res Commun 1:239–246, 1974.

47. Habener JF, Kemper B, Potts JT Jr: Calcium-dependent intracellular degradation of parathyroid hormone: A possible mechanism for the regulation of hormone stores. Endocrinology 97:431–441, 1975.

48. Flueck JA, DiBella FP, Edis AJ, Kehrwald JM, Arnaud CD: Immunoheterogeneity of parathyroid hormone in venous effluent serum from hyperfunctioning parathyroid glands. J Clin Invest 60:1367–1375, 1977.

49. Marx SJ, Sharp ME, Krudy A, Rosenblatt M, Mallette LE: RIA for the middle region of human PTH—studies with a radioiodinated synthetic peptide. J Clin Endocrinol Metab 53:76–84, 1981.

50. Roos BA, Lindall AW, Aron DC, Orf JW, Yoon M, Huber MB, Pensky J, Ells J, Lambert PW: Detection and characterization of small midregion parathyroid hormone fragment(s) in normal and hyperparathyroid glands and sera by immunoextraction and region-specific radioimmunoassays. J Clin Endocrinol Metab 53:709–721, 1981.

51. Catherwood, B.D., Friedler, R.M., Singer, F.R.: Sites of clearance of endogenous parathyroid hormone in the vitamin D-deficient dog. Endocrinology 98:228-232, 1976.

52. Corvilain J, Manderlier T, Struyven J, Fuss M, Bergans A, Nijs N, Brauman H: Metabolism of human PTH by the kidney and the liver. Horm Metab Res 9:239–242, 1977.

53. Segre GV, Niall HD, Sauer RT, Potts JT Jr: Edman degradation of radioiodinated parathyroid hormone: application to sequence analysis and hormone metabolism in vivo. Biochemistry 16:2417–2427, 1977.

54. Canterbury JM, Bricker LA, Levey GS, Kozlovskis PL, Ruiz E, Zull JE, Reiss E: Metabolism of bovine parathyroid hormone. Immunological and biological characteristics of fragments generated by liver perfusion. J Clin Invest 55:1245–1253, 1975.

55. Segre GV, Perkins AS, Witters LA, Potts JT Jr: Metabolism of parathyroid hormone by isolated rat Kupffer cells and hepatocytes. J Clin Invest 67:449–457, 1981.

56. Hruska KA, Martin K, Mennes P, Greenwalt A, Anderson C, Klahr S, Slatopolsky E: Degradation of parathyroid hormone and fragment production by isolated perfused dog kidney—effect of glomerular filtration rate and perfusate Ca^{++} concentrations. J Clin Invest 60:501–510, 1977.

57. Hunziker W, Blum JW, Fischer JA: Plasma kinetics of exogenous bovine parathyroid hormone in calves. Pflugers Arch 371:185–200, 1977.

58. Hruska KA, Kopelman R, Rutherford WE, Klahr S, Slatopolsky E, Greenwalt A, Bascom T, Markham JA: Metabolism of immunoreactive parathyroid hormone in the dog. J Clin Invest 56:39–48, 1975.

59. Berson SA, Yalow RS: Immunochemical heterogeneity of parathyroid hormone in plasma. J Clin Endocrinol Metab 28:1037, 1968.

60. DiBella FP, Gilkinson JB, Flueck J, Arnaud CD: Carboxy-terminal fragments of human parathyroid hormone in parathyroid tumors. J Clin Endocrinol Metab 46:604–612, 1978.

61. DiBella FP, Kehrwald JM, Laakso K, Zitzner L: Parathyrin radioimmunoassay: Diagnostic utility of antisera produced against carboxyl-terminal fragments of the hormone from the human. Clin Chem 24:451–454, 1978.

62. Manning RM, Hendy GN, Papapoulos SE, O'Riordan JLH: Development of homologous immunological assays for human parathyroid hormone. J Endocrinology 85:161–170, 1980.

63. Papapoulos SE, Manning RM, Hendy GN, Lewin IG, O'Riordan JLH: Studies of circulating parathyroid hormone in man using a homologous amino-terminal specific immunoradiometric assay. Clin Endocrinol 13:57–68, 1980.

64. Stewart AF, Horst R, Deftos LJ, Cadman EC, Lang R, Rasmussen H, Broadus AE: Biochemical evaluation of patients with malignancy-associated hypercalcemia: Evidence for humoral and non-humoral groups. N Engl J Med 303:1377–1383, 1980.

65. Desplan C, Jullienne A, Moukhtar MS, Milhaud G: Sensitive assay for biologically active fragments of human parathyroid hormone. Lancet 2:198–199, 1977.

66. Segre GV, Harris ST, Tully G, Neer R: Amino-terminal radioimmunoassay for human parathyroid hormone. Proceedings of the 63rd Annual Meeting of The Endocrine Society, Abstract #584, 1981.

67. Mallette LE, Renfro M, Lemoncelli J, Rosenblatt M: RIAs for the 28-48 region of PTH detect intact hormone but not hormone fragments. Calcif Tissue Int 33:375–380, 1981.

68. Segre GV, Tregear GW, Potts JT Jr: Development and application of sequence-specific radioimmunoassays for analysis of the metabolism of parathyroid hormone. In Colowick SP and Kaplan NO (Eds.): Methods of Enzymology vol. 37, Academic Press, New York, pp 38–66, 1975.

69. Woodhead JS, Davies SJ, Lister D: Two-site assay of bovine parathyroid hormone. J Endocrinol 73:279–288, 1977.

70. Zanelli JM, Gainesdas RE: International collaborative study of N.I.B.S.C. research standard for human parathyroid hormone for immunoassay. J Endocrinol 86:391–304, 1980.

71. Tower BB, Clark BR, Rubin RT: Preparation of ^{125}I polypeptide hormones for radioimmunoassay using glucose oxidase with lactoperoxidase. Life Sci 21:959–966, 1977.

72. Lindall AW: New sensitive parathyroid radioimmunoassays with sequence-specific antibodies. Ligand Rev 2:7–14, Fall, 1980.

73. Arnaud CD, Tsao HS, Littledike T: Radioimmunoassay of human parathyroid hormone in serum. J Clin Invest 50:21–34, 1971.

74. Barrett PQ, Neuman WF: The cleavage and adsorption of parathyroid hormone at high dilution. Implications for receptor binding studies. Biochim Biophys Acta 541:223–233, 1978.

75. Manning RM, Adami A, Papapoulos E, Gleed JH, Hendy GN, Rosenblatt M,

O'Riordan JLH: A carboxy-terminal specific assay for human parathyroid hormone. Clin Endocrinol 15:439–449, 1981.

76. Drezner MK, Lebovitz HE: Primary hyperparathyroidism in paraneoplastic hypercalcemia. Lancet 1:1004–1005, 1978.

77. Raisz LG, Yajnik CH, Bockman RS, Bower BF: Comparison of commercially available parathyroid hormone immunoassays in the differential diagnosis of hypercalcemia due to primary hyperparathyroidism or malignancy. Ann Intern Med 91:739–740, 1979.

78. Habener JF, Segre GV: Parathyroid hormone radioimmunoassay. Ann Intern Med 91:782–785, 1979.

79. Dunlop DAB, Papapoulos SE, Lodge RW, Fulton AJ, Kendall BE, O'Riordan JLH: Parathyroid venous sampling—anatomic considerations and results in 95 patients with primary hyperparathyroidism. Br J Radiol 53:183–191, 1980.

80. Heath H: Biogenic amines and the secretion of parathyroid hormone and calcitonin. Endocr Rev 2:319, 1980.

81. Craig RK, Hall L, Edbrooke MR, Allison J, MacIntyre I: Partial nucleotide sequence of human calcitonin precursor mRNA identifies flanking cryptic peptides. Nature 295:345–347, 1982.

Chapter 10

General Techniques for Raising Antisera Against Parathyroid Hormone and Calcitonin

L. E. Mallette

Antisera against parathyroid hormone (PTH) or calcitonin (CT) have many uses. They are the critical reagents for radioimmunoassay (RIA). They can also be used for immunohistologic studies or as neutralizing reagents in studies of hormone action in vivo or in vitro. This chapter will discuss the important steps necessary for generation of antisera, including selection of immunogen and its preparation for injection, choice of animal species, techniques of injection, harvesting of antisera, and techniques for screening the antisera. I will begin by reviewing some points of immunologic knowledge that are especially pertinent to the problem of antiserum production.

I. Immunological Knowledge Relating to Antibody Production

Interaction of the immunogen with both lymphocytes and macrophages is important in the immune antibody response. Although the lymphocyte produces the immunoglobulin, the macrophage plays an important role in regulating the immune response. The macrophage binds, ingests, and degrades the antigen, and somehow processes or presents the antigen to the lymphocyte. Interaction of the antigen with the macrophage seems to stimulate T helper cell activity rather than T suppressor cells, and thus enhances the immune response.[1] If antigen bypasses the macrophage and interacts directly with lymphocytes, T suppressor cells are stimulated. The goal of immunogen injection is actually to introduce the antigen into an environment potentially rich in macrophages. The intramuscular, intradermal, and subcutaneous routes each meet this objective. Intravenous injection of an antigen, on the other hand, minimizes its exposure to macrophages and tends to enhance the induction of T suppressor cells and thereby produce immunological tolerance to the antigen.[1]

Emulsification of the antigen before injection has an adjuvant effect and serves two purposes. First, it may prevent rapid catabolism of antigen. A more prolonged exposure to antigen will enhance the antibody response. Second, the emulsion may enhance exposure of macrophages to the antigen by eliciting a foreign body reaction. Adjuvants may also play a role other than these physical roles; chemical constituents of such adjuvants as killed tubercle bacillus or *Bordetella pertussis* may enhance the macrophage response and thereby promote T helper cell activity.[1] The use of adjuvant may decrease 10–100-fold the dose of immunogen needed to elicit the antibody response.[1]

The size of the antigen dose also has an important influence on the antibody response. At the time of the initial antigen injection, a large number of lymphocytes are capable of producing immunoglobulins recognizing the antigenic site of interest. Only a small number of these lymphocytes are statistically likely to be destined to produce a high-affinity antibody. To obtain a high-affinity antiserum, these high-affinity cells must be selectively activated to replicate. Replication is initiated when a receptor globulin on the lymphocyte surface binds the antigen. The receptor globulin has the same antigenic recognition site (and affinity) as the globulin eventually secreted by the lymphocyte clone. A small dose of antigen will provide enough peptide to saturate the surface receptor globulin only of those lymphocytes destined to make higher-affinity globulins. With larger doses of antigen, both high- and low-affinity cells will have enough surface receptors occupied to cause activation.[1] Thus, it is recommended that a small dose of antigen be used for the first dose or two. Even if the primary dose is too small to induce detectable titers of antibodies, it may trigger selectively the initial proliferation of high-affinity clones, which will then respond vigorously to the next antigen dose.[1]

The degree of purity of the immunogen may also be an important consideration. The concomitant administration of several antigens may blunt the antibody response to the immunogen of interest.[1] This "antigenic competition" is poorly understood, but probably involves a suppressive effect of cells on cells, that is, T suppressor cells elicited by antigen X can apparently block proliferation of anti-Y B cells.[1] Antigenic competition is not seen in thymus-deprived mice, unless they are given thymus cells.[1] An example of antigenic competition is shown below.

With repeated exposures to immunogen, the antibody response tends to occur earlier and to lead to higher titers, a phenomenon called the "anamnestic response." In addition, the character of the antiserum tends to change with time. I observed that the affinity of the midregion-specific PTH antibodies from a given animal will tend to increase steadily over the course of the first three to five booster doses[2] (Table 1). This phenomenon is well recognized in the immunology literature and is termed "antibody maturation"[1]; it has been less well appreciated in the hormone RIA literature. The increase in affinity occurs most sharply at the time of the booster dose, but seems to continue during the interval between doses; it also seems to occur in a fashion that is relatively independent of titer.[2] This maturation of antibody affinity represents a selection of clones of cells producing higher-affinity globulins. As the depot of antigen slowly becomes depleted with time, the immune system is exposed to a progressively lower dose of antigen, and the higher-affinity clones therefore are selectively stimulated.[1] The maturation of affinity may be prevented or retarded by the use of larger antigen doses.[1] The practical result of the phe-

Table 1. Maturation of Antibody Affinity During the Course of Several Booster Doses of bPTH or hPTH Given to Goat 1

Weeks After First Dose	ID-50[a]	Immunogen
22	400	bPTH
34	316	bPTH
106	67	bPTH
127	47	bPTH
161	48	bPTH
166	52	bPTH
188	49	hPTH
198	33	hPTH
203	25	hPTH
208	13	hPTH
216	10	hPTH
229	8	hPTH

[a]ID-50 (inhibitory dose-50) is the concentration of hPTH standard (in nlEq/tube) that reduces B/F by 50%. Thus, a decreasing ID-50 reflects an increase in affinity. The antiserum's affinity for the hPTH standard remained stable for many weeks after the fourth booster dose of bPTH, but increased further when the immunogen was switched from bPTH to hPTH. Data were obtained with a midregion-specific assay using labeled bPTH(41–84) as tracer. The normal range for midregion PTH in human serum is up to 150 nlEq/ml, or 15 nlEq/tube in the usual assay.[3]

nomenon of antibody maturation is that the best antisera are likely to be produced after several booster doses have been given.

The interval between booster doses may also have an effect on antibody maturation, but this has not been studied systematically for hormone antisera. The optimum interval might in fact depend on the route of administration and adjuvant employed. The general practice is to give booster doses only after the titer has peaked and begun to fall, or usually at intervals of 4 or more weeks.

II. Choice of Immunogen

Antisera have been prepared against many different forms of PTH or CT. Highly purified native hormones may be used as immunogens, but are quite expensive. Antisera for use in RIA can be prepared more economically by using a crude or partially purified hormone preparation. Assay specificity is then assured by use of a highly purified or synthetic radioligand. For use as immunohistologic reagents, however, antisera should not contain antibodies against other cellular components, since high background staining may result. Antisera against highly purified forms of the hormone would be more likely to prove useful for this purpose. Alternately, the globulins specific for PTH or CT could be purified from an antiserum against crude hormone by affinity chromatography, using synthetic or highly purified PTH or CT as the adsorbing ligand. The remainder of this chapter will deal with antisera for use in RIA.

A potential objection to the use of partially purified hormone preparations as immunogens is the presence of other protein components that might produce "antigenic competition" and thereby limit the rise in titer. Whether this effect would be important in the production of PTH or CT antisera might depend on the animal species being used, as well as the purity of the particular batch of hormone employed. This potential objection does not seem to have prevented the production of useful antisera using partially purified preparations. The question is, however, whether even higher titer and affinity might be achieved by use of highly purified hormones as immunogen. Their expense at the present time limits our ability to answer this question.

A second potential objection to the use of partially purified hormone preparations as immunogen is that they may contain hormone fragments as well as intact hormone. The concern is that these fragments might somehow alter the type of antibodies produced, or otherwise interfere in the immune response to the antigen of interest, which is generally considered to be the intact hormone. This concern is probably of little consequence, for the following reasons. First, solutions known to be enriched in hormone fragments have been used deliberately as immunogens and have produced useful antisera.[4] Second, the use of highly purified PTH as immunogen would not avoid the presence of hormone fragments, since the hormone is catabolized to various fragments in the recipient. Third, it is possible that fragments rather than the intact hormone may serve as the species that interact with lymphocytes and macrophages to induce the antibody response. This possibility is suggested by the observation that goat antibodies directed against the midregion of hPTH show a two- to fourfold greater molar affinity for the midregion PTH(44–68) fragment than to intact hormone.[3]

Despite the potential objections cited above, PTH that is partially purified to the TCA precipitate stage (hPTH–TCA) is pure enough to elicit an adequate antibody response, at least in some species. For example, the second injection of hPTH–TCA produced measurable anti-PTH titers in each of seven goats and rabbits (Table 2). In contrast, two injections of less pure hPTH

Table 2. PTH-binding Activity (B/F Value) After Two Injections of PTH Given 4 Weeks Apart

	Form of PTH Used as Immunogen	
Animal	Crude Extract of Parathyroid[a]	hPTH–TCA[b]
Goats	0.01	0.74, 0.86, 1.41
Rabbits	0.01, 0.01, 0.01, 0.01	0.13, 0.12, 0.46, 0.06

Note: Data are B/F at final dilution 1:50 after 48-h incubation of antiserum with [^{125}I]bPTH(41–84) at 4°C. Dextran–charcoal separation.
[a]*Extract:* supernatant (50,000 g, 30 min) of a urea–HCl homogenate of hyperplastic human parathyroid tissue (about 30 μg immunoreactivity per dose).
[b]*hPTH–TCA:* hPTH partially purified from the supernatant (see Appendix A) (about 40 μg immunoreactivity per dose). Titers were markedly higher after the third and subsequent injections of hPTH–TCA.

(crude parathyroid extract) failed to elicit PTH-binding activity in any of five animals, probably as a result of antigenic competition produced by the numerous other cellular components in the extract (Table 2).

In contrast to this lack of success with crude parathyroid extracts, calcitonin antisera have been prepared successfully in guinea pigs and rabbits using a crude extract of human medullary thyroid carcinoma.[5,6]

Synthetic PTH peptides or synthetic CT may also be used as immunogen. Disadvantages include the fact that they are more expensive than partially purified hormone preparations such as hPTH–TCA. Also, they may not necessarily be homogeneous. The synthetic process is not perfect, and amino acid deletions can occur. Purification of synthetic peptides by high-performance liquid chromatography may be necessary to prepare a homogeneous peptide with the amino acid sequence of native hormone. The synthetic peptides, because of their smaller size, may also be less antigenic than the native hormone.

The planned use of the antiserum, specifically which animal species is to be studied by RIA, may be the most important consideration in the choice of immunogen. Where possible, the immunogen should be obtained from the same species of animal that is to be studied. Homologous antisera may be expected to have higher affinity than heterologous ones. For example, antisera raised against bovine PTH usually crossreact to some extent with hPTH, but their affinity for hPTH is less than for bPTH. It may, therefore, be necessary to inject a larger number of animals with bPTH than with hPTH to obtain an antiserum useful for measuring hPTH. Of four goats I injected with bovine PTH, only one made an antiserum with useful affinity for hPTH.[2,7] In contrast, two of four goats injected with human PTH made antisera with an affinity for human PTH that exceeded that of the best anti-bPTH serum by three to five times.[3,8] To illustrate this concept further, it was found that one goat injected with bPTH had produced an antiserum with stable affinity to hPTH for more than 2 years. When the injected immunogen was switched from bPTH to hPTH, antiserum from this animal further increased its affinity for hPTH several fold (Table 1).

III. Preparation of Immunogen for Injection

Native parathyroid hormone may be purified from parathyroid glands for use as immunogen by a modification of the method described by Rasmussen et al.[9] Their scheme used a defatted powder of parathyroid tissue as the starting material. We prefer to start instead with a parathyroid homogenate. The following procedure is convenient.

Parathyroid glands from the desired species are collected as they become available and are stored at $-80°C$. We avoid glands from subjects with a history of hepatitis or with positive results of serologic tests for hepatitis. Each gland is placed in a cold Petri dish and, with a small forceps and No. 10 scalpel, is meticulously trimmed of fibrous tissue. Trimming will greatly facilitate homogenization and should probably be done before freezing and thawing have caused activation of lysosomal enzymes. When sufficient tissue has been accumulated, the parathyroid glands are thawed, minced at 4°C into 5-mm fragments, quickly weighed, and homogenized. The supernatant is carried through a scheme (Appendix A) in which other proteins are salted out, PTH is precipitated with ether and then redissolved, and other proteins again salted out. The material enriched in PTH is finally precipitated with trichloroacetic acid (TCA). The TCA precipitate contains a small percentage of PTH by weight. Material in the TCA supernatant of some batches can be recovered by ether precipitation and can also be used as immunogen; it is enriched in PTH fragments. These immunogens are dissolved in dilute acetic acid and stored at −80°C.

Human CT likewise can be purified from extracts of medullary carcinoma for use as immunogen.[6]

Synthetic PTH peptides or CT can be dissolved in dilute acetic acid and emulsified for use as immunogen. Since there is concern that their smaller molecular weight might limit antigenicity, smaller peptides are sometimes coupled covalently to a larger protein molecule.[10] Useful antisera, however, have been generated against PTH peptides or CT without prior coupling.[11,12] Unfortunately, there has been no systematic study to determine which method is preferable. The optimum method might differ for various peptides in different animal species. We have attempted to use a compromise method, namely, to adsorb the peptide onto a larger molecule. Solutions of methylated bovine serum albumin and the peptide are mixed at room temperature for about 30 min and then emulsified with Freund's adjuvant for injection. I have obtained some PTH-binding activity with this technique, but titers have not been very high after three booster doses. Dr. R. Gagel in our Section has recently generated a high-titer goat antiserum against synthetic human CT in this manner (personal communication).

It is important that the antigen emulsion be quite stable. Various devices are available for preparation of emulsions, but I have relied on the older and less expensive method of cross-injection between syringes (Appendix B). Complete Freund's adjuvant (Difco Laboratories, Detroit, Michigan) contains killed tubercle bacillus and is generally used only for the primary immunization. Further injections of complete Freund's adjuvant may produce an intense granulomatous reaction at the injection sites, causing crusted, indurated lesions.* Incomplete Freund's adjuvant is used for subsequent injections; it contains the emulsion oil, but lacks the tubercle bacillus component. As an additional adjuvant I have used DPT vaccine, given a few days before the primary antigen dose. Vaitukaitis et al.[13] had recommended crude *Bordetella pertussis* vaccine as an adjuvant for hormone antisera; this material has recognized adjuvant activity.[1] There are no data, however, on how well these adjuvants

*I noted a severe granulomatous reaction in one goat after the first antigen dose in complete Freund's adjuvant. I was unable to trace the history of this animal and do not know whether it might previously have had an infection with an animal strain of tuberculosis or may have received the complete adjuvant previously.

enhance the titer or affinity of antisera against PTH or CT, or whether they might actually cause lower titers by antigenic competition, especially in the case of the DPT vaccine.

The optimum dose of PTH or CT is not established. Bovine PTH doses as low as 2–3 µg produced significant, but small, rises in titer in goats.[2] I have used hPTH–TCA doses that contained 20–60 µg of immunoreactivity, judged against the WHO hPTH Reference Standard for Immunoassay, 75/549. Human CT doses of 100–500 µg have been used.[6,10] Doses above 100 µg are probably not necessary and might be more likely to inhibit development of high-affinity antisera. Systematic study is needed in this area.

IV. Choice of Animal Species

Generation of high-affinity, high-titer antisera for use in hormone radioimmunoassay may depend partly on luck in choosing an animal strain genetically capable of producing high-affinity antibodies against the particular antigenic region of interest, and in finding the individual animal that will produce a very-high-affinity antiserum. Most animals will produce some binding activity against any injected foreign protein. Titer and average affinity vary widely between individuals, however, and only an occasional animal will generate an optimal antiserum.

Useful antisera against PTH have been generated in a number of species, including guinea pig,[14] rooster,[15,16] goat,[17,18] and sheep.[19] There may be a species-specific tendency for immunological recognition of certain regions of PTH. Thus, all eight goats I injected with either bPTH or hPTH generated antibodies specific for the 1–34 and 44–68 regions, but only one made significant titers of 53–84-specific antibodies. Also, the 44–68-specific antibody species tended to be of much higher affinity than the 1–34-specific species.[2] Useful antisera against the 1–34 and 53–84 regions have been generated by others in roosters and guinea pigs. CT antisera have usually been generated in either guinea pigs or rabbits,[6,10–12] but recently the goat has also provided good CT antisera (Drs. R. Gagel and H. Heath, personal communications).

In the choice of animal species there are also practical considerations, such as initial and maintenance costs and longevity. Initial purchase cost is a minor factor, since the animals will be maintained long-term, and since maintenance per diem costs tend to multiply quickly. Maintenance costs are usually considerably higher for large animal species such as goats and sheep, and not all facilities are equipped to house these animals. Larger animals have the advantage that large volumes of serum can be obtained, but in the case of the goat[2] this potential advantage may be offset by the fact that titers tend to be somewhat lower than have been reported in other species such as chicken or guinea pig (1:10,000 versus 1:100,000–1,000,000). Larger animals generally have longer life expectancy than smaller ones. It is frustrating to obtain a useful antiserum in a guinea pig or chicken only to have the animal die a short time later. Two of my goats have produced useful antisera of stable affinity for more than 3 years.

V. Techniques of Injection and Harvesting Antisera

There are few data to indicate the optimum route of antigen injection for PTH and CT. I have used the intradermal route successfully in goats. I had less success with intradermal injections in the rabbit, guinea pig, and chicken, but this could relate to species more than to route. Intramuscular and subcutaneous injections have also been recommended.[1] The foot pad as an injection site in the chicken is said to be superior to other routes, and has also been used in the guinea pig.[6] Repeated use of the same site may lead to higher titers.

For intradermal injections, the viscous emulsion is drawn into a 1-cc syringe* through a No. 19 or 21 needle. The needle is then changed to No. 25 for the injections. Small injections (0.1–0.2 ml) are made at a minimum of six sites on the animal's flank by the intradermal skin test technique. Animal restraint and bleeding methods are discussed in Appendix C.

Once a detectable titer has been obtained and some maturation of antibody affinity has occurred, a considerable rise in titer may be produced by intravenous injection of very small antigen doses (below 5 μg). The potential effects of this technique on antiserum affinity have not been reported for hormone antisera, however.

To document the rise in titer caused by each boost (should this be of interest), a small bleed can be made at the time the antigen injection is given. Antiserum is then harvested at intervals after each booster dose. Maximum titers tend to occur 10–14 days after the first two or three booster doses and at earlier times after subsequent doses.[1] Each antiserum harvest is allowed to clot at room temperature for 3–4 h to allow time for clot retraction. Clear serum is then removed with an aspirating bulb or pipette. The remaining serum, surrounding the clot and admixed with red and white cells, is next removed and recentrifuged to harvest an additional 10–30% of serum. The clot may be allowed to stand at 4°C overnight for further clot retraction; the small amount of serum thus recovered is usually hemolyzed, but may be valuable if the animal is producing a good antiserum. Sera are labeled with date and name of donor animal (a different-colored label for each animal) and are stored at −20°C. Portions of the best antisera are also placed in a second freezer at −80°C. I store a typical 100-ml harvest in six plastic counting vials (15 ml serum each) and in two or three screw-capped serum vials (1–2 ml each). The smaller vials are useful during antiserum screening to avoid thawing and refreezing of larger amounts of the antiserum.

---

*Glass syringes are preferable. Disposable plastic can also be used, but the rubber plunger will become swollen and sticky after contact with the oily emulsion, so that it cannot be retracted again once it is advanced. For this reason care should be taken to avoid introducing air while drawing up the emulsion, and air bubbles should be expelled only after the entire dose is loaded.

VI. Evaluation of Univalent Antisera

Evaluation of antisera for use in RIA involves determination of three basic characteristics: (1) titer, (2) sensitivity, and (3) regional specificity. Titer is a function of the number of immunoglobulin molecules present, their average affinity for the antigen, and the specific activity of the tracer. Sensitivity is a function of affinity and tracer specific activity. Titer and sensitivity are influenced also by the conditions under which the assay is run. Regional specificity will be discussed below.

Let us first consider the simple case of a univalent antiserum. We will assume that a small synthetic peptide is used as immunogen, and that the antiserum contains only antibodies directed against a single antigenic site in the peptide. Despite the seeming simplicity of this system, these antibodies are the products of a large number of lymphocyte clones, each with its own affinity for the antigen. The average affinity of the antiserum for a given animal at any time is determined by the relative abundance of the immunoglobulins secreted by the various clones. The average affinity and the titer of antiserum from a given animal will tend to vary independently with time,[2] so that each antiserum bleed is unique.

The first step in characterizing an antiserum lot is to estimate its titer by incubating radiolabeled hormone with varying dilutions of antiserum under conditions similar to those planned for the RIA (Appendix D). For initial screening, the dilutions might range from 1:50 to 1:50,000. For PTH antisera, the tracer may be either intact bPTH(1−84) or the synthetic fragment used as immunogen (if it contains a tyrosine or histidine residue and can therefore be radiolabeled). Titer is usually defined as the dilution that yields a B/F value of 1.0, i.e., where half the tracer is bound. Figure 1 shows an experiment in which six antiserum lots were titered using labeled synthetic midregion hPTH as tracer. The titer of three of the lots is indicated by vertical dotted arrows. After titer is estimated, standard curves may be constructed to estimate antiserum sensitivity.

When several antiserum bleeds have been obtained, one may wish to compare the sensitivity of the various bleeds to select the one with greatest sensitivity. In order to compare different lots of antiserum as to sensitivity, it is necessary to construct each standard curve with the same B_0/F_0* value. The need for equal B_0/F_0 values arises because the apparent sensitivity of the standard curve with a given antiserum lot may depend on B_0/F_0. The standard curve usually becomes more sensitive at lower B_0/F_0 values.† The following steps will allow the construction of a set of standard curves with different anti-

---

*B_0/F_0 is the B/F value for the zero point on the standard curve.
†Diluting out the antiserum to produce a lower B/F value does, however, reach a point of diminishing return, where the benefit from improved sensitivity is negated by increased statistical error. The error arises from the fact that as B/F falls, the nonspecific binding becomes an increasingly greater percentage of the total bound counts. A B_0/F_0 value of 0.4–0.7 seems to be optimal for the PTH assay systems I have investigated.

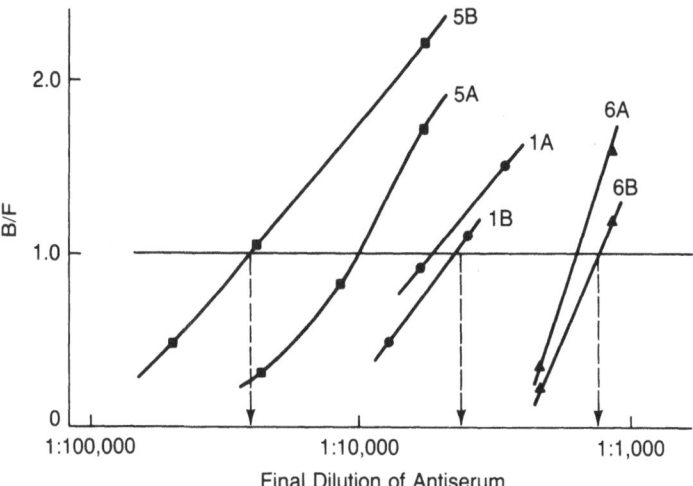

Figure 1. Determination of titer by serial dilution of midregion-specific antibodies in antisera from goats 1, 5, and 6. Midregion specificity was assured by use of labeled [tyr[43]]hPTH(43–84) as tracer. The samples were incubated 24 h with tracer, followed by dextran–charcoal phase separation. From the curve of each antiserum, a dilution was chosen (vertical arrows) at which a standard curve was constructed (see Figure 2).

serum lots in such a way that each curve will have approximately the same B_0/F_0 value.

First, as in Figure 1, a serial dilution in the absence of added PTH or CT standard is performed simultaneously with each antiserum lot to be tested.[2] For each lot, a dilution value can then be interpolated which should give the desired B_0/F_0 value (Figure 1). Next, a family of standard curves is constructed, one with each antiserum lot at its proper dilution (Figure 2). Under these conditions, B_0/F_0 should be similar for each lot. From the normalized standard curves (Figure 3), it will be possible to select the antiserum lot that gives the most sensitive assay.

VII. Evaluation of Multivalent Antisera: Regional Specificity

When intact hormone is used as immunogen, the degree of complexity of the antiserum increases. PTH and CT are large enough peptides to possess numerous potential antigenic determinants. The reslting antisera will be multivalent: they will contain several families of immunoglobulins, each recognizing a different antigenic determinant. Dr. Deftos has discussed regional specificity of CT antisera (Chapter 9). This discussion will, therefore, focus on PTH antisera. The regional specificity of an antiserum is a function of the number of antigenic sites recognized by the animal's immune system and the relative abundance of immunoglobulins directed to each of the sites. The average affin-

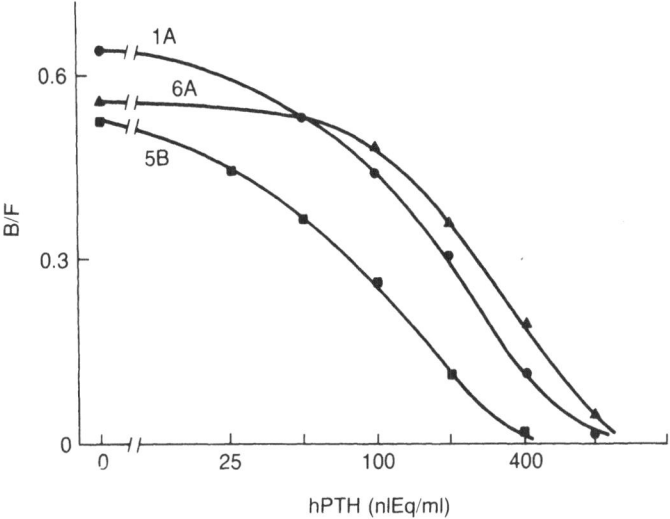

Figure 2. Midregion-specific standard curves with three of the antiserum lots from Figure 1. Data are B/F values versus hPTH concentration; the same tracer as for Figure 1 was used, but was added after 3 days preincubation of standard with antiserum; phase separation occurred 18 h after tracer addition. Note that B_0/F_0 values for each antiserum are similar, but not identical. This represents the usual amount of scatter of B_0/F_0 values in such titration experiments.

ity and titer of the immunoglobulins directed against each antigenic site will vary independently with time in each animal. Each of these antisera will, therefore, have more distinctive properties than univalent antisera. The regional specificity of these more complex antisera may be established in three different ways.

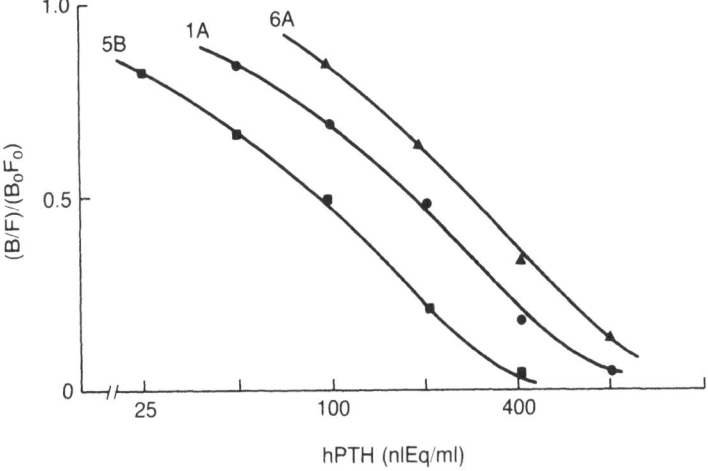

Figure 3. Normalized standard curves from Figure 2. B_0/F_0 is set equal to 1.0.

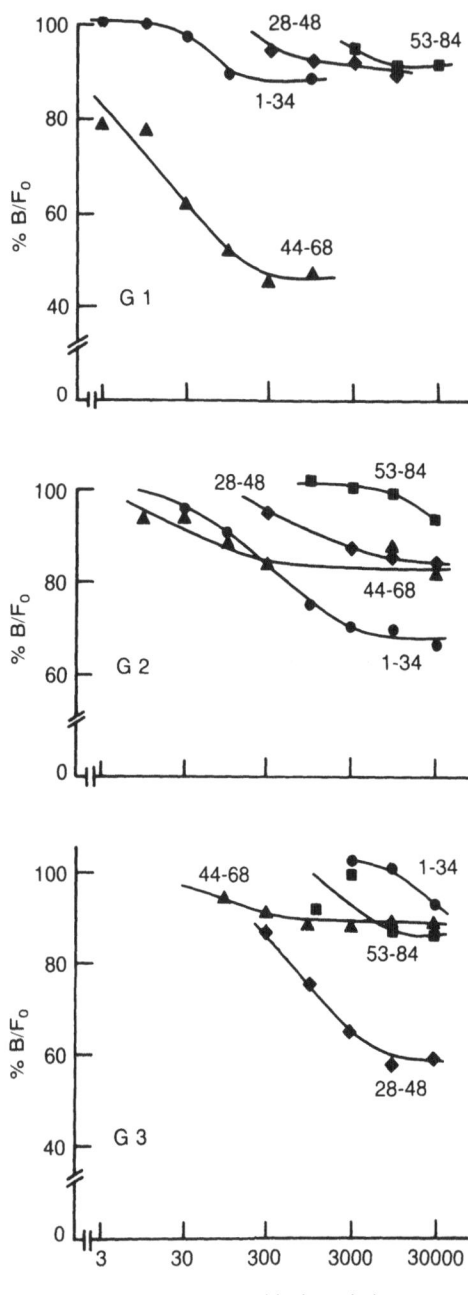

Figure 4. Regional specificity of three goat anti-bPTH sera. Each antiserum was incubated 4 days with varying concentrations of each of four PTH fragments, after which [^{125}I]bPTH(1–84) was added, with phase separation done 2 days later. The four fragments studied were hPTH(1–34), bPTH(28–48), hPTH(44–68), and hPTH(53–84). (Reprinted from Ref. 2 with permission.)

First, each individual region can be considered singly by using a synthetic peptide as tracer, analogously to that discussed above for a univalent antiserum. In fact, the examples in Figures 1–3 actually employed antisera against native hPTH. For antisera against native PTH, the following fragments are now available commercially for use as radioligands to characterize the immunoglobulins specific for the amino-terminal region, cleavage region, midregion, and carboxyterminal region, respectively: hPTH(1–34), [tyr^{27}]hPTH(27–48), [tyr^{43}]hPTH(43–68), and [tyr^{52}]hPTH(52–84). With the use of these synthetic peptides, the overall characteristics of the antiserum can then be summarized in terms of its titer and affinity for each region.

The second approach to characterizing the properties of antisera against native PTH is to use intact bPTH(1–84) as radioligand, testing the ability of various synthetic peptides to inhibit its binding (Figure 4). At a high dosage, each peptide fragment will cause only partial inhibition of binding of the intact hormone, in proportion to the relative titer of globulins directed against it. If high levels of the peptide have caused enough inhibition of tracer binding, it may be possible to estimate an ID-50 value from the standard curve. In the examples (Figure 4) the relative abundance of the three different antibody species was different for each of the three antisera, but there was a trend for the antibodies directed against the midregion (44–68) to show the highest affinity, as demonstrated by the position of their standard curves further to the left.

The technique of characterizing antisera using intact PTH as tracer is rapid and simple, but has three potential disadvantages. First, immunoglobulin species present in relatively low titer may not be detected. These antibodies might be useful for certain purposes, despite their limited abundance. For example, the 28–48 region-specific antibodies accounted for only a small fraction of the PTH binding in two of the antisera against native bPTH (Figure 4), but were quite useful for the study of the steric configuration of the 28–48 region of bovine vs. human PTH.[20] The second potential disadvantage is that the apparent sensitivity of the antibodies against a given region may be different with the intact hormone as tracer from that observed with a synthetic peptide tracer. The third disadvantage is that labeled intact PTH has a relatively high nonspecific binding value, which averages 12–20% and adds a considerable "noise factor" to the assay.

A third technique can be used to characterize antisera generated against native PTH, but is more cumbersome. Intact hormone again is used as radioligand, but individual regions are examined separately by blocking the other regions with an excess of unlabeled synthetic peptide. For example, one of my goat antisera is essentially bivalent, recognizing chiefly the 1–34 and 44–68 regions. Blocking the low-affinity amino-terminal-specific antibody species with an excess of unlabeled hPTH(1–34) left only midregion-specific species able to bind the tracer bPTH(1–84), and produced a midregion-specific assay almost as sensitive as that using the synthetic midregion peptide as tracer (Mallette, unpublished observations). This blocking technique may be useful if the avail-

able synthetic peptide representing the region to be studied does not contain
tyrosine or histidine and thus cannot be labeled easily. It is, however, wasteful
of synthetic peptide, since high levels of peptide are necessary to block these
low-affinity unwanted antibody species. Furthermore, polyvalent antisera must
be preincubated with a different set of unlabeled peptides for each region to be
screened.

VIII. Miscellaneous Additional Considerations in Antibody Screening

In general, the affinity of antiserum from a given animal remains relatively
stable after the initial period of antibody maturation (see Table 1). Unfortunately, an occasional animal will, after an initial period of increasing affinity,
show a marked decrease in affinity (Mallette, unpublished observations). The
factors that trigger the proliferation of the low-affinity lymphocyte clones in
such a case are unknown. It is, nevertheless, prudent to screen antiserum sensitivity promptly and to obtain a larger harvest at an early date from animals
producing promising antiserum.

For screening of antiserum sensitivity, the peptide used as standard should
resemble as closely as possible that which the investigator desires eventually to
measure. Thus, in order to select an antiserum for measurement of human
PTH, the human hormone should be used as standard, rather than bovine
PTH. We observed quite marked differences in the relative order of sensitivity
of three different anti-bPTH sera to the bovine versus the human hormone.[2]
The relative order of sensitivity of several anti-human PTH sera to canine PTH
also differed considerably from that to human PTH (Mallette, unpublished
observations).

Standard curves done to screen antisera may be run under either equilibrium or nonequilibrium conditions. When the curve is run at equilibrium, the
affinity constant of the antiserum can be estimated, providing a physiochemical
parameter that may be compared between laboratories. If, however, the curve
is run under the same conditions as will eventually prevail in the assay, one
obtains an immediate prediction of the ultimate assay sensitivity. For adequate
sensitivity of PTH assays, it is necessary to use nonequilibrium conditions.
Antiserum is incubated first with the samples to be measured, and tracer is
added at a later time. The assay then undergoes phase separation before equilibrium has time to be reestablished. Although sensitivity is enhanced under
these conditions, affinity constants cannot be estimated.

IX. Conclusion

Using the techniques and information given here, any investigator, given a bit
of luck, should be able successfully to generate useful antisera. We urge that

future protocols for antiserum generation carry an experimental design that might answer some of the questions mentioned above regarding adjuvant effects, optimum timing and routes of administration, and effects of dose size.

Acknowledgments This work was supported by the Veterans Administration. Ms. Melissa Renfro skillfully performed the radioimmunoassays and assisted in obtaining the antisera.

References

1. Eisen HN: Immunology. An Introduction to Molecular and Cellular Principles of the Immune Responses. Harper & Row, Hagerstown, MD, pp. 400–438, 1980.
2. Mallette LE: Goat antisera to bovine parathyroid hormone: characterization of regional specificity, and evolution of titer and affinity of the 44–68 region specific antibody species. Horm Metab Res 13:523–528, 1981.
3. Mallette LE, Tuma SN, Berger RE, Kirkland JL: Radioimmunoassay for the middle region of human parathyroid hormone using an homologous antiserum with a carboxyterminal fragment of bovine parathyroid hormone as radioligand. J Clin Endocrinol Metab 54:1017–1024, 1982.
4. DiBella FP, Gilkinson JD, Flueck J, Arnaud, CD: Carboxyl-terminal fragments of human parathyroid hormone in parathyroid tumors: unique new source of immunogens for the production of antisera potentially useful in the radioimmunoassay of parathyroid hormone in human serum. J Clin Endocrinol Metab 46:604–612, 1978.
5. Tashjian AH Jr, Howland BG, Melvin KEW, Hill CS Jr: Immunoassay of human calcitonin. Clinical measurement, relation to serum calcium and studies in patients with medullary carcinoma. N Eng J Med 283:890–895, 1970.
6. Wright CR, Voelkel EF, Tashjian AH Jr: Measurement of human calcitonin by affinity chromatography and radioimmunoassay. In Abraham GE (Ed.): Handbook of Radioimmunoassay. Marcel Dekker, New York, pp. 391–423, 1977.
7. Mallette LE: A radioimmunoassay for human parathyroid hormone utilizing a goat anti-bovine PTH serum. Acta Endocrinol 96:215–221, 1981.
8. Marx SJ, Sharp ME, Krudy A, Rosenblatt M, Mallette LE: Radioimmunoassay for the middle region of human parathyroid hormone: studies with a radio-iodinated synthetic peptide. J Clin Endocrinol Metab 53:76–84, 1981.
9. Rasmussen H, Sze YL, Young R: Further studies of the isolation and characterization of parathyroid polypeptides. J Biol Chem 239:2852–2857, 1964.
10. Silva OM, Snider RH, Becker KL: Radioimmunoassay of calcitonin in human plasma. Clin Chem 20:337–339, 1974.
11. Parthemore J, Deftos LJ: The regulation of calcitonin in normal human plasma as assessed by immunoprecipitation and immunoextraction. J Clin Invest 56:835–841, 1975.
12. Hennessy JF, Wells SA Jr, Ontjes DA, Cooper CW: A comparison of pentagastrin injection and calcium infusion as provocative agents for the detection of medullary carcinoma of the thyroid. J Clin Endocrinol Metab 39:487–495, 1974.
13. Vaitukaitis J, Robbins JB, Nieschlag E, Ross GT: A method for producing spe-

cific antisera with small doses of immunogen. J Clin Endocrinol Metab 33:988–991, 1971.

14. Berson SA, Yalow RS, Aurbach GD, Potts JT Jr: Immunoassay of bovine and human parathyroid hormone. Proc Natl Acad Sci USA 49:613–617, 1963.
15. Reiss E, Canterbury JM: A radioimmunoassay for parathyroid hormone in man. Proc Soc Exp Biol Med 128:501–504, 1968.
16. Chase LR, Slatopolsky E: Secretion and metabolic efficacy of parathyroid hormone in patients with severe hypomagnesemia. J Clin Endocrinol Metab 38:363–371, 1974.
17. Fisher JA, Binswanger V, Dietrich FM: Human parathyroid hormone. J Clin Invest 54:1382–1394, 1974.
18. Manning RM, Hendy GN, O'Riordan JL: Characterization of antibodies against intact human parathyroid hormone. J Endocrinol 73:38P–39P, 1977.
19. Wood WG, Butz R, Casaretto M, Hehrmann R, Juppner H, Marschner CWI, Sahn H, Hesch RD: Preliminary results on the use of an antiserum to human parathyrin in a homologous radioimmunoassay. J Clin Chem Clin Biochem 18:789–795, 1980.
20. Mallette LE, Refro M, Lemoncelli J, Rosenblatt M: Radioimmunoassays for the 28–48 region of parathyroid hormone detect intact hormone but not hormone fragments. Calcif Tiss Int 33:375–380, 1981.

Appendix A: Purification of PTH from Parathyroid Tissue

Material

1. Parathyroid tissue, trimmed of fibrous tissue, minced
2. Glassware, glass homogenizer, high-speed centrifuge, small centrifuge (flameproof or in a well-vented hood)

Reagents

1. Acetic acid, glacial and 0.1 N
2. Acetone
3. Ether
4. NaCl, dry and 1.0 N
5. Trichloroacetic acid, 40% and 5% aqueous solutions
6. Urea
7. HCl, 2.0 M
8. Alphamercaptoglycerol (αMG)
9. MB-3 exchange resin

Solutions

1. Urea–HCl: 8 M urea, 0.2 N HCl, 0.1% αMG. Stir 10 M urea solution with MB-3 resin (about 100 mg/100 ml) to remove carbamate ion. Start with 2 vol of 1 N HCl, add 0.01 vol αMG and make up to 10 vol with 10 M urea. Use same day.
2. Acetone–ether (1:1, v/v).
3. 20% glacial acetic acid with 0.1% αMG.

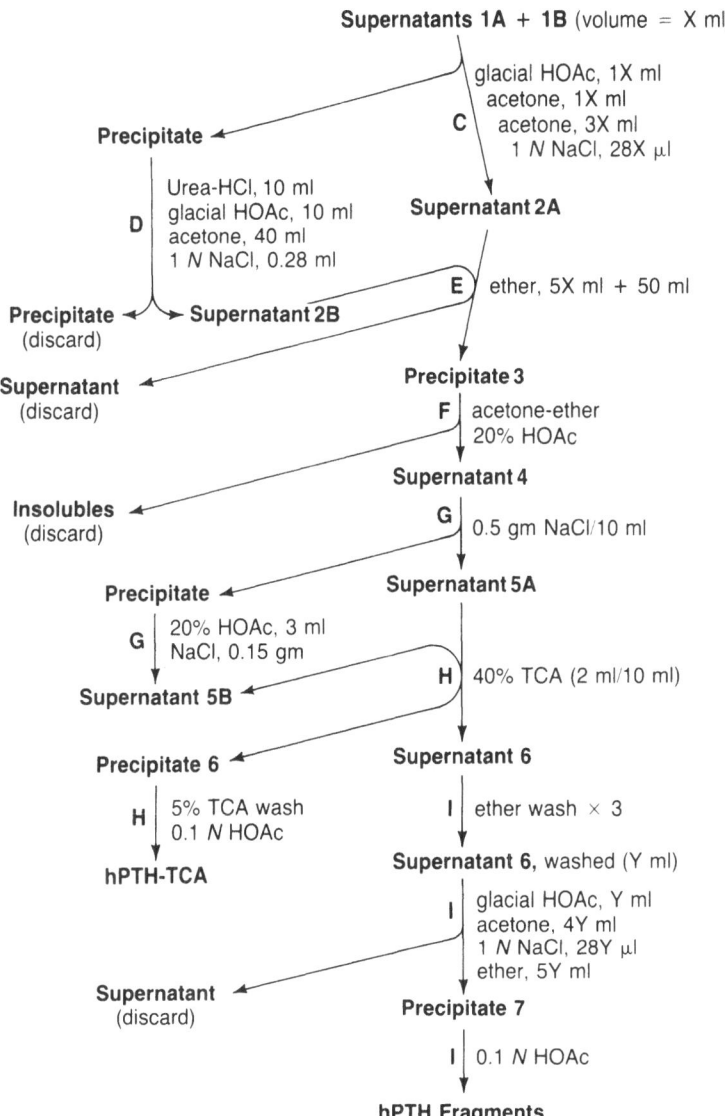

Figure A.1. Flow scheme for purification of PTH.

Procedure (Caution, Flammable Reagents!)

Refer to Figure A.1 for flow diagram. The volumes given here are appropriate for a starting mass of about 2 g of tissue; the procedure can be scaled up proportionally for a larger mass of tissue.

Note: At each stage, the total volume of supernatant or solution should be recorded, and small aliquots of solution taken for determination of yield. I suggest 100-μl aliquots, which are mixed with 900 μl of PTH assay buffer containing 30% hypoparathyroid serum and stored at $-20°$ C for future RIA. Supernatants to be aliquoted are indicated below with "AQ."

Homogenize Parathyroid

Place 1 g tissue fragments/7 ml urea–HCl in glass homogenizer and homogenize thoroughly. Centrifuge (50,000 g, 30 min). A thin fat layer will float to the top; remove with spatula and discard. Pour off supernatant into a graduated cylinder and take sample for yield ("AQ 1A").

Reextract Pellet

Transfer pellet back to homogenizer and rehomogenize in urea–HCl (3.5 ml/g tissue). Centrifuge as above, in same tube. Sample this supernatant ("AQ 1B") and pool remainder with supernatant 1A in the cylinder; record pooled volume (X ml) and transfer to beaker that will hold 7X ml.

Salt Out Non-PTH Proteins

To the pooled supernatants add with constant stirring X ml of glacial acetic acid, then X ml of acetone. Let stand covered 1 h at room temperature to allow flocculation. Then add 3X ml of acetone and 28X μl of 1 N NaCl (i.e., 28 μl/ml original volume X). Let stand at room temperature another hour, then centrifuge (1000 g, 15 min). Sample supernatant ("AQ 2A") and transfer the remainder to an Erlenmeyer flask that will hold 10X ml. Use pellet for next step.

Back-extract the Salted Out Material

Dissolve pellet in 10 ml urea–HCl, add 10 ml glacial acetic acid, 40 ml acetone, and 0.28 ml 1 N NaCl. Let stand 1 h, centrifuge (1000 g, 15 min). Sample supernatant ("AQ 2B") and add to the Erlenmeyer flask that contains supernatant 2A. Discard pellet.

Precipitate PTH with Ether

To the pooled supernatant 2A and 2B add (5X + 50) ml ether. Allow precipitate to settle, then decant and discard supernatant. (Alternatively, the precipitate may be collected in a small number of 15-ml glass centrifuge tubes. Several portions of the ether solution are centrifuged in each tube; each portion of supernatant is discarded, resulting in accumulation of a growing pellet in each tube. This step should be done in a sparkproof centrifuge or at least in a well-vented hood.)

Wash and Redissolve Ether Precipitate

Wash the precipitate once in acetone–ether and dissolve in about 10–20 ml 20% acetic acid–aMG. Solvation may require 1–3 h of gentle mixing. Centrifuge (1000 g) to pellet any insoluble material for disposal. Sample supernatant ("AQ 4") and save for next step.

Salt Out Other Proteins

To each 10 ml of supernatant 4, add 0.5 g NaCl crystals and mix until the NaCl dissolves. Centrifuge (1000 g, 10 min). Sample supernatant ("AQ 5A") and save.

Back-extract the pellet by dissolving it in about 3 ml of 20% acetic acid; add 0.15 g NaCl, allow it to dissolve and centrifuge. Sample supernatant ("AQ 5B") and pool superantant 5A and 5B. Discard pellet.

Precipitate PTH with TCA

Add 2.0 ml of 40% TCA for each 10 ml of pooled supernatant. PTH will precipitate. Centrifuge (1000 g, 10 min). Save supernatant (supernatant 6) for next step. Wash precipitate in 5% TCA, dissolve in as small a volume of 0.1 N acetic acid as possible. Sample this solution ("AQ 7A") and store remainder at $-70°$ C. This solution (hPTH–TCA) will contain 20–40 μg hPTH/ml.

Recover Fragments of PTH

Extract supernatant 6 with ether 3 times to remove TCA: shake equal volumes of supernatant 6 and ether in a stoppered cylinder, with frequent venting. Discard ether washes. Estimate residual aqueous volume (Y ml). Add Y ml glacial acetic acid, 4Y ml acetone, 28Y μl 1 N NaCl, and 5Y ml ether. A visible precipitate may form at this stage. If so, it may be recovered by centrifugation for use as immunogen. The supernatant is discarded and the pellet dissolved in 0.1 N acetic acid. Sample this solution ("AQ 7B") and store remainder at $-70°$ C. This solution will be enriched in hPTH fragments and will contain a highly variable amount of PTH immunoreactivity. Store at $-70°$ C.

Appendix B: A Method for Preparation of Antigen Emulsion

Antigen solution (e.g., 2.0 ml) is loaded into one Luerlock syringe (5 ml) and Freund's adjuvant (2.0 ml) into a second syringe (5 ml). A stopcock or double-hubbed needle is then attached to the second syringe (5 ml), and all the air is expelled from the syringe-stopcock, which is then fitted carefully to the first syringe (air also carefully excluded). The solutions can then be mixed and emulsified by cross-injection. After about 100 strokes, the resistance to cross-injection will increase markedly as the more viscous emulsion forms. At least 20 further strokes should be performed in order to assure a stable emulsion. As a test, a small portion of the emulsion (or a test emulsion with water rather than antigen solution) can be set aside. A satisfactory emulsion will not separate into layers after several hours of standing.

Appendix C: Some "Restraint" Methods

Goats

Secure control of the animal by grabbing the handle nature provided (a horn). The animal is then placed against a wall with an assistant securing the chin and a horn (to rotate the neck slightly away from the side to be punctured). The animal is fixed in place by the assistant who holds a knee or leg gently but firmly against the animal's flank just ahead of the hind leg. The external jugular is easily located by compressing the base of one side of the neck gently with a finger. Blood may be withdrawn into 20-

ml Vacutainer tubes or into a vacuum bottle. We generally take from 10 to 15 20-ml tubes per bleed. Brief pressure after venipuncture will secure hemostasis. Injections may be made intradermally on the animal's flank and back. We clean the skin with an alcohol swab before venipuncture or injection, but shaving is not necessary. A few slices of carrot after the procedure will produce a more cooperative animal in the future. Nanny goats are recommended. Billy goats can be used but should be made more docile by neutering.

Guinea Pigs

Guinea pigs should not be picked up by their rib cage and should have their feet supported while being handled. Pulmonary contusions occur easily and can be fatal. Blood may be obtained by one of four routes. We have used cardiac puncture in the past, but the mortality rate may approach 5%. Others more skillful than I have reported successful femoral venipuncture or have obtained blood from the retroorbital venous plexus. Our preferred method now is to perform a venisection (with a scalpel or sharp razor blade) of the vein on the instep of the hind limb. The hind limb is shaved and coated with silicone grease or petroleum jelly. An incision is then made rapidly across the instep, and the hindlimb is quickly inserted into a vacuum bleeding apparatus (see section on the rabbit, below). All of the above procedures are done under general anaesthesia; methoxyfluorane (from a veterinary supply house) placed in an ether jar works well and does not induce bronchospasm as ether tends to do.

Chickens

The bravest assistant gains control of the animal by grabbing (with gloved hand) the neck and legs. The animal will become relaxed when inverted by the legs. With practice the animal can be secured supine with a wing extended. A large central vein is visible near the joint on the undersurface of the wing and with practice can be punctured. Do the first venipuncture as distally as possible, and alternate wings to allow tissue repair between bleeds.

Rabbits

The animal should be placed in a specially constructed restraining cage from which the ears protrude and which allows access to the animal's flank and back.

Method A

Xylol is applied to the skin over the central ear artery. After several seconds, vasodilation will facilitate arterial puncture, which will provide several milliliters of blood.

Method B

A specially constructed vacuum bottle with seal can be purchased (Rabbit bleeding apparatus, Bellco Glass Inc., Vineland, New Jersey) to fit over the animal's ear. Blood is then pumped from a small nick in the marginal ear vein by applying vacuum to the bottle. This method is less sterile, but larger volumes are obtained more reliably.

Appendix D: Usual Conditions for PTH RIA

	Usual Assay	Reduced Volume
Buffer, with or without antiserum	300 μl	120 μl
Serum	100 μl	40 μl
Tracer in buffer	100 μl	40 μl
Charcoal–dextran	200 μl	400 μl

Note: Buffer is sodium barbital, 0.05 M, pH 8.6.

Serum aliquot may be hypoparathyroid serum, with or without standard, or known (internal standard) or unknown serum. For purposes of titering the various antiserum lots, a 20% solution of hypoparathyroid serum in buffer may be substituted for the serum, or (what amounts to the same thing) the antiserum may be diluted in 5% hypoparathyroid serum, with 400 μl dispensed per tube.

Chapter 11

Preparing the Tracer:
Iodination Techniques

Anne P. Teitelbaum

Radiolabeled hormones have a wide variety of uses, for example in radioimmunoassay procedures, in pharmacokinetic analyses, as tracers for receptors, and for autoradiography studies. In general, it is not difficult to prepare radioiodinated hormones in the laboratory. However, iodination of parathyroid hormone (PTH) and calcitonin (CT) has been particularly troublesome because radiolabeling procedures often damage these hormones. With the development of several new techniques during the past decade, it is possible to choose the method best suited to the needs of the investigator.

The method of choice for radioiodination of PTH or CT depends on the proposed use of the radioligand and on the amino acid sequence of the peptide. The major factors determining the selection are the requirements for biologic activity and specific activity. If the peptide contains methionine residues in a region of the molecule critical for biologic activity or binding activity, the strongly oxidizing conditions of many iodination procedures will probably inactivate the hormone.[1,2] The native sequences of bovine, human, and porcine PTH and human CT contain at least one methionine residue in their respective biologically active regions. However, if iodination conditions are mild enough to preclude harmful oxidation, a radioligand with low specific activity could result. Thus, it may be necessary to sacrifice high specific activity in order to retain biologic activity.

The iodination procedure used may depend on the amino acid residue to be iodinated. Tyrosine is labeled in preference to histidine when both are present in the peptide especially at pH > 7.5, that is, under conditions in which some tyrosines are negatively charged.[3] It is generally more difficult to iodinate histidine than tyrosine, even when histidine is in a peptide containing no tyrosine, presumably because tyrosine is the stronger acid. It has been shown that bPTH(1–84) is labeled predominantly at amino acid 43, a tyrosine, despite the presence of four histidines in the molecule.[4,5]

The problem of choosing between high specific activity and retention of biologic activity can be circumvented by the use of analogs of PTH or peptides which contain no methionine. Rosenblatt et al.[6] have synthesized an analog of the biologically active amino-terminal fragment of bovine PTH, [8Nle, 18Nle, 34Tyr]bPTH(1–34) amide with norleucine residues substituting for methionines and a tyrosine–amide residue in place of phenylalanine. Since norleucine is similar in size to methionine, the substitutions do not affect the biological activity of the hormone. Highly oxidizing conditions can be used to iodinate this peptide with no loss of biological activity.[6] Furthermore, the presence of tyrosine allows labeling of only one residue. Two of the three species of salmon CT contain tyrosine residues and lack methionine, whereas the human sequence has a methionine at position 8. Thus, salmon CT is generally iodinated in preference to human CT.

I. Theory

To introduce radioactive iodide into tyrosine or histidine residues, the isotope must be converted from iodide (I^-) to iodine (I_2). The I_2 is then polarized by its association with the oxygen of H_2O:

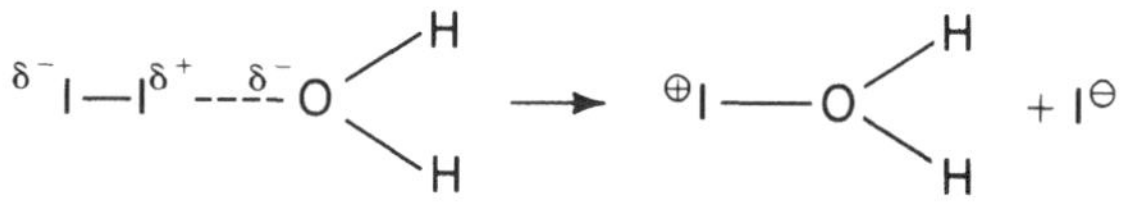

The positively charged iodide molecule is an electrophile which attacks tyrosine on the carbon ortho to the negatively charged oxygen:

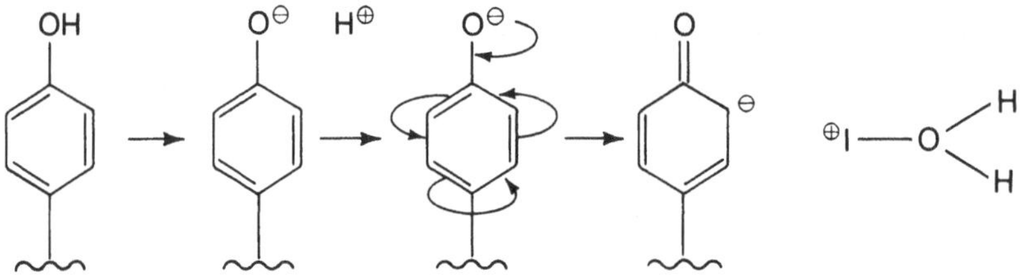

or on the carbon adjacent to the negatively charged nitrogen:

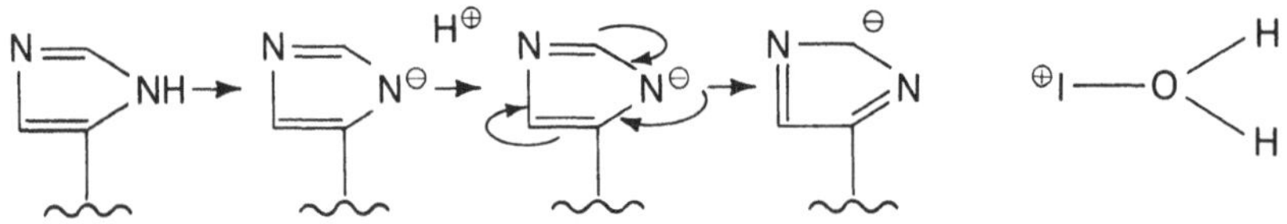

Monoiodotyrosine is more acidic than free tyrosine and so is more readily iodinated.[3] Thus, even if the molar ratio of iodide to tyrosine is unity, some fraction of the tyrosine residues [approximately 20% for bPTH(1–84)[4,5]] are diiodinated. The pH of the iodination reaction mixture should be slightly alkaline to partially ionize tyrosine or histidine. At pH > 8, other species of iodide such as hypoiodous acid are formed reducing the effective concentration of I^+.[3]

Table 1. Iodination Techniques

Method	Advantages	Disadvantages	References
Chloramine-T	Easy, high S.A.[a]	Highly oxidizing, damages methionines	7,8,9,13,14
Iodine monochloride	Easy	Low S.A., highly oxidizing	10
Sodium hypochlorite	Easy	Highly oxidizing	11
Lactoperoxidase (LP)	Fairly easy, gentle	Must separate LP from labeled hormone	12,13,14,29
Solid-phase LP	Easy, gentle	Low S.A.	15
Electrolytic	Gentle	Requires special equipment and purification of ^{125}I-labeled hormone by receptor adsorption	4,16,17,18,19
Iodogen (1,3,4,6-tetrachloro-3a,6a-diphenylglycoluril)	Easy, gentle	Low S.A.	20
Iodo beads	Easy, gentle, solid phase support	Low S.A.	—
Conjugation, Bolton–Hunter[b]	Hormone not exposed to oxidant	Low S.A.	21
Methyl *p*-hydroxybenzimidate	Hormone not exposed to oxidant	Low S.A.	22

[a]S.A. = specific activity.
[b]N-succinimidyl-3-(4-hydroxy 5-[^{125}I]iodophenyl)propionate

The reaction $2I^- \rightarrow I_2 + 2e^-$ requires an oxidation potential of 592 mV. Side reactions, which can convert iodine to iodate and other products likely to oxidize methionines to sulfoxides, require slightly more than 1000 mV. Thus, the oxidation potential must be confined to a very narrow range to avoid damage to methionine-containing peptides. Iodination procedures using Chloramine-T or other strong oxidants are uncontrolled with respect to potential difference. Methods that limit the concentration of oxidant or control oxidation potential include lactoperoxidase labeling, electrolytic iodination, and the iodogen method. Although other methods for radioiodinating peptides are available, this chapter will be limited to the above-mentioned procedures. Two alternative methods of iodination involve conjugation of radioiodinated N-succinyl propionate (Bolton and Hunter[21]) or methyl-*p*-hydroxybenzimidate to the amino groups of lysine residues. In these procedures, the hormone is not exposed to oxidizing conditions. However, introduction of these compounds into PTH or CT could impair receptor binding and/or biologic activity. Table 1 summarizes the advantages and disadvantages of using various iodination techniques.

PTH has also been radiolabeled with [^{3}H]-, [^{14}C]-, or [^{35}S]-labeled amino acids using parathyroid gland slices and cell-free systems.[23] Indirect labeling of PTH has been accomplished by [³H]-acetamidation of lysine and free amino groups with no loss of biologic activity.[24]

II. Methods

A. General

The radioisotope most widely used for radioiodination is ^{125}I, chosen for its stability and counting efficiency. The half-life for ^{125}I is 60 days, compared with 8 days for ^{131}I, the other generally available radioisotope of iodide. ^{125}I and ^{131}I are manufactured as the sodium salts. The isotope is stable for several days after production at room temperature[25] and at alkaline pH.

Radioiodide is generally shipped in NaOH, pH = 7–11. Since iodination reaction volumes are small (<300 μl), addition of the isotope could change the pH of the reaction mixture. Thus, the iodination reaction should be strongly buffered.

It is advisable to use a small reaction volume to maximize the concentrations of the reactants. Column chromatography usually follows iodination, and application of a small volume ensures good separation of unreacted from incorporated iodide.

Thorough preparation of reagents and equipment for radioiodination is necessary to minimize the time required to perform the reaction. Precise timing and rapid transfer of reagents are important both to increase the success of the procedure and to decrease the dose of radiation received by the operator. Safety precautions for handling ^{125}I can be found in "Radioiodination Techniques."[25] This handbook contains useful and important information, not directly related to iodination of PTH or CT, covered only briefly here.

The protocols for iodination that follow are designed to yield radioligands with less than one iodide per protein molecule. Thus, the molar ratio of iodide to PTH or CT is generally given as close to unity. If this ratio is altered, it may be necessary to change the concentration of oxidant used and/or the reaction time. To convert millicuries of carrier-free iodide to millimoles, use this formula (see Ref. 26):

$$21.7 \times 10^5 \text{ mCi} = 1 \text{ mmol } ^{125}\text{I}$$

Most of the procedures given here have been used previously to label PTH or CT, and, in general, the interassay variability of these techniques is low. However, it is recommended that changes in a protocol be made only after simultaneous comparison with the original. Table 1 includes references for alternative protocols.

B. Iodination Techniques

Chloramine-T

This method is used most often for preparation of tracers for radioimmunoassay.

PTH[27]

1. Add to 12 × 75-mm disposable glass test tube
    70 μl potassium phosphate buffer (0.5 *M*, pH 7.5)
    1 mCi ^{125}I (1–10 μl) (adjust molar ratio of ^{125}I to protein for other PTH
      peptides based on molecular weight of hormone)
    5 μg bPTH(1–84) in 5 μl 10 m*M* acetic acid
    10 μg Chloramine-T (20 mg/ml, in 0.5 *M* PO₄ buffer, pH 7.5)
    Total volume: 100 μl
2. Shake gently for 30 s at room temperature.
3. Stop reaction by adding sodium metabisulfite, a reducing agent, 50 μl of 4.8
    mg/ml 0.5 *M* PO₄, pH 7.5. Place on ice until purification.
    Optional: before purification by cellulose chromatography (see below), 50
    μl of hypoparathyroid serum can be added to minimize adsorption to
    glass.
    Chloramine-T and sodium metabisulfite should be prepared on the day
    of iodination.

[⁸Nle,¹⁸Nle,³⁴Tyr]bPTH(1–34) amide[27]

1. Incubate 30 s at room temperature
    0.2 *M* sodium phosphate, pH 7.4
    10 μg hormone (for example only; molar ratio is given in reference)
    5 mCi ^{125}I
    500 nmol Chloramine-T
2. Stop reaction with 1 μmol sodium metabisulfite.
3. Purify
    a. Using QUSO (see below).
    b. Add urea to tracer to a final concentration of 8 *M*.
    c. Gel filter on BioGel P-10 using 0.1 *M* ammonium acetate containing
      2% bovine serum albumin, pH 5.0 as eluant.

Calcitonin[28]

1. Add to a 12 × 75-mm disposable glass test tube
    25 μl 0.3 *M* phosphate buffer, pH 7.5 (Na⁺ or K⁺)
    1 mCi ^{125}I (1–10 μl)
    1 μg calcitonin (1–5 μl in 10 m*M* acetic acid)
    10 μl Chloramine-T (1.6 mg/ml 0.3 *M* PO₄, pH 7.5)
    Total volume: 50 μl
2. Shake gently for 10 s.
3. Stop reaction by adding sodium metabisulfite as above (50 μl of 4.8 mg/ml
    0.3 *M* PO₄ buffer, pH 7.5) or 80 μg cysteamine-HCL in 100 μl 0.3 *M* phos-
    phate buffer. Place on ice until purification on BioGel P-10, BioGel P-60 or
    Sephadex G-50 (see below).

Lactoperoxidase Method

The enzyme lactoperoxidase catalyzes the production of I⁺ in the presence of
low levels of H_2O_2.

PTH[29]
1. To a 12 × 75-mm disposable glass test tube, add
     10 µl 0.1 M potassium phosphate buffer, pH 7.4
     1 mCi ^{125}I (∼10 µl)
     5 µg bPTH(1–84) (∼5 µl) in 10 mM acetic acid
     1 µg lactoperoxidase in 25 µl 0.1 M phosphate buffer, pH 5.6
     10 µl hydrogen peroxide (0.8 mM)
2. Incubate at 37°C for 10 min.
3. Add 10 µl H_2O_2 solution; incubate 3–7 min.
4. Add 10 µl (30 µl total) H_2O_2 and incubate a final 3–7 min.
5. Stop the reaction by adding 0.5 ml of 0.1 M phosphate buffer, 0.05 M mercaptoethanol, pH 7.4. Place on ice until purification.
     The purification procedure chosen (see below) must separate the radioligand from iodinated lactoperoxidase and [^{125}I]tyrosine. If albumin or serum is added before purification, [^{125}I]tyrosine will bind to these proteins.

Calcitonin[14]
To a 12 × 75-mm disposable glass test tube, add
     0.4 M sodium acetate, pH 5.0
     4 µg human calcitonin
     3 µg lactoperoxidase
     1 µCi ^{125}I
     Total volume 40 µl
Add 10 µl of 0.6% H_2O_2 (v/v in H_2O) twice at 5-min intervals
Stop reaction by adding sodium azide (200 µl, 25 mM in H_2O)

Solid-phase Lactoperoxidase[15]
Lactoperoxidase immobilized on acrylic resin "beads" has been used to iodinate bPTH(1–34). The beads also contain glucose oxidase which produces small quantities of H_2O_2 as it metabolizes glucose.
1. To a 12 × 75-mm disposable glass test tube, add
     5 mCi ^{125}I (10–20 µl)
     50 µl 0.2 M phosphate buffer, pH 7.2
     25 µl β-D-glucose (1% w/v in H_2O)
     10 µg bPTH(1–34) in 10 µl of 10 mM acetic acid
     25 µl lactoperoxidase beads, rehydrated in distilled H_2O according to directions in package insert
2. The mixture is incubated at room temperature for 20 min
3. The reaction is stopped by centrifuging the reaction mixture at 1000 g for 1 min at 4°C, which separates the beads from the radioligand

Electrolytic Method 17

Initially described by Pennisi and Rosa,[16], this method was adapted for PTH by Sammon et al.[4]. In this procedure, the oxidation potential is maintained at a level high enough to permit conversation of iodide to iodine, but lower than

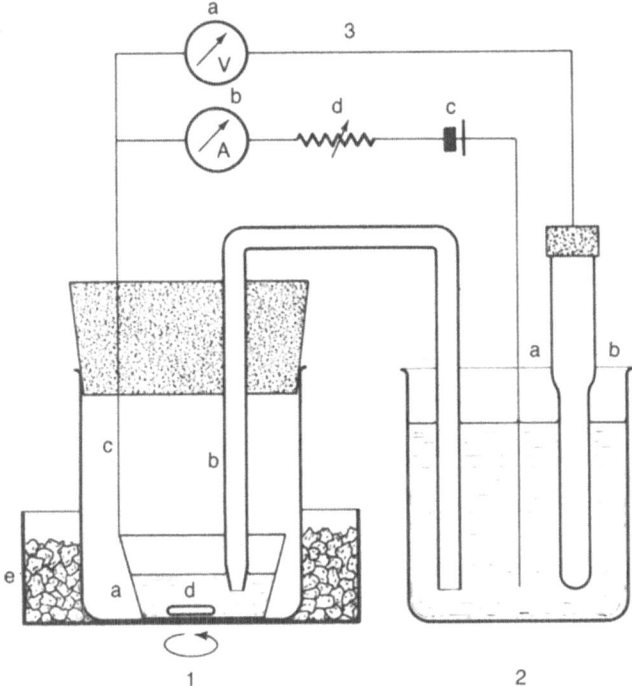

Figure 1. Electrolytic iodination apparatus. *1:* Reaction vessel; *a:* Platinum crucible (1-cm diameter) anode; *b:* Agar/salt bridge [Preparation: gradually warm 30 mg agar in 10 ml 90% saturated KCl (commercially available), with stirring, keep covered. Fill U-shaped glass tube by suction. Avoid bubbles. Seal ends with agar. Store in saturated KCl.]; *c:* 10-ml beaker and close-fitting rubber stopper with holes for platinum wire and salt bridge; *d:* Magnetic stirring bar; *e:* Ice bath. *2a:* Cathode, and *b:* Calomel reference electrode, both in saturated KCl. *3a:* Potentiometer; *b:* Microammeter; *c, d:* Variable potential source.

that required to oxidize methionine residues. This is accomplished by using the apparatus shown in Figure 1. In our laboratory, we use a BK precision 280 digital voltmeter (Dynascan Corp.) and a Simpson ammeter which also serves as the variable potential source.

Assembling the equipment for this reaction takes only a few minutes, if the components are used exclusively for this purpose. The wiring can be permanent and color coded. The agar bridge (and at least one spare) can be stored, airtight, in saturated KCl. The platinum crucible must be made to hold 500 μl and is cleaned by flaming after the reaction.

PTH

To the platinum dish, add

200 μl sodium phosphate buffer (0.5 *M*, pH 7.5)

25 μl bPTH (1–34), containing 0.1 *M* NaCl in 25 μl 10 m*M* acetic acid

6–8 mCi ^{125}I (10–20 μl)

Mix with magnetic stirrer. Surround reaction vessel with ice. Apply sufficient current to raise the potential difference to 680 mV, but do not exceed 20 μA. This may take 2–5 min. Decrease current, without changing voltage, to 5 μA (10–15 min). Turn off battery. Add 2 μl of 1 mM mercaptoethanol to reduce free iodide.

Calcitonin

Human CT has been iodinated using the electrolytic method,[19] as follows:

    80 μl 0.5 M sodium phosphate buffer (pH 7.5) containing 0.1 M NaCl
    10 μg human calcitonin
    10 μCi [125]I

The same procedure as for labeling bPTH (1–34) was used, except that the potential difference was maintained at 740 mV.

Iodogen and Iodobeads

I know of no published reports using these organic, chlorinated substances as oxidants for radioiodination of PTH or CT. The structures of these compounds are:

Iodogen   Iodobeads

The following procedure is used for iodination of glucagon[30]:

1. Iodogen is insoluble in water and must be coated on the walls of the reaction tube. Several test tubes can be plated at once and stored in a dessicator at 4°C in the dark for up to 6 months. A 20 μg/ml solution of Iodogen in chloroform is added (25–50 μl) to a 10 $\times$ 75-mm test tube. The liquid is swirled over the bottom of the tube and then evaporated with a mild stream of nitrogen at room temperature.
2. The hormone (1–2 nmol), [125]I (2–4 mCi in 10 μl NaOH (0.1 N) and 0.5 M phosphate buffer (pH 7.2) are mixed in a final volume of 50–100 μl and then transferred to an Iodogen-coated tube for 1.5 min with repeated vortexing.

The reaction is terminated by removing the reaction mixture from the test tube.

Iodobeads have been available only recently. The information given here was taken from the package insert. A phosphate or Tris buffer (100 mM, pH > 7) is used; the volume can range from 100 to 1000 μl/bead. The number of beads and the molar ratio of iodide to protein will determine specific activity. The reaction is carried out at room temperature or on ice; no reaction times are suggested.

C. Assessment of Specific Activity

The specific activity (S.A.) of the radioligand is calculated as the amount of ^{125}I incorporated into the protein divided by the amount of protein used, and is generally expressed as Ci/mmol or μCi/μg. To obtain the numerator, the radioactivity in a portion (5–10 μl) of the reaction mixture is determined after stopping the reaction; the fraction of radioactivity which is incorporated is determined by paper chromatoelectrophoresis (CME), precipitation by trichloracetic acid or by column chromatography. The latter technique can also be part of the purification procedure and will be discussed below.

Paper CME is described elsewhere in detail.[31] Separation of labeled hormone from free iodide is achieved because of the latter's greater electophoretic mobility. It is essential to add ⌐1 mM unlabeled iodide as carrier.

A CME profile of the postiodination reaction mixture will usually show [^{125}I]PTH or [^{125}I]CT ("hormone" peak) at the origin, a peak of ^{125}I closest to the cathode, and one or two relatively small peaks in between. The small peak closest to the hormone is considered to be "damaged" hormone in our laboratory and is included in the estimation of the percentage of radioactivity incorporated into protein. Thus,

$$\text{S.A.} = \frac{\text{mCi }^{125}\text{I} \times (\text{cpm in "hormone" and "damage" peaks} \div \text{total cpm applied to CME strip})}{\text{mg protein}}$$

A more rapid, but less reliable, method for separating radiolabeled PTH or CT from free iodide is precipitation of a portion of the postiodination reaction mixture with ⌐1 ml 10% TCA in the presence of 0.1% BSA. Incubate on ice for 10 min, centrifuge 10 min at 2000 g. Intact hormone will precipitate while free iodide will remain in the supernatant. Damaged hormone may or may not precipitate.

D. Purification of Tracer

As soon as possible after the iodination, the labeled hormone should be separated from free ^{125}I. In general, bovine serum albumin (0.1% w/v) (BSA) or hypoparathyroid serum (10% v/v) is added before purification to prevent adsorption to glassware. The number of technqiues described here is a small proportion of those described in the literature.

Ion–Exchange Chromatography[32]

Free iodide adsorbs to the resin and ^{125}I-labeled hormone is eluted. The column is prepared on the day of iodination:

1. A 5-ml glass pipette is broken at the 1.8-ml mark and plugged with glass wool.
2. Dowex 1–4X resin (washed twice in distilled H_2O to remove fines) is added in distilled H_2O until column is full.
3. Wash column with 4.0 ml of 1 M KI in 10 mM HCl, 1 mM mercaptoethanol (MCE)
4. Wash column with 1.0 ml of 0.1 M KI in 10 mM HCl, 1 mM MCE.
5. Wash column with 1.0 ml of 0.1 M KI in 10 mM HCl, 1 mM MCE 0.25% BSA.
6. Wash column with 1.0 ml of 10 mM HCl, 1 mM MCE + 0.25% BSA.
7. Cover bottom of column with parafilm until use.
8. Apply iodination reaction mixture to column, collect six fractions (0.5 ml) using, as eluant, 10 mM HCl containing 1 mM MCE and 0.25% BSA. [^{125}I]protein usually elutes in the second and third fractions. Free ^{125}I remains bound to the column.

Cellulose Chromatography[27]

Free iodide and PTH fragments do not absorb to cellulose. This allows separation of these compounds from [^{125}I]PTH(1–84), which can be eluted from cellulose with hypoparathyroid serum.

1. Break a 5-ml glass pipette at the 2.5-ml level, plug with glass wool.
2. Fill with gently stamped cellulose powder (Whatman) to 3.5-ml level.
3. Transfer the postiodination reaction mixture to the column.
4. Wash column with 7 ml of 0.05 M sodium barbital, pH 8.6; discard eluted radioactivity. A stream of air can be blown onto the column for more rapid elution ($\sim$1 ml/min).
5. Rinse top of column with sodium barbital buffer and wipe carefully.
6. Elute PTH with five consecutive 0.5-ml washes of undiluted hypoparathyroid serum into five test tubes containing 10 ml of sodium barbital buffer, 0.1 M, pH 8.6.
7. The peak of radioactivity will be in the second and third fractions.

QUSO[27]

Microfine silica (QUSO, Philadelphia Quartz Co.) adsorbs PTH and CT, but not free iodide or damaged hormone fragments. This rapid procedure is often used to repurify radiolabeled hormone after storage.

1. 10–60 mg QUSO is added to 1 ml of ^{125}I-labeled hormone in the presence of 5% hypoparathyroid serum or 0.1% BSA.
2. Vortex, centrifuge 3 min at 2000 g, at 4°C.
3. Discard supernatant.
4. Elute radioactivity with 1 ml 20% acetone, 1% glacial acetic acid, 0.05 M mercaptoethanol.

BioGel P-2 for Purification of ^{125}I-labeled PTH or CT

Column chromatography of the postiodination reaction mixture on BioGel P-2 (Bio Rad) separates ^{125}I-labeled PTH or CT from free iodide and other salts. The dimensions of the column should be approximately 1 × 20 cm. In our laboratory, ^{125}I-labeled peptides are eluted with 0.1 M acetic acid containing 0.1% BSA and come out in the void volume. If this procedure is used without prior purification of the ^{125}I peptides, it will be necessary to discard the column after one use since the free ^{125}I does not wash out easily.

BioGel P-60 for Purification of ^{125}I-labeled CT

Gel filtration of ^{125}I-CT can be performed after the postiodination mixture has been purified by adsorption to QUSO G32.[29] ^{125}I-CT is eluted from a 1 × 17 cm column of BioGel P-60 using 50 mM ammonium acetate, pH 4.0 containing 1% BSA. The Sephadex procedure is as follows:

Sephadex A-25 (QAE) for purification of ^{125}I-labeled PTH

[^{125}I]bPTH(1–84) can be separated from unlabeled PTH on QAE Sephadex A-25 at pH 9.4 in the presence of 8 M urea in 20 mM Tris–HCl.[5] PTH is eluted from a 1.2 × 30-cm column using a linear gradient of NaCl (0–15 mM); 4-ml fractions are collected. The peak of unlabeled PTH elutes in fraction 25; [^{125}I]PTH elutes in fraction 35.

Sephadex G-50 for purification of ^{125}I-labeled CT

Sephadex G-50 chromatography separates [^{125}I]CT from aggregated hormone and free iodide.[33] The column dimensions should be 0.5 × 50 cm and the elution buffer is 0.01 M sodium acetate/acetic acid, pH 4.2, containing 50 μg/ml sodium azide, 10 mg/ml BSA and 0.7% mercaptoethanol; 1.4 ml fractions are collected and [^{125}I]CT elutes in fraction 8.

Receptor Purification[34]

This procedure is used in our laboratory to purify electrolytically labeled [^{125}I]bPTH(1–34). The hormone is adsorbed to partially purified chicken renal plasma membranes; the membranes are washed several times and bound hormone is eluted with acid. Selection of biologically active [^{125}I]bPTH(1–34) molecules by renal receptors results in a preparation of labeled hormone with greatly improved binding activity.

Preparation of Crude Chicken Renal Plasma Membranes

Ten female chickens (8–12 weeks old) are decapitated. Their kidneys are excised and placed in ice-cold 0.9% NaCl, 1 mM EDTA, 10 mM Tris–HCl, pH 7.5. Kidneys are dissected free of fat and connective tissue and are washed several times with ice-cold buffer. All subsequent steps are performed at 4°C. Kidneys are homogenized in 3 vols (relative to wet weight) of ice-cold 0.25 M sucrose, 1 mM EDTA, 10 mM Tris–HCl, pH 7.5 (SET buffer), using 10 strokes of a motor-driven, loose Teflon pestle. After centrifugation at 1475 g for 10 min, the pellets are washed once with SET buffer, then resuspended in 1 vol (relative to initial wet weight) of 2.0 M sucrose, 1 mM EDTA, and 10 mM Tris–HCl, pH 7.5. After centrifugation at 13,300 g for 10 min, the supernatant is saved and the pellet resuspended by vortexing in 2.0 M sucrose/ EDTA/Tris and recentrifuged at 13,300 g for 10 min. The supernatant, containing most of the particulate material, is combined with the previous supernatant and diluted eightfold with 1 mM EDTA and 10 mM Tris–HCl, pH 7.5. This suspension is then centrifuged at 30,000 g for 15 min. The fluffy layer of the pellet is resuspended in SET buffer and recentrifuged at 30,000 g for 15 min. The fluffy layer of the pellet is then resuspended in 40–50 ml SET buffer to yield a protein concentration of 8 to 10 mg/ml. This suspension is stored in 0.5-ml aliquots at −80°C.

Receptor Purification

"Receptor purification" is performed by incubating 15–25 μCi of electrolytically labeled [^{125}I]bPTH(1–34) with 0.5 ml of the crude chicken renal membrane suspension in a 2.0-ml final volume containing 25 mM Tris–HCl (pH 7.5), 2 mM MgCl$_2$, and 0.1% bovine serum albumin. Eight tubes are generally processed together. After a 2-h incubation at 4°C, the tubes are centrifuged at 2000 g for 5 min. The supernatants are discarded, and the pellets are washed 3 times by resuspension in 5 ml of ice-cold 25 mM Tris–HCl (pH 7.5), 2.0 mM MgCl$_2$, followed by centrifugation at 2000 g for 5 min. Bound [^{125}I]bPTH(1–34) is then eluted from the pellets by resuspension in 3 ml of ice-cold 0.1 M acetic acid. After centrifugation at 2000 g for 5 min, supernatants containing receptor-purified [^{125}I]bPTH(1–34) are frozen and lyophilized overnight. The residues after lyophilization are resuspended in 10 mM acetic acid and centrifuged at 2000 g for 15 min to remove residual particulate material. Supernatants are pooled and stored at a concentration of $1–2 \times 10^6$ cpm/ml at 4°C. No detectable loss of binding activity is observed over a 4-week storage period. A typical yield of receptor-purified [^{125}I]bPTH(1–34) is 5% of the original electrolytically labeled hormone added.

High-pressure Liquid Chromatography (HPLC)

Purification of radiolabeled PTH or CT has not yet been described, to my knowledge. Reversed-phased HPLC has been employed in the isolation of PTH from parathyroid glands.[35,36] Thus, this technique offers the potential for separation of the iodinated from noniodinated hormone.

Identifying[32] Amino Acids Labeled with [125]I

Enzyme digestion of ^{125}I-labeled PTH or CT followed by chromatography on Sephadex G-15 will determine the proportions of monoiodinated and diiodinated tyrosine and histidine. Radiolabeled hormone (10 μg) is incubated in 0.5 μl of 0.1 M NaCl and 0.05 M Tris, pH 7.4, successively with chymotrypsin (118 U) and trypsin (24 $\times$ 10 U) for 4 h and 8 h respectively and boiled for 2 min. Carboxyl–peptidase A (173 U) is added and the mixture is incubated overnight at 37°C (0.5 ml 0.5 M Tris, pH 7.4, containing 0.25 mM CaCl$_2$, 0.25 mM MgCl$_2$, 50 mM PO$_4$, 0.1 M NaCl). Gel filtration on Sephadex G-15 (9 $\times$ 500 mm) using the 0.05 M Tris, pH 7.5, containing 0.1 M NaCl as eluant resolves the mono- and diiodinated amino acids which elute after the excluded volume of the column.

E. Assessment of Biological Activity

Using a sensitive bioassay, it is not difficult to determine whether a preparation of radiolabeled hormone has retained or lost biological activity. However, this approach does not, by itself, prove that the ^{125}I-labeled peptide is biologically active, unless the preparation consists solely of monoiodinated hormone. As stated above, separation of iodinated bPTH(1–84) from the noniodinated species can be achieved using a Sephadex column.[5]

The most convenient bioassay for PTH is the activation of renal adenylate cyclase activity in the presence of 5'guanyl-imidodiphosphate[37] (see Chapter 14 for a discussion of the in vitro bioassay of PTH). In this assay, as little as 30 pg-Eq/ml of bPTH(1–34) can be detected. Thus, very low amounts or radioactivity (less than 2000 cpm for a peptide with SA $=$ 100 μCi/μg) can be assayed with no interference with [32P] or [3H] counting. Several dilutions of the labeled hormone preparation should result in a dose-response curve identical to those obtained with the unlabeled peptide and with hormone which has been sham-iodinated.

Specific, high-affinity binding of an ^{125}I-labeled peptide to target tissue receptors[34,38] is further evidence that the hormone has not been damaged. This test is insufficient by itself because antagonists possess binding activity but lack biologic activity.

In vitro systems which have been used to measure the biologic activity of PTH include PTH stimulation of liver mitochondrial enzymes,[39] PTH-stimulated [45Ca] release from cultured embryonic bone,[40] and a cytochemical bioassay (see Chapter 15) which is sensitive to extremely low levels of PTH.[41]

In vivo bioassays require large amounts of hormone, even when using small animals. One such assay, previously used to demonstrate the biologic activity of $[^{125}I]bPTH(1-84)$[18] is the chick hypercalcemic bioassay of Parsons et al.[42] An alternative method is the Munson bioassay[43] in rats, which measures the ability of PTH to prevent the fall in serum calcium following thyroparathyroidectomy.

Several in vivo bioassays for calcitonin, using rats, have been described, which are based on the hormone's hypocalcemic action.[44,45] The method described in[44] has been used to assay the biologic activity of electrolytically labeled $[^{125}I]$human calcitonin.[46] The response to CT in vitro (i.e., adenylate cyclase stimulation[47] or inhibiton of bone resorption[48]) is generally too insensitive for use as a bioassay.

F. Storage of Radioactive Tracers

The shelf-life of a radioligand is usually not limited by the half-life of the radioisotope, but by the loss of biological activity which occurs on storage. The concentrations of free iodide and breakdown products resulting from radiolysis increase with time. These substances can sometimes be removed by the purification procedures described above. In general, radioiodinated PTH and CT can be stored at $-20°C$ in dilute acidic solutions for 2–4 weeks.

References

1. Rasmussen H: Effect of oxidation and reduction upon the biological activity of PTH. Science 128:1347, 1958.
2. Tashjian AH Jr, Ontjes DA, Munson PL: Alkylation and oxidation of methionine in bovine PTH: effects on hormonal activity and antigenicity. Biochemistry 3:1175, 1964.
3. Hughes WL: The chemistry of iodination. Ann NY Acad Sci 70:3, 1957.
4. Sammon PJ, Brand JS, Neuman WF, Raisz LG: Metabolism of labeled parathyroid hormone. I. Preparation of biologically active ^{125}I-labeled parathyroid hormone. Endocrinology 92:1596, 1973.
5. Christie DL, Barling PM: Isolation of iodinated bPTH using ion exchange: demonstration of its immunological characteristics and biological activity. Endocrinology 103:204, 1978.
6. Rosenblatt M, Goltzman D, Keutmann HT, Tregear GW, Potts JT Jr: Chemical and biological properties of synthetic, sulfur-free analogues of PTH. J Biol Chem 251:159, 1976.
7. Hunter WM, Greenwood FC: Preparation of iodine-131 labeled human growth hormone of high specific activity. Nature 194:495, 1962.
8. Arnaud CD, Tsao HS, Littledike T: Radioimmunoassay of human parathyroid hormone in serum. J Clin Invest 50:21, 1971.
9. Habener JF, Potts JT Jr: Modified method of CT iodination. In Antoniades HN (Ed.): Hormones in Human Blood. Harvard University Press, Cambridge, MA, p. 551, 1976.

10. Glover JS, Salter DN, Sheperd BP: A study of some factors that influence the iodination of ox insulin. Biochem J 103:120, 1967.

11. Redshaw MR, Lynch SS: An improved method for the preparation of iodinated antigens for radioimmunoassay. J Endocrinol 60:527, 1974.

12. Marchalonis JJ: An enzymic method for the trace iodination of immunoglobulins and other proteins. Biochem J 113:299, 1969.

13. Robinson CJ, Reit B, Martin TJ: Effect of iodination by the chloramine-T and lactoperoxidase methods upon the biological activity of PTH. Proc Soc Endocrinol. 27P, 1975.

14. Dermody WC, Levy AG, Davis PE, Plowman JK: Heterogeneity of chloramine-T and lactoperoxidase-radioiodinated human calcitonin. Clin Chem 25:989, 1979.

15. Teitelbaum AP, Nissenson RA, Abbott SR, Arnaud CD: Radioiodination of parathyroid hormone using solid-phase lactoperoxidase. Clin Res 28:54A, 1980.

16. Pennisi F, Rosa U: Preparation of radioiodinated insulin by constant current electrolysis. J Nucl Biol Med 13:64, 1969.

17. DiBella FP, Dousa TP, Miller SS, Arnaud CD: Parathyroid receptors of renal cortex: Specific binding of biologically active, ^{125}I-labeled hormone and relationship to adenyl cyclase activation. Proc Natl Acad Sci 71:723, 1974.

18. Nielsen ST, Barrett PQ, Neuman MW, Neuman WF: The electrolytic preparation of bioactive radioiodinated parathyroid hormone of high specific activity. Anal Biochem 92:67, 1979.

19. Scarpace PJ, Deftos LJ: Preparation and immunological characteristics of biologically active radioiodinated human calcitonin. Endocrinology 101:1398, 1977.

20. Fraker PJ, Speck JC, Jr: Protein and cell membrane iodinations with a sparingly soluble chloramide, 1,3,4,5-tetrachloro-3d,6a-diphenylglycoluril. Biochem Biophys Res Commun 80:849, 1978.

21. Bolton AE, Hunter WM: The labelling of proteins to high specific radioactivities by conjugation to a ^{125}I-containing acylating agent. Biochem J 133:529, 1973.

22. Wood FT, Wu MM, Gerhart JC: The radioactive labeling of proteins with an iodinated amidination reagent. Anal Biochem 69:339, 1975.

23. Habener JF, Kronenberg HM: Parathyroid hormone biosynthesis: structure and function of biosynthetic precursors. Fed Proc 37:2561, 1978.

24. Zull JE, Chuang J: Further studies on acetamidination as a technique for preparation of biologically valid ^{3}H labeled tracer for PTH. J Biol Chem 250:1668, 1975.

25. Bolton AE: Radioiodination Techniques. Amersham Corporation, Arlington Heights, IL, 1978.

26. Weast RC (Ed.): Handbook of Chemistry and Physics, 59th ed. CRC Press, West Palm Beach, Fla., 1978.

27. Segre GV, Rosenblatt M, Reiner BL, Mahaffey JE, Potts JT Jr: Characterization of parathyroid hormone receptors in canine renal cortical plasma membranes using a radioiodinated sulfur-free hormone analogue. J Biol Chem 254:6980, 1979.

28. As performed in the laboratories of CD Arnaud and SB Arnaud, personal communication.

29. Sutcliffe HS, Martin TJ, Eisman JA, Pikczyk R: Binding of PTH to bovine kidney cortex plasma membranes. Biochem J 134:913, 1973.

30. Schrader WT, O'Malley BW (Eds.): Laboratory Methods Manual for Hormone Action and Molecular Endocrinology, 6th edition. Houston Biological Association, Houston, 1981.

31. Yalow RS, Berson SA: Immunoassay of plasma insulin. In Glick D (Ed.): Methods of Biochemical Analysis, vol. 12. Interscience Publishers, New York, p. 69, 1964.
32. Neuman WF, Neuman MW, Sammon PJ, Lake K: The metabolism of labeled parathyroid hormone. II. Methodological studies. Calcif Tissue Res 18:241, 1975.
33. Abbott SR: Doctoral dissertation, University of Leeds, England, 1980.
34. Nissenson RA, Arnaud CD: Properties of the parathyroid hormone receptor adenylate cyclase system in chicken renal plasma membranes. J Biol Chem 254:1469, 1978.
35. Zanelli JM, O'Hare MJ, Nice EC, Corran PH: Purification and assay of bovine parathyroid hormone by reversed-phase high performance liquid chromatography. J Chromatogr 223:59, 1981.
36. Bennett HPJ, Solomon S, Goltzman D: Isolation and analysis of human parathyrin in parathyroid tissue and plasma. Biochem J 197:391, 1981.
37. Nissenson RA, Abbott SR, Teitelbaum AP, Clark O, Arnaud CD: Endogenous biologically active human parathyroid hormone: measured by a guanyl nucleotide-amplified renal adenylate cyclase assay. J Clin Endocrinol Metab 52:840, 1981.
38. Segre GV, Rosenblatt M, Reiner BL, Mahaffey JE, Potts JT Jr: Characterization of parathyroid hormone receptors in canine renal cortical plasma membranes using a radioiodinated sulfur-free hormone analogue. J Biol Chem 254:6980, 1979.
39. Rasmussen H, Shirasu H, Ogata E, Hawker C: Parathyroid hormone and mitochondrial metabolism. J Biol Chem 242:4669, 1967.
40. Raisz LG, Niemann FI: Effect of phosphate, calcium and magnesium on bone resorption and hormonal responses in tissue culture. Endocrinology 85:446, 1969.
41. Goltzman D, Henderson B, Loveridge N: Cytochemical bioassay of parathyroid hormone. J Clin Invest 65:1309, 1980.
42. Parsons JT: A bioassay for PTH using chicks. Endocrinology 92:454, 1973.
43. Munson PL: Biological assay of PTH. In Greco RO, Talmage RV (Eds.): The Parathyroids. Charles C. Thomas, Springfield, IL, p. 94, 1961.
44. Hirsch PF, Voelkel EF, Munson PL: Thyrocalcitonin: hypocalcemic hypophosphatemic principle of the thyroid gland. Science 146:412, 1964.
45. Arnaud CD, Tsao HS: Porcine calcitonin. Simple procedure for isolation in high yield. Biochem 8:449, 1969.
46. Scarpace PJ, Neuman WF, Raisz LG: Metabilism of radioiodinated calcitonin in rats. Endocrinology 100:1260, 1977.
47. Loreau N, LaJotte C, Wahbe F, Ardaillou R: Effects of guanyl nucleotides on calcitonin-sensitive adenylate cyclase and calcitonin binding in rat renal cortex. J Endocrinol 76:533, 1978.
48. Tashjian AH Jr, Wright DR, Ivey JL, Pont A: Calcitonin binding sites in bone: relationships to biological response and "escape." Recent Prog Horm Res 34:285, 1978.

Chapter 12

Radioimmunoassay Procedure for Parathyroid Hormone

Charles D. Hawker

A radioimmunoassay (RIA) procedure for parathyroid hormone (PTH) was first described in 1963 by Berson et al.[1] However, this assay could differentiate only half of patients with primary hyperparathyroidism (1°HPT) from normal subjects, and many clinicians believed that the assay's value would be in research rather than patient evaluation. This belief persisted during the next 8–10 years for several reasons: First, Berson and Yalow[2] published their classic report on the phenomenon of immunoheterogeneity which showed that there were at least two and probably several different immunoreactive forms of PTH that existed. Second, probably related to the question of immunoheterogeneity, there were numerous conflicting published reports of PTH assays with considerable variation in clinical performance. Third, the reagents necessary for a PTH radioimmunoassay (purified PTH and antiserum) were extremely hard to obtain. And fourth, because of the low circulating levels of PTH in normal humans, the radioimmunoassay "art" had to be extended nearly to its practical limits in order to have the sensitivity required for the assay.

In the past decade, however, several developments have made the measurement of PTH much more routine. As a result, the PTH test is now one of the more frequently ordered radioimmunoassays and, although technically demanding, the procedures are much simpler than in prior years.

This chapter is not intended to be an absolute reference method for radioimmunoassay of PTH. No single chapter could ever serve that role. Each antiserum that might be used in the assay will have different characteristics of specificity, affinity, etc., and different conditions of temperature, pH, salt concentration, protein concentration, etc. may be used that will depend on the nature of the antiserum and labeled ligand employed.

Instead, this chapter describes standard procedures that have worked well in the author's laboratory for many years, and highlights the key points in the procedures that need to be considered by anyone interested in developing or evaluating an assay. Numerous recent publications describe improvements in the PTH radioimmunoassay that can be incorporated as methods are implemented. Some of these improvements are being used by the author in newer PTH methods now being validated. However, we believed that this chapter should consist primarily of the standard methods used so that it could be a starting point from which laboratorians could design assays they are establishing regardless of the source of their reagents.

Our laboratory has routinely performed assays for intact PTH and C-terminal PTH. Each assay has its own value in clinical settings, with the C-terminal assay primarily serving the interests of the endocrinologist and the intact assay being used in patients with chronic renal failure and in selective venous catheterization studies. The procedures that follow cover both of these tests, their similarities and differences, and identify the areas that need attention in evaluating and implementing a new assay. Moreover, although other chapters have covered antiserum preparation, tracer preparation, and interpretation of results in detail, these aspects are also covered here in limited fashion so that the complete details of these procedures are provided.

I. Materials and Methods

A. Reagents

Antisera

The fundamental difference between the two PTH assays described is the antiserum used.

1. Antiserum against C-terminal PTH fragments. This antiserum was produced in a guinea pig by intradermal immunization with the C-terminal human PTH fragment preparation described below. Details of the immunization protocol have been described.[3]
2. Antiserum against intact PTH. This antiserum was produced by subcutaneous immunization of a guinea pig with a Sephadex G-100 preparation of bovine PTH (see below) conjugated to bovine serum albumin. Use of the conjugate in primary or booster immunizations did not lead to useful antiserum, but subsequent booster immunizations with unconjugated hormone resulted in rapid development of a high-titer, high-affinity antiserum to PTH. Complete details of the immunization protocol have been reported elsewhere.[4]

Parathyroid Hormone Preparations

1. Bovine hormone was purified from parathyroid glands by the methods of Hawker et al.[5] The product of the G-100 gel filtration step possessed a specific biological activity[6] of 300–500 USP units/mg and was used as the immunogen for the intact PTH antiserum without further purification. For use as primary RIA standard and for preparation of the radioiodinated hormone (see below), this material was further purified by gel filtration and ion-exchange chromatography.[5] The final product is salt-free and has a spe-

cific biological activity of 3000 USP units/mg. Note: Some commercial PTH preparations may not be salt-free.

2. A human C-terminal PTH fragment preparation, for use as the immunogen for the C-terminal antiserum, was obtained from pooled human parathyroid adenoma tissue, as described.[3]

Reagents for Radioiodination

1. Cellulose powder: CF-1 (Whatman, H. Reeve Angel and Co., Inc., Clifton, New Jersey)
2. Chloramine T (Eastman Kodak Co., Rochester, New York)
3. Sodium [^{125}I]iodide: IMS-300 (Amersham Corp., Arlington Heights, Illinois)

Other Reagents for RIA Procedures

1. Charcoal: Norit A, neutral activated charcoal, pharmaceutical grade (Amend Drug and Chemical Co., Irvington, New Jersey)
2. Dextran T-70 (Pharmacia Fine Chemicals, Inc., Piscataway, New Jersey)
3. Thimerosal, sodium salt (Sigma Chemical Co., St. Louis, Missouri)
4. Aprotinin (10,000 inhibitor units/ml): Trasylol (FBA Pharmaceuticals, Mobay Chemical Corp., New York, New York)
5. Normal horse and guinea pig sera (Pel-Freez Biologicals, Rogers, Arkansas).
6. Normal human sera for quality-control pools and C-terminal diluent. Outdated blood-bank plasma, previously tested for low PTH content and minimal ability to produce "damaged" ^{125}I-labeled PTH (see below for details) is first converted to serum, as follows:

    To 1 l of plasma, add 6 ml of thrombin (Fibrindex; Ortho Pharmaceutical Corp., Raritan, New Jersey) and 6 ml of protamine sulfate, USP (Eli Lilly & Co., Indianapolis, Indiana). Stir. Let mixture stand in 37°C water bath for 1 h. Centrifuge for 20 min at 10,000 g. After filtering the supernate through glass wool, dispense it into capped 12 × 75-mm polystyrene tubes and store at −76°C.

7. Hyperparathyroid serum, for quality-control pools. Previously assayed sera from patients with surgically proven primary hyperparathyroidism is used after first testing for hepatitis B surface antigen. Pool those found to be negative, filter through glass wool, and store as for normal human serum (above).

Common Reagents

All other chemicals are of reagent grade. Glass-distilled, deionized water is used throughout.

B. Solutions

Buffers

1. Sodium barbital buffer, 0.1 M, pH 8.6. Dissolve 44.39 g of barbital and 197.71 g of sodium barbital in about 3 l of H_2O by stirring magnetically overnight. Titrate to pH 8.6 with NaOH solution (5 M). Add H_2O to a final volume of 12 l; check the pH. Store in a polyethylene container at 4°C. Note: Do not store this solution in glass.
2. Sodium barbital buffer, 0.02 M, pH 8.6. Make from the above by diluting fivefold. Check pH.
3. Sodium phosphate buffer, 0.5 M, pH 7.45. Dissolve 17.25 g of sodium phosphate (monobasic, $NaH_2PO_4 \cdot H_2O$) in 200 ml of H_2O. Titrate to pH 7.45 with NaOH solution (5 M); dilute to 250 ml. Check pH. Store at room temperature.
4. Sodium phosphate buffer, 0.05 M, pH 7.45. Make from the above by diluting 10-fold. Check the pH and store at 4°C.

For Radioiodination

1. Chloramine T. Just before use, weigh 35 mg of Chloramine T into a large test tube. Add 10 ml of 0.05 M sodium phosphate buffer, pH 7.45, to dissolve the solid.
2. Sodium metabisulfite. Prepare fresh daily. Dissolve 24 mg in 10 ml of 0.05 M sodium phosphate buffer, pH 7.45.
3. PTH. Weigh the purified bovine hormone on a Cahn Electrobalance (or equivalent) and dissolve in 0.01 M acetic acid to a concentration of 0.5 μg/μl. Accurately dispense 5-μl aliquots with an automatic pipettor into the tips of 0.4-ml polyethylene conical centrifuge tubes; cap the tubes and store at −76°C for radioiodination at a later date. Generally, 65 μg of purified PTH, when diluted as described, will supply enough hormone for 1 year of radioiodinations performed every other week.

For RIA Procedures

1. Dextran-coated charcoal suspension. To 13 g of charcoal and 1.3 g of Dextran T-70 in an Erlenmeyer flask add 78 ml of 0.1 M barbital buffer, pH 8.6; 7.8 ml of normal human serum; and 304 ml of water. Stir the suspension at 4°C overnight (with magnetic stirrer). Centrifuge the suspension (10 min, 4°C, 275 g) and decant to remove fine charcoal particles and uncoated dextran. Resuspend the precipitate in 400 ml of 0.02 M sodium barbital buffer, pH 8.6, containing 20 ml of normal horse serum/l. Stir magnetically at 4°C overnight; use within 3 days.
2. Diluents. For C-terminal PTH assay, use a diluent solution of 0.085 M sodium barbital buffer, pH 8.6, containing, per liter, 500,000 inhibitor units of Trasylol, 100 ml of normal human serum (see above), and 0.2 g of thi-

merosal, to dilute standard PTH, antiserum, labeled PTH, and assay tube contents to final volume.

For intact PTH assay, use a diluent solution of 0.02 *M* sodium barbital buffer, pH 8.6, containing, per liter, 250,000 inhibitor units of Trasylol, 20 ml of normal guinea pig serum, and 0.2 g of thimerosal, to dilute standard PTH, antiserum, labeled PTH, and assay tube contents to final volume. Note: The diluents for the two assays must not be interchanged; this seriously hampers the performance of each assay. It is apparently a characteristic of each antiserum that certain constituents or concentrations of buffer may work successfully for one antiserum, but may inhibit the performance of the other.[7] Other antisera used by the reader may have different characteristics and require different diluents.

3. PTH RIA standards. For C-terminal PTH assay, dilute 2 μl of the PTH solution (1 μg of PTH) for radioiodination to a concentration of 11 ng/ml with acetic acid (0.01 *M*). Dispense in 200-μl aliquots in 12 × 75-mm polystyrene tubes, cap, and store at $-76°$C.

For intact PTH assay, dilute 5 μl of the above PTH solution (2.5 μg of PTH) for radioiodination to a concentration of 1 ng/ml with intact-assay diluent. Dispense diluted solution in 400-μl aliquots and store as above.

4. Antisera. For the C-terminal PTH assay, dilute the antiserum (GP75U) 1:1500; add 0.1 ml of this solution to each assay tube requiring antiserum. The final assay volume is 0.5 ml so the final dilution is 1:7500. Prepare this just before use in C-terminal diluent from aliquots of stock antiserum stored at $-76°$C.

For the intact PTH assay dilute antiserum (GP204) 1:40,000 with intact-assay diluent; final dilution in the assay itself is 1:400,000, because the final assay volume is 1 ml.

C. Special Apparatus

For Radioiodination

1. Chromatoelectrophoresis apparatus: Model 3-1200 CE (Buchler Instruments, Inc., Fort Lee, New Jersey).
2. Power supply (for the above), adjustable to 1000 V, with constant-voltage regulation: Buchler Instruments, Model 3-1014A.
3. Chromatography paper: Toyo No. 21-689 (Nuclear Associates, Inc., Carle Place, New York). This paper is tested by supplier's consultant for ability to absorb PTH and other peptide hormones.
4. Radiochromatogram scanner: Model 7201 (Packard Instrument Co., Inc., Downers Grove, Illinois).
5. Microcolumns. Columns for purification of ^{125}I-labeled PTH by adsorption to cellulose are made from 10-ml polystyrene disposable pipettes (Falcon) cut in half with a sharp knife. The tip of the lower half is fitted with a small glass wool plug and filled with dry cellulose powder, which, when tamped firmly with a small rod, occupies a volume of 2 ml.

For RIA Procedures

1. Automatic pipettor: Model 25004 with Model I 200-μl pump and Model B 1000-μl pump (Mircomedic Systems, Inc., Horsham, Pennsylvania).
2. Freezer, ultracold (to $-76°$C): Model UC-105 (Kelvinator, Inc., Manitowoc, Wisconsin), or similar.
3. Reciprocating shaker: Model 6000 (Eberbach Corp., Ann Arbor, Michigan).
4. Automatic gamma-scintillation spectrometer: Model 1285, 1008-tube capacity (Tracor Analytic, Inc., Chicago, Illinois), or similar.
5. Refrigerated centrifuge: Beckman J-6 (Beckman Instruments, Inc., Fullerton, California). Must be able to generate a centrifugal force of 2000 g and hold at least 200 10 $\times$ 75-mm tubes.
6. Incubation tubes, soda-lime glass: Neutrex (Chase Instruments, Corp., Poultney, Vermont). Note: Avoid use of tubes made of borosilicate glass or plastic. PTH is notorious for its ability to bind to almost any surface. This binding has been found to be lowest in soda-lime (or flint) glass.

D. Preparation of ^{125}I-labeled PTH

Highly purified bovine PTH is labeled with ^{125}I every 2 weeks by modification of the procedures of Hunter and Greenwood.[7,8]

1. Thaw one small tube containing 2.5 μg of pure bovine PTH in 5 μl of acetic acid.
2. Add to this 1.2 mCi of Na[125]I (diluted to 20 μl with sodium phosphate buffer, 0.5 M).
3. At time zero add 24 μl of freshly prepared Chloramine T solution; vortex–mix.
4. After 25 s stop the reaction by adding 96 μl of sodium metabisulfite solution; vortex–mix.
5. Add 600 μl of normal human serum.
6. Transfer the contents of reaction tube to the top of a dry column of cellulose CF-1.
7. Wash column 5 times with 1.5 ml of sodium barbital buffer (0.02 M, pH 8.6).
8. Elute with five 1.0 ml volumes of acetone/acetic acid/water (20:0.8:79.2, by volume). Collect separate 1.0-ml fractions.
9. Divide the most highly radioactive fraction of the acetone/acetic acid/water eluate into aliquots of about 0.1 ml each. Store at $-76°$C.
10. Analyze droplets of the acetone/acetic acid/water eluate and the original reaction mixture by paper chromatoelectrophoresis and radiochromatogram scanning[9] to determine the purity and specific activity of the labeled PTH preparation. The ^{125}I-labeled PTH prepared as described should be essentially completely free of unreacted Na[125]I and "damaged" [125]I-

labeled PTH components; its specific activity should range from 250 to 350 Ci/g.

E. Collection and Handling of Specimens

Obtain blood samples between 0700 and 0900 hours from patients after an overnight fast and allow to clot at room temperature for 1 h. Separate the serum by centrifugation and store frozen ($-20\,^{\circ}$C) until assay. Although we have no conclusive data, it seems best to avoid assaying sera more than 12 months old. Plasma samples are acceptable for assay; we prefer serum, however, because it seems to produce fewer problems in the RIA. Avoid specimens collected in Corvac or similar tubes, which may absorb some of the PTH. Severely lipemic or badly hemolyzed specimens will produce unacceptable interference and are best avoided.

F. Radioimmunoassay Procedure

The basic procedures for assay of C-terminal or intact PTH are nearly identical. Both RIAs use the same bovine PTH for standards and radioiodination, the same dextran-coated charcoal preparation (for separation of antibody-bound and free ^{125}I-labeled PTH), and the same types of equipment, such as soft glass incubation tubes, pipettors, etc. The only major difference between the assays is the antiserum. We will first describe the C-terminal PTH assay in detail, then the assay for the intact hormone. For the latter we will point out only the significant areas of difference from the C-terminal assay.

Procedure for C-terminal PTH Assay

The day before the assay, consecutively number disposable 12×75-mm glass culture tubes and place all frozen serum specimens to be assayed and an appropriate number of standards (see below) in the refrigerator ($4\,^{\circ}$C) to thaw overnight. The size of an assay run is governed by the number of tubes that can be centrifuged at one time. Refer to the assay protocol.

On the day assay is started (day 0), place all assay tubes and solutions in an ice bath (keep reagents cold throughout all procedures). Wear disposable gloves at all times when patients' sera are being handled.

Using the automatic pipettor, deliver the appropriate amount of diluent into the assay tubes shown in the protocol (Table 1). Mix each serum specimen thoroughly before pipetting, in succession, the 50-, 100-, and 200-μl aliquots into the tubes shown on the protocol. Note that two 100-μl and 200-μl aliquots of each specimens are pipetted. Only one of each volume will receive antiserum; the other will become a "serum damage" tube (see Section I.G) for that patient.

Pipette the quality-control serum pools the same way as the samples.

Table 1. Protocol for C-terminal PTH Assay (volume added, μl)

Tube No.	Diluent	Antibody	Standard	Serum	Description
1–4	300	—	—	—	Total counts tubes
5–8	300	—	—	—	Buffer damage tubes
9–12	200	100	—	—	Zero standards
13–14	—	100	200	—	400-pg standard
15–16	—	100	200	—	240-pg standard
17–18	—	100	200	—	144-pg standard
19–20	—	100	200	—	86.4-pg standard
21–22	—	100	200	—	51.8-pg standard
23–24	—	100	200	—	31.1-pg standard
25–26	—	100	200	—	18.7-pg standard
27[a]	150	100	—	50	50-μl serum tube
28[a]	100	100	—	100	100-μl serum tube
29[a]	0	100	—	200	200-μl serum tube
30[a]	200	0	—	100	100-μl serum damage tube
31[a]	100	0	—	200	200-μl serum damage tube

[a]Repeat this five-tube sequence for each unknown serum or quality-control pool.

For every 7 assays runs (approximately), use one 200-μl tube of the 100 ng/ ml PTH standard. Add 0.8 ml of diluent directly to the tube and vortex-mix. This is now a 20 ng/ml solution of PTH. For the first RIA standard, dilute 0.2 ml of this solution to 2.0 ml by adding to 1.8 ml of diluent in a tube numbered "standard No. 1"; vortex–mix. Prepare 6 other consecutively numbered tubes and add 0.6 ml of diluent to each. Then carefully remove 0.9 ml of standard No. 1 and add to standard tube No. 2. Vortex–mix, then remove 0.9 ml of tube No. 2 and add to standard tube No. 3. Continue this procedure of adding 0.9 ml of diluted standard to the next numbered tube until all seven standards are prepared. Each successive tube will have 0.6 the concentration of PTH of the previous tube (i.e., 2000 pg/ml for standard No. 1, 1200 pg/ml for No. 2, etc., for the 7 tubes). This entire procedure must be done with utmost care to ensure the greatest possible accuracy and precision in the standard curve. Dispense 200-μl aliquots of each standard PTH concentration, as shown in the protocol (Table 1). Use a Hamilton repeating syringe or equivalent.

When all diluent, standard, patients' sera, and quality-control pools have been pipetted, the final step is addition of 100 μl of diluted antiserum to the appropriate tubes; use an automated pipettor for the addition. Vortex–mix all tubes and cover with clear plastic wrap secured with a rubber band. Place all racks in shakers in a cold (4°C) room for 3 days. In our experience, shaking increases the binding of labeled PTH to antibody by 10–20%. Thus, a bit less antibody can be used to achieve a satisfactory degree of binding of the tracer, which in turn, enhances the sensitivity of the assay.

After the 3 days' incubation, add the ^{125}I-labeled PTH. First prepare the solution of labeled hormone, as follows: Allow 0.2 ml of C-terminal diluent per assay tube plus at least 20% extra. Add to this volume of diluent thawed ^{125}I-labeled PTH stock (stored at −76°C) diluted to a count rate of 3000–3250 dpm/200 μl. The amount of labeled hormone stock to be added varies somewhat with each iodination lot. Mix the ^{125}I-labeled PTH solution well by gentle swirling. Keep the solution on ice. Add the label (200 μl of correctly diluted ^{125}I-labeled PTH), using a Micromedic or other similar device, to all tubes in the run while they are in an ice bath. Vortex–mix the contents of the tubes and return them to the shaker in the refrigerated (4°C) room for an additional 3 days.

After the total of 6 days' incubation, separate "bound" and "free" ^{125}I-labeled PTH as follows. Remove the charcoal–dextran suspension from the refrigerator and place it in an ice bath, on a magnetic mixer. Mix for at least 10 min to ensure that all of the sedimented charcoal is resuspended. Keep mixing the suspension throughout the pipetting step. Prepare a second set of numbered tubes for the supernates. Number these tubes consecutively. Add 1.5 ml of buffered horse serum to tubes 1–4 of the run. Vortex, mix, cap, and let stand. These will be the "total counts" tubes for that run. Add 1.5 ml of dextran-coated charcoal to the rest of the tubes of the run, beginning with No. 5, and vortex–mix. Centrifuge all tubes from No. 5 to end of run (See Table 1) at 2000 g for 10 min. Note: Each centrifuge run must contain a complete set of controls and standards because of interassay variation in timing of charcoal addition and centrifugation. Therefore, use the centrifuge that can hold the largest number of tubes, and determine the total run size based on this capacity.

Pour the supernate carefully from the assay tube into the corresponding numbered tube in the second set. Cork the tubes, then count the radioactivity of the supernates for 5 min.

Procedure for Intact PTH Assay

The procedure for this assay (Table 2) includes tubes for later calculation of the following parameters: binding of ^{125}I-labeled PTH to glass, ^{125}I-labeled PTH "damaged" by incubation, and interference of serum (50-, 100-, and 200-μl volumes) with adsorption of ^{125}I-labeled PTH to dextran-coated charcoal.

For every two assay runs, thaw one 400-μl tube of the 100 ng/ml PTH standard. Dilute each 400-μl aliquot with 1.6 ml of diluent to produce a 20 ng/ml solution of PTH; mix thoroughly. Into each of nine 15-ml vials, pipette 1 ml of diluent per tube of standard thawed. With a serological pipette withdraw that same volume of 20 ng/ml standard from the vial of diluted standard and transfer quantitatively to the first of the series of vials; mix thoroughly. Withdraw the same volume from this second vial and transfer quantitatively to the third vial, using the same pipette. Mix thoroughly. By repeating this sequence through all of the vials, a series of 10 concentrations of PTH standard ranging from 20 ng/ml to 39 pg/ml will have been prepared, each concentration being exactly half of that in the preceding vial. This entire procedure should be done

Table 2. Protocol for Assay of Intact PTH (volume added, μl)

Tube No.	Diluent	Antibody	Standard	Serum	Description
1–4	300	—	—	—	Total counts tubes
5–8	(These tubes receive the contents of tubes 9–12)				Glass binding controls
9–12	300	—	—	—	Buffer instead of charcoal
13–16	(These tubes receive the supernates from tubes 17–20 after addition of dextran-coated charcoal)				
17–20	300	—	—	—	Buffer damage controls
21–24	(These tubes receive the supernates from tubes 25–28 after addition of dextran-coated charcoal)				
25–28	250	—	—	50	Serum interference with charcoal
29–32	(These tubes receive the supernates from tubes 33–36 after addition of dextran-coated charcoal)				
33–36	200	—	—	100	Serum interference with charcoal
37–40	(These tubes receive the supernates from tubes 41–44 after addition of dextran-coated charcoal)				
41–44	100	—	—	200	Serum interference with charcoal
45–48	200	100	—	—	Zero standards
49–50	—	100	200	—	4000-pg standard
51–52	—	100	200	—	2000-pg standard
53–54	—	100	200	—	1000-pg standard
55–56	—	100	200	—	500-pg standard
57–58	—	100	200	—	250-pg standard
59–60	—	100	200	—	125-pg standard
61–62	—	100	200	—	62.5-pg standard
63–64	—	100	200	—	31.25-pg standard
65–66	—	100	200	—	15.6-pg standard
67–68	—	100	200	—	7.8-pg standard
69[a]	150	100	—	50	Patient's serum
70[a]	100	100	—	100	Patient's serum
71[a]	0	100	—	200	Patient's serum

[a]Repeat this tube sequence for each unknown serum or quality-control pool

with maximum care to ensure the greatest accuracy and precision in preparing the standard curve.

Diluent (for intact assay), standards, patient sera, and quality-control serum pools are pipetted as described for the C-terminal assay. Note that three tubes (containing 50, 100, or 200 μl of serum) are run for each patient. After adding antiserum, incubate the gently mixed tubes on a shaker at 4°C for 2 days. Then add the ^{125}I-labeled PTH, prepared as for the C-terminal assay, except that intact PTH assay diluent is used and a count rate of 2500–2750 dpm/200 μl of diluted tracer is required. After addition of the labeled PTH, add an additional 500 μl of diluent to each tube. Vortex–mix the tubes' contents and return them to the cold room to shake for another 3 days.

Add 1.5 ml of well-mixed charcoal suspension to each tube in the protocol after tube No. 12 (see Table 2). Add 1.5 ml of 0.02 M barbital buffer, pH 8.6, containing 20 ml horse serum/l, to tubes 9–12. After vortex–mixing, centrifuge all tubes after No. 12 at 2000 g for 10 min. Transfer the supernates from tubes 9–12 to tubes 5–8 with a Pasteur pipette. Wash tubes 9–12 with 1.5 ml of water, vortex–mix, and discard the supernates. Save other supernates as indicated on the protocol. Remove the supernates from all other tubes by aspiration. Note: Take care not to disturb the charcoal pellets.

Add 2.5 ml of water to each tube. Vortex–mix again to resuspend charcoal pellets. Recentrifuge as before. Aspirate supernates. Cap the tubes and count the radioactivity of the pellets for 5 min.

G. Calculation of Results

C-terminal PTH

The values for antibody-bound counts (supernates) are counted directly. However, these counts must be corrected for incubation damage and serum interference with the binding of ^{125}I-labeled PTH to charcoal as follows: the average counts for the incubation damage controls (tubes 5–8, Table 1) are subtracted from the counts for each standard or zero standard tube, from the counts for serum tubes in which the serum speciman was diluted before assay, and from the counts for all tubes containing 50 μl of serum. The counts for the individual serum specimen controls containing 100 or 200 μl of each serum specimen are substracted from the count values for the corresponding tubes that also contained antiserum (see Table 1).

Intact PTH

The values for antibody-bound counts are computed rather than being counted directly as in the C-terminal assay. The average number of counts in the glass-binding controls is subtracted from the counts for each tube containing charcoal, because this portion of the ^{125}I-labeled PTH is adsorbed to glass rather than charcoal. The counts for each standard or zero-standard tube and the average incubation damage counts are subtracted from the average total counts to obtain the values for antibody-bound counts for the standards. The counts for all tubes containing unknown or control serum and the counts for the serum interference controls for the corresponding serum volumes are subtracted from the average total counts to obtain the values for antibody-bound counts for all the serum tubes.

The rest of the calculation procedure is the same for both assays. The average antibody-bound counts in the zero standards are divided by the average total counts to determine "assay binding" for each assay run. All of the antibody-bound counts for standards and unknown or control serum samples are divided by the antibody-bound counts for the zero standards and multiplied by 100 to obtain the percentage of ^{125}I-labeled PTH bound for each standard or serum tube. The standards are plotted by the logit procedure,[10] in which the logit of the percent bound is plotted as a function of the log of the picograms of standard PTH in each tube. The amount of PTH in each unknown or control serum tube is then obtained from the standard curve by using the logit of the percent bound for each serum tube. This amount of PTH is then divided by the volume of serum sample assayed to give the final value. PTH values for both assays are expressed as pg Eq/ml. The term "equivalent" is used because the PTH assay standard is bovine hormone, not human.

II. Results and Discussion

A. Performance Characteristics

The performance characteristics of PTH assays can vary considerably. In reviewing the performance characteristics of the assays described in this chapter, the author emphasizes the importance of these characteristics in the overall clinical performance of the assay. These characteristics included specificity, sensitivity, accuracy and precision, and clinical validation and are discussed in their order of importance to the assay.

Specificity

Of all the performance characteristics of a PTH assay being evaluated, specificity is the most critical. This question of specificity is important because of its relationship to clinical interpretation of results. As discussed in other chapters and below, assays with specificity for the C-terminal region of PTH have the broadest applicability to evaluation of hyper- and hypocalcemia. However, due to the role of the kidney in the removal of these fragments, patients with chronic renal failure can accumulate the C-terminal fragments to very high levels, and the C-terminal assay frequently fails to correlate with clinical improvement in dialysis patients. An assay for intact PTH, on the other hand, does show strong clinical correlation in dialysis patients even though its discriminatory ability for primary hyperparathyroidism (1°HPT) is not as good as a C-terminal assay. The key consideration is whether or not the assay being evaluated measures C-terminal fragments, since nearly all PTH antisera will detect intact PTH. If the assay is specific for the C-terminal region of PTH with good sensitivity for fragments, then that assay will be useful in general endocrine settings. But, in addition, an assay that measures intact PTH but not C-terminal fragments will be needed for use in dialysis patients and in venous catheterization studies.

   As previously reported,[7] the two assays discussed here have different reactivities toward pure peptide fragments of bovine PTH. The antiserum used for the intact PTH assay (GP204) does not react with a fragment from the C-terminal region (amino acids 53–84) of intact PTH, or with a fragment from the middle region (27–44) of intact PTH. Reactivity with a fragment from the N-terminal region (1–34) of intact PTH is only about 0.007% of the reactivity with intact PTH.[7] In sharp contrast, antiserum GP75U, used in the C-terminal PTH assay, reacts with the 53–84 C-terminal fragment on an equimolar basis with the intact PTH; however, it does not recognize large excesses of either the 1–34 or 27–44 fragments.[7]

The antisera have been further characterized by gel filtration chromatography studies[11,12] using sera from patients documented 1°HPT and secondary hyperparathyroidism (2°HPT) due to chronic renal failure. This technique separates the forms of PTH in serum on the basis of their molecular sizes, thus permitting direct evaluation of the reactivity of each antiserum with these various forms. These studies confirm that the C-terminal PTH antiserum recognizes primarily C-terminal fragments and intact PTH. The intact PTH antiserum also apparently recognizes fragments which seem to encompass the N-terminal region.

Sensitivity

Sensitivity is as important as specificity. An antiserum may have C-terminal specificity, but such poor sensitivity that it would be of little clinical value. Both analytical and clinical sensitivity must be considered. The former refers to the actual mass of PTH that can be measured by the assay, whereas the latter refers to the percentage of patients with known disease (for example, 1°HPT) that can be differentiated from normal. The former depends on the K value (binding constant) of the antiserum and the conditions selected for its use; the latter depends on specificity as well as sensitivity.

It is not really possible with the current state of the art to define analytical sensitivity in terms of an absolute mass of PTH that an assay should be able to detect, because there is no accepted standard for human PTH available. Use of synthetic fragments is questionable because they may not have the same structural characteristics as natural fragments and intact hormone. Use of PTH standards other than human may give variable cross reactivities with different antisera so that the absolute mass of human PTH being measured cannot be determined.

Finally, the very nature of PTH in human blood (the existence of multiple immunoreactive forms) means that the assay result, no matter what standard it is based on, really represents the sum of several different forms of PTH which are measured to varying degrees because of differing cross reactivities of the antiserum for these forms. Therefore, an absolute number (for the lowest mass of PTH detected), no matter how small, would be difficult to interpret since the forms of PTH contributing to that number might not be known with any certainty.

A more useful approach in defining analytical sensitivity is the evaluation of the assay in normal subjects. The PTH assay should be able to measure PTH in at least 90% of normal individuals tested. If the sensitivity for PTH in normals is less than 90%, patients with hypoparathyroidism will be difficult to differentiate from normals because too many normals will have nondetectable results.

Finally, even if the assay detects PTH in more than 90% of normals, it could still have inadequate clinical sensitivity. To be clinically useful, the PTH assay should also be able to differentiate at least 90–95% of all proven 1°HPT patients tested from normal subjects or patients with other causes of hypercalcemia. This level of sensitivity will in part depend on the assay's specificity as well as the antiserum's binding constant for the C-terminal region of PTH, since C-terminal fragments are predominant in 1°HPT patient serum.[13]

Precision and Accuracy

Accuracy is difficult to assess objectively in a PTH assay, because of immunoheterogeneity and lack of a valid standard. Therefore, accuracy must be evaluated clinically in terms of how often the assay gives the expected results. It is important to use patients with proven rather than suspected diagnoses, because some patients in the latter category may actually have a different disorder. Moreover, it is important to take all patients of a given diagnosis randomly, rather than just the ones with obvious disease. Excessive use of the latter patients will bias the assay's performance to make it appear better than it really is.

An alternative method for assessing one aspect of accuracy, however, is to determine analytical recovery of added intact PTH. In analytical recovery studies done for the C-terminal antiserum used here, recovery of PTH ranged from 85 to 107% (mean, 96.7%). The correlation coefficient (picograms of pure human PTH added to hypoparathyroid serum versus picograms of PTH measured by RIA) was 0.98.[14]

A separate section discusses quality control and describes our specific efforts to control PTH assays and our results. However, the following more general discussion is intended to illustrate the impact that poor precision can have on the clinical utility of an assay.

The coefficient of variation (CV) is defined as 1 standard deviation (SD) divided by the means times 100 (giving a percent basis). Since the 95% confidence interval (CI) about any given result is ± 2 SD, this interval is really $\pm$ 2 CV. Therefore, when the CV for a given assay at a given concentration is 10%, the 95% confidence interval for a result at that level is ± 20%. At a 15% CV, the 95% CI would be ± 30%, and at a 20% CV the 95% CI would be $\pm$ 40%.

A CV of 10% is about the best that has ever been reported for a PTH assay, and some laboratories have had a CV of 15–25%. If one is comparing a given result with normal, then obviously that result has to be 2 CV above the upper limit so that it can reliably be considered abnormal. Therefore, an assay with a CV of 15–20% that requires a 30–40% increase above normal to be statistically significant will probably not be clinically very reliable, since a large percentage of abnormal results may be expected in the range immediately above normal.

This point becomes even more dramatic when one is comparing two or more results, perhaps a pre- and posttreatment, or specimens collected from different veins in a catheterization study. Since each result can have its own 95% CI, two results must be more than 4 CV apart in order to be statistically significantly different. If the assay CV is 25%, the results would have to be 100% different in order to be significant.

Clearly any step that can be taken to improve precision is of great value. A laboratory that runs many specimens will have an advantage over a laboratory that runs few specimens. A laboratory that can use automated pipetting has an advantage over laboratories with manual pipetting. In the author's laboratory, the CV decreased to half of its prior value when an automatic pipetting station was implemented. Triplicates are recommended over duplicates because of the additional precision that results. Also, multiple dilutions of the patient serum as a variation of triplicates give the same benefit as triplicates on precision, but have the additional benefit of an assessment of parallelism between the PTH in the patient specimen and the standard. Some patient specimens occasionally do not dilute parallel to the standard because of nonspecific interference or presence of an unusual profile of PTH forms. Reporting of a result when only one volume of serum was tested could be clinically misleading.

Quality Control

Three quality-control serum pools representing one high and two normal PTH concentrations are used in every assay. One normal pool is run in triplicate at different locations throughout the run to monitor "drift" (see note below). Only freshly thawed aliquots of these serum pools are used; all pools are stored at $-76\,^{\circ}$C and thawed only once. Thus any variations observed in the PTH values for these pools should be due to test factors only and not to deterioration. At least 100 determinations are required on each pool to establish the acceptable limits for results. These limits are determined by computer for each run, and the overall means and 95% confidence intervals (mean $\pm$ 2 SD) are continually updated as more values for each serum pool are accumulated. If no more than one pool in a run has values outside these 95% confidence intervals, the run may be considered acceptable.

The following performance characteristics should be met for every assay run to verify consistent performance:

1. Percent binding for the zero standard tubes (counts bound corrected for nonspecific binding) should be between 25 and 35% of the counts added, for both assays.
2. Nonspecific binding (i.e., "damage" counts in the supernates of tubes con-

taining no antibody, representing the ^{125}I-labeled PTH not bound to the dextran-coated charcoal) should be less than 15% of the total counts.

3. Poor agreement (CV > 30%) in the PTH values determined for the multiple dilutions of any patient's serum requires that such specimens be reassayed. A large number (>15%) of specimens with such characteristics in any one run may indicate problems in the assay itself. Each patient's serum must be run at multiple dilutions to verify "parallelism" between the PTH in the assay standard and the PTH in the specimen. A few patients' sera (perhaps as many as 2%) produce nonspecific interference in the assay; PTH cannot be validly quantitated in such specimens, of course.[15]

4. Consistent bias in quality-control pools in a series of runs (i.e., all high or all low) may reflect deterioration of the assay standard of the ^{125}I-labeled PTH.

5. Any assay is valid only if it provides results consistent with other clinical information on the patients tested.

The quality-control serum pools used in the author's PTH assays consistently give coefficients of variation for between-run precision of 7–11%. The other parameters mentioned above are consistently within their stated limits; if not, corrective action is instituted by the supervisor.

Clinical Validation

The diagnostic performance and reliability of any laboratory test can be no better than its clinical validation. In the case of a PTH assay we and many others believe that the following must be done for such validation:[15] (1) identification of the region of the PTH molecule recognized by the PTH assay in question, through use of pure PTH fragments (commercially available) and by gel-filtration studies; (2) examination of at least 300 serum specimens from normal subjects and patients with disorders of parathyroid gland function and calcium homeostasis.

C-terminal PTH Assay

Normal Subjects   The 183 normal subjects evaluated with this included approximately equal numbers of men and women. These age-matched individuals all had physical examinations and normal results for several conventional laboratory tests. The mean C-terminal PTH (±SD) in this group was 256 (± 59.5) ng Eq/l, and the mean ± 2 SD range would therefore be 137–375 ng Eq/l. Because the lower limit of detection in this assay is considered to be 150 ng Eq/l, the normal range is expressed as <150–375 ng Eq/l. Only 7 (4%) of the normal subjects had C-terminal PTH values <150 ng Eq/l. There were no statistically significant correlations between concentrations of C-terminal PTH and calcium in this group of normal subjects.

Patients with Calcium–Parathyroid Disorders   With this C-terminal PTH assay, approximately 95% of patients with surgically confirmed primary hyperparathyroidism have above-normal values (>375 ng Eq/1), as have about 20% of hypercalcemic patients with cancer.[15] All patients with chronic renal failure have increases in C-terminal PTH concentration. In hypoparathyroid patients, C-terminal PTH is generally nondetectable.[16] Correct interpretation of a patient's PTH value (whether measured with the C-terminal or the intact assay) can only be made by also considering the serum calcium value (as measured by atomic absorption spectroscopy).

Intact PTH Assay
Normal Subjects Mean intact PTH ($\pm$SD) in 93 normal subjects was 255 ($\pm$46) ng Eq/1. The mean $\pm$ 2 SD range is 163–347 ng Eq/1. This assay demonstrates an inverse relationship between PTH and serum calcium concentrations in normal individuals:[4] a plot of intact PTH versus calcium shows a highly significant negative correlation ($r = -0.412$, $P < .001$).

Patients with Calcium–Parathyroid Disorders In hyperparathyroid patients, only about 55% have intact PTH values higher than 347 ng Eq/1. However, as previously reported, many of the apparently normal values in these patients are, in fact, inappropriately high for the serum calcium values; 97% of these patients are separated from normals and other hypercalcemic individuals on the basis of computer-assisted formal discriminant analysis of calcium and intact PTH.[17] In a group of 160 patients with chronic renal failure, 86% had increased concentrations of intact PTH. The interested reader is referred to recent articles for detailed discussions of the clinical interpretation of results for both of these PTH assays in several large groups of patients.[7,16]

In order to have enough clinical data with a PTH assay for the assay to be useful the following minimum numbers of patients are suggested:

Normals	100
Primary hyperparathyroidism	100
Other hypercalcemia	100*
Chronic renal failure	100
Surgical hypoparathyroidism	5

This quantity of specimens is a minimum for proper evaluation of the clinical performance of an assay. Larger numbers would be desirable, but the numbers shown above should ensure that the assay results can be reasonably well interpreted and that the laboratory or physician will know what to expect. Strong communication with physicians is to be encouraged. Their feedback is very important in assessing the reliability of the assay.

As mentioned above, the use of randomly collected patients in each category as opposed to selecting on the basis of overt disease is extremely important. Evaluation of an assay's performance will be biased high if only the patients with severe disease are used in the data base, and the physician will be unable to assess the assay's performance in equivocal cases.

*Includes at least 40 with nonparathyroid malignancy.

Other Performance Considerations

Several important factors must be controlled if an RIA procedure for PTH is to be useful and reliable in routine clinical diagnosis. The preparations of ^{125}I-labeled PTH used must have relatively uniform specific activities and high purity, to prevent large variations in assay binding and other problems. Key reagents such as antisera, PTH standards, ^{125}I-labeled PTH, and quality-control serum pools should be stored in small aliquots at $-76\,^{\circ}$C to protect against losses of activity. The PTH assays described in this report have achieved a consistent record of performance, reliability, and clinical usefulness because of attention to details such as these, and strict adherence to the quality control and performance criteria discussed.

Assay Speed

Hypercalcemia is seldom life threatening. Therefore, speed should be the last concern when a PTH assay is being evaluated. Most of the pressures for faster turnaround time have been artificially developed by competition or exist for political or financial reasons. Only after the basic criteria for specificity, sensitivity, accuracy, etc., have been satisfied should speed of the assay become a criterion.

B. Clinical Interpretation

Other chapters have addressed this aspect in detail. The reader can be referred to several articles from the author's work for specific clinical interpretation information on the assays described here.[4,7,15,16,18,19]

In general, however, it is recommended that a sensitive C-terminal-specific assay for PTH be used in essentially all endocrine evaluations, with an N-terminal or intact type of assay used only in a follow-up or supplemental manner as necessary. In selective venous catheterization studies the intact or N-terminal assay is clearly superior to the C-terminal assay.[20] In patients with chronic renal failure, both types of assays have a definite clinical value and both should be performed on the same specimen at intervals of 3–4 months to obtain the best assessment of such a patient's response to treatment.[19]

C. Recent Developments

Several recent publications have described newer approaches and method enhancements for the PTH radioimmunoassay. Some of these will undoubtedly be of interest to the laboratorian setting up a new PTH assay. Indeed, some of these techniques are now being used in the author's laboratory.

Although assay speed (incubation time) is largely a function of the antiserum binding constant and the requisite sensitivity, the reaction time can be significantly shortened by using polyethylene glycol in the assay diluent.[21] Moreover, although double antibody techniques apparently would not work with our original antiserum, such techniques are now in place in our laboratory with a newer antiserum. Use of a double antibody method eliminates the need for charcoal to separate antibody-bound from free-labeled hormone. This in turn solves several problems inherent with charcoal systems: drift due to variations in charcoal particle size from the beginning of the run to the end, interference with the binding of labeled hormone to charcoal due to serum proteins, and high nonspecific binding components. Charcoal also requires more handling steps than double antibody and leads to poorer assay precision.

Other improvements recently described include so called "midregion" assays,[22,23] and the use of synthetic human PTH peptides for radiolabeling, antibody production, standards, etc.[24-28] However, only one report has been published with any clinical data on the use of midregion assays; the ability to discriminate hyperparathyroid patients from normal subjects was only about 80% as opposed to 90-95% with C-terminal assays.[23,29] There are also potential problems in the use of synthetic PTH peptides. Since the molecular structures of endogenous PTH fragments are not yet known, it is not possible to know what relative portions of intact PTH and various fragments will be measured. The final measure of the utility of such assays remains their clinical performance in the separation of 1°HPT patients from normal and in evaluation of chronic renal failure patients. These data are as yet too preliminary for such assays.

References

1. Berson SA, Yalow RS, Aurbach GD, Potts JT Jr: Immunoassay of bovine and human parathyroid hormone. Proc Natl Acad Sci USA 49:613–617, 1963.
2. Berson SA, Yalow RS: Immunochemical heterogeneity of parathyroid hormone in plasma. J Clin Endocrinol Metab 28:1037–1047, 1968.
3. Di Bella FP, Gilkinson JB, Flueck J, Arnaud CD: Carboxyl-terminal fragments of human parathyroid hormone in parathyroid tumors: unique new source of immunogens for the production of antisera potentially useful in the radioimmunoassay of parathyroid hormone in human serum. J Clin Endocrinol Metab 46:604–612, 1978.
4. Hawker, CD: Parathyroid hormone: radioimmunoassay and clinical interpretation. Ann Clin Lab Sci 5:383–398, 1975.
5. Hawker CD, Glass JD, Rasmussen H: Further studies on the isolation and characterization of parathyroid polypeptides. Biochemistry 5:344–352, 1966.
6. Munson PL, Biological assay of parathyroid hormone. In Greep RO, Talmage RV (Eds.): The Parathyroids. Charles C Thomas, Springfield, Il, pp 94–107, 1961.
7. Hawker CD, Di Bella FP: Human parathyroid hormone: a review of the radioimmunoassay procedure and clinical interpretation. In Natelson, S, Pesce AJ, Dietz AA (Eds.): Clinical Immunochemistry. American Association for Clinical Chemistry, Washington, DC, pp 329–343, 1978.
8. Hunter WM, Greenwood FC: Preparation of iodine 131-labeled human growth hormone of high specific activity. Nature 194:495–496, 1962.
9. Yalow RS, Berson SA, Immunoassay of plasma insulin. Methods Biochem Anal 12:69–96, 1964.

10. Rodbard D, Bridson W, Rayford PL: Rapid calculation of radioimmunoassay results. J Lab Clin Med 74:770–781, 1969.
11. Di Bella FP, Hawker CD: Parathyrin (parathyroid hormone): radioimmunoassays for intact and carboxyl-terminal moieties (proposed selected method). Clin Chem 28:226–235, 1982.
12. Hawker CD, Di Bella FP, Parathyroid hormone in chronic renal failure: studies with two different parathyroid hormone radioimmunoassays. Contrib Nephrol 20:21–37, 1980.
13. Flueck JA, Di Bella FP, Edis AJ, Kehrwald JM, Arnaud CD: Immunoheterogeneity of parathyroid hormone in venous effluent serum from hyperfunctioning parathyroid glands. J Clin Invest 60:1367–1375, 1977.
14. Di Bella FP, Kehrwald JM, Laakso K, Zitzner L: Parathyrin radioimmunoassay: diagnostic utility of antisera produced against carboxy-terminal fragments of the hormone from the human. Clin Chem 24:451–454, 1978.
15. Di Bella FP, Hawker CD: Radioimmunoassay for parathyroid hormone: Bane or laboratory test *par excellence?* I. What to look for and why. Ligand Q 3(4):29–31, 1980.
16. Hawker CD, Di Bella FP: Radioimmunoassays for intact and carboxyl-terminal parathyroid hormone: clinical interpretation and diagnostic significance. Ann Clin Lab Sci 10:76–88, 1980.
17. Hawker CD, Brunden MN, Sutter ML, Spencer TW: Immunoreactive parathyroid hormone (iPTH) radioimmunoassay: Enhancement of clinical utility with bivariate classification analysis (abstract). Clin Chem 23:1151, 1977.
18. Di Bella FP, Hawker CD: Radioimmunoassay for parathyroid hormone: Bane or laboratory test *par excellence?* II. Endocrine applications. Ligand Q 4(1):26–29, 1981.
19. Hawker CD, Di Bella FP: Parathyroid hormone: a laboratory test not just for endocrinologists (editorial). Int J Artif Organs 4:112–115, 1981.
20. Di Bella FP, Hawker CD: Use of an "intact" parathyroid hormone assay increases "step-up" gradients in samples obtained by selective venous catheterization (abstract). Calcif Tissue Int 33:320, 1981.
21. Kao PC, Jiang N-S, Klee GG, Purnell DC: Development and validation of a new radioimmunoassay for parathyrin (PTH). Clin Chem 28:69–74, 1982.
22. Marx SJ, Sharp ME, Krudy A, Rosenblatt M, Mallette LE: Radioimmunoassay for the middle region of human parathyroid hormone: studies with a radioiodinated synthetic peptide J Clin Endocrinol Metab 53:76–84, 1981.
23. Roos BA, Lindall AW, Aron DC, Orf JW, Yoon M, Huber MB, Pensky J, Ells J, Lambert PW: Detection and characterization of small midregion parathyroid hormone fragment(s) in normal and hyperparathyroid glands and sera by immunoextraction and region-specific radioimmunoassays. J Clin Endocrinol Metab 53:709–721, 1981.
24. Shiraki M, Kawada N, Akiguchi I, Yamauchi H, Kojima A, Orimo H: 1–34 Human parathyroid hormone radioimmunoassay: properties of antiserum against synthetic 1–34 human parathyroid hormone and its clinical application. Endocrinol Jpn 28:239–244, 1981.
25. Mallette LE, Renfro M, Lemoncelli J, Rosenblatt M: Radioimmunoassays for the 28-48 region of parathyroid hormone detect intact hormone but not hormone fragments. Calcif Tissue Int 33:375–380, 1981.
26. Mallette LE, Bradley WA: Bovine parathyroid hormone- (41–84), a hormone

fragment with desirable properties for use as radioligand. J Lab Clin Med 98:886–895, 1981.

27. Mallette LE: Goat antisera to bovine parathyroid hormone: characterization of regional specificity, and evolution of titer and affinity of the 44-68 region-specific antibody species. Horm Metab Res 13:523–528, 1981.

28. Manning RM, Adami S, Papapoulos SE, Gleed JH, Hendy GN, Rosenblatt M, O'Riordan JLH: A carboxyl-terminal specific assay for human parathyroid hormone. Clin Endocrinol 15:439–449, 1981.

29. Martin KJ, Hruska K, Freitag J, Bellorin-Font E, Klahr S, Slatopolsky E: Clinical utility of radioimmunoassays for parathyroid hormone. Min Elect Metab 3:283–290, 1980.

Chapter 13

Radioimmunoassay for Calcitonin

Hunter Heath III and
Glen W. Sizemore

Calcitonin (CT) is one of the most recently discovered calcium-regulating hormones; the first brief allusion to its existence appeared in 1961.[1] The first assays for CT were based on its hypocalcemic property upon injection into rats.[2,3] Radioimmunoassays for animal calcitonins soon followed, but the availability of sensitive radioimmunoassays for human CT (hCT) awaited determination of the amino acid sequence and synthesis of the hormone.[4] There are now well over 20 published radioimmunoassays for human CT that vary in such basic elements as antisera, buffer constituents, incubation conditions, and tracer preparation; sensitivity and final results also vary.

This chapter describes a particular homologous radioimmunoassay[5] with the requisite sensitivity and specificity to detect the hormone in normal human plasma, and to measure increases in immunoreactive CT (iCT) concentration after appropriate stimuli.[6-8] The greatest usefulness of the human CT assay is in research, but CT measurements are important in preoperative diagnosis of familial medullary thyroid carcinoma[9-13] and posttreatment evaluation of all types of medullary thyroid carcinoma.[14,15] The assay is difficult, and susceptible to variation and error from many sources.[16-19] It will be cost-effective only in laboratories with large research needs, commercial reference laboratories processing many samples, or clinical laboratories supporting a large clinical oncology practice.

I. Principle

This assay for human CT is a conventional, homologous radioimmunossay. The same synthetic CT monomer is used as antigen for the antiserum, as radioiodinated tracer, and as standard. To improve sensitivity, incubation conditions are nonequilibrium; that is, tracer is added only after a delay of 4–5 days. Certain practical benefits are realized by counting both phases of the charcoal separation, and expressing binding as a ratio.[5,20] The assay method described here may not provide optimal conditions for all antisera; one must be prepared to explore alternative buffers, pH, incubation intervals, and methods of phase separation.

II. Materials and Methods

A. Reagents

For Production of Antiserum

1. Synthetic human CT monomer (hCT), 1 µg/µl in 0.01 mol/l acetic acid. Note: We presently use hormone from Ciba-Geigy, Basel; however, other sources may be used, such as from Beckman Bioproducts (Palo Alto, California); or Organon Pharmaceuticals (Oss, The Netherlands).
2. Complete Freund's adjuvant (No. 344289, Calbiochem–Behring Corp., La Jolla, California).
3. Mycobacterial adjuvant (Perrigen-SW, No. 869098. Calbiochem–Behring Corp., La Jolla, California).
4. Pertussis vaccine *(Eli Lilly and Co., Indianapolis, Indiana)*, supplied in 1:10,000 dilution, 7.5 ml vial).

For Radioiodination of Human CT

1. ^{125}I (Na^{125}I, carrier-free, for protein iodination, 100 mCi/ml, No. IMS-30, Amersham Corp., Arlington Heights, Illinois).
2. Phosphate buffer, pH 7.5. Mix 3 vol 0.3 mol/l sodium phosphate dibasic (Cat. No. S-374, Fisher Scientific Co., Fair Lawn, New Jersey) with 1 vol 0.3 mol/l sodium phosphate monobasic (Fisher Cat. No. S-369).
3. Chloramine-T (Cat. No. 1022, Eastman Kodak Co., Rochester, New York). Mix 1 g (3.55 mmol) per liter of phosphate buffer. Add buffer to powder immediately before use.
4. Sodium metabisulfite, anhydrous (ACS grade, No. S-224, Fisher Scientific Co.), 0.9 g (4.73 mmol) per liter of phosphate buffer. Add buffer immediately before use.
5. Egg albumin (Cat. No. A-388, Fisher Scientific Co.), 50 g/l in 1 mol/l tris(hydroxymethyl)aminomethane (Tris) buffer (Trizma base, Cat. No. T-1503, Sigma Chemical Co., St. Louis, Missouri) with 0.05 mol/l 2-mercaptoethanol (No. M-6250, Sigma Chemical Co.). Allow this solution to mix by passive diffusion at 4°C; stirring mechanically results in excessive foaming. Stored at 4°C, this is usable for at least 6 months.
6. Quso G-32, microfine precipitated silica (Philadelphia Quartz, Co., Chester, Pennsylvania), 6 mg of dry powder in 12 × 75 = mm glass radioimmunoassay tubes; flint glass is preferred (slightly less adsorption of hormone).
7. 2-Mercaptoethanol (see step 5). 0.05 mmol/l in distilled or deionized water.
8. Acetone–acetic acid: Add 200 ml of ACS grade acetone (No. A-18, Fisher

Scientific Co.) and 10 ml of glacial acetic acid (No. 3-9508, Baker Chemical Co., Phillipsburg, New Jersey) to 790 ml of 0.05 mol/l mercaptoethanol in distilled or deionized water.

9. Chromatographic column (1 × 50 cm) of Bio-Gel P-10, 100–200 mesh (Cat. No. 150-1040, Bio-Rad, Inc., Richmond, California) equilibrated in 0.05 mol/l tris buffer, pH 7.8.

For Radioimmunoassay

1. Tris buffer, 0.05 mol/l. Solution A (0.1 mol/l HCl):16.7 ml of concentrated hydrochloric acid diluted to 2 l with distilled water. Solution B (0.1 mol/l Tris):30.28 g Tris (Trizma base, see step 5) diluted to 2.5 l with distilled water. For 5 l of final volume, mix 2500 ml of solution B, 1725 ml of Solution A, and 775 ml of distilled or deionized water; add 0.5 g of thimerosal (ethylmercurithiosalicylic acid, sodium salt, Eastman Kodak), and adjust pH to 7.8 (20 °C) with NaOH.

2. Hypocalcitoninemic plasma, obtained by plasmapheresis from a person who has had total thyroidectomy (for a disease other than medullary thyroid carcinoma) and preferably postsurgical thyroid ablation with radioiodine. The plasma should contain no detectable CT when tested with one or more established calcitonin radioimmunoassays. Repeated use of a single donor provides improved long-term assay reliability. Some laboratories use outdated blood-bank plasma, on the unproved assumption that all calcitonin has been destroyed by endogenous proteases; others "strip" normal plasma with adsorbants such as charcoal or Quso. The latter methods may not remove all authentic calcitonin reactivity[21], and certainly remove many other plasma constituents.

3. Assay diluent (used to dilute antiserum and tracer, and as a replacement volume in assay tubes). To make 1 l, mix 900 ml of 0.05 mol/l Tris, 50 ml of hypocalcitoninemic plasma, and 50 ml of aprotinin (Trasylol, 1000 KIU/ml; FBA Pharmaceuticals, New York, New York). Store frozen at −20 °C in aliquots sufficient for approximately 1 week's use; after thawing, centrifuge if necessary to clear flocculent protein precipitate.

4. Tracer CT. Radiolabeled CT made up in assay diluent to 2000–3000 cpm 200 μl. Dilute to this concentration from the column-purified tracer (see iodination procedures, below) no more than 24 h before use.

5. Antiserum. Antiserum G-1701 used in the submitters'[5] laboratory is stored at −80 °C, undiluted. Aliquots are diluted 100-fold in assay diluent ("working stock"), and also stored at −80 °C. The 100-fold dilution is relatively unstable, so the aliquots are made to last no more than 6 months. For the assay, dilute this "working stock" 120-fold with assay diluent (12,000-fold dilution) for a final dilution of 60,000-fold when pipetted with the rest of the assay components. (The dilution obviously varies widely for different antisera.) At this dilution, under described conditions, the bound/free ratio is 0.4–0.6 after correction for nonspecific binding.

6. Quality control. Internal reference standard plasma. A low-concentration standard can be made from a pool of normal plasma. If available, pooled plasma from patients with medullary thyroid carcinoma may be diluted as needed with hypocalcitoninemic plasma to form conveniently internal reference standards of any desired concentration. Such plasmas should be stored in small aliquots at $-80°C$ and thawed as few times as possible to avoid a slow change in standard concentration.

7. Phase separation buffer. Make to 20l with distilled water: 29.4 g of sodium barbital (sodium diethylbarbiturate), 19.42 g of sodium acetate (ACS grade), 2.0 g of thimerosal, 153 g of sodium chloride (ACS grade), and 8.33 ml of HCl (ACS grade). Adjust pH to 7.4 $\pm$ 0.05 at room temperature. Store at 4°C.

8. Charcoal. (Carbon, decolorizing, Norit, Cat. No. C-170, Fisher Scientific Co.).

9. Dextran T-70 (Pharmacia, Inc., Uppsala, Sweden).

10. Human albumin (human albumin, fraction V, Sigma No. A-2386).

11. Standard diluent: 2.5 g of bovine serum albumin (Sigma No. A-4378) per liter of 0.05 mol/l Tris buffer, pH 7.8.

12. Standards. Accurately weigh out an aliquot of lyophilized human CT powder. After correcting for contamination with water, bogus peptides, and acetate salt (as detailed by the manufacturer), dilute to 1 $\mu g/\mu l$ in 0.01 mol/l acetic acid. Progressively dilute with standard diluent to obtain 12 standards ranging in concentration from 1.25 pg/100 μl to 200 pg/100 μl (12.5 to 2000 ng/L). Note: The concentrations of standards needed depend upon sensitivity characteristics of the antiserum available.

Store at $-20°C$ or colder, and freeze–thaw as little as possible. Note: Do not use longer than 3–6 months, or as determined by stability testing; sudden decreases in assay quality may follow use of old standards. It may be advisable to store standards in small aliquots, to minimize freeze–thaw cycles. Some authorities recommend dilution of standards in hypocalcitoninemic plasma; although theoretically ideal, this procedure results in intolerably rapid degradation of the standard.[5] If the antiserum used is very sensitive to nonspecific effects of human plasma, however, plasma-based standards may be required.

B. Apparatus

Aside from items commonly available in radioimmunoassay laboratories, the following equipment is especially useful.

1. Incubation boards. Bottom, 28–30-mm-thick wood, drilled to receive 80 12 × 75-mm assay tubes, with two nails or screws on front and back; painted or sealed. Top, 13-mm-thick board with two nails or screws protruding from front and back; 13–19-mm-thick closed-cell foam rubber sheet glued to underside; entire top then covered with polyethylene plastic sheet. The

cushioned top seals the assay tubes and is fastened to the bottom by rubber bands hooked over the nails or screws.

2. Vortex-type mixers (SMI Model 2600 Multi-tube vortexer, Scientific Manufacturing Industries, Emeryville, California; and Vortex-Genie mixer, Fisher Cat. No. 12-812-V1).

3. One or two tabletop centrifuges with four or eight carriers for 12 × 75-mm tubes (Model GLC-2B, Dupont/Sorvall, Newton, Connecticut).

4. Counting tube racks: 28–30-mm-thick boards, 232 × 232 mm, drilled to receive 100 120 × 16-mm plastic counting tubes. Paint or seal for water resistance and washability.

5. Repipet Dispensers, 1-ml capacity with conical flasks (Fisher Scientific) for dispersing diluent, antiserum, and CT tracer.

C. Collection and Handling of Specimens

Collecting blood after an overnight fast is recommended, although most observers find no effect of feeding on human plasma iCT.[6] Draw 10–20 ml of venous blood into heparinized evacuated collection tubes, and chill immediately after mixing. Transport on ice, centrifuge within 1 h of collection, and store plasma at −20°C or colder. Screw-cap polypropylene 7-ml vials are highly satisfactory for this purpose. Hemolysis must be assiduously avoided because of unpredictable nonspecific interference with tracer binding. Plasma is preferred in this assay because serum gives variably and erratically higher CT values. Plasma specimens may require centrifugation after thawing, to avoid aspiration of minor protein precipitates. Keep specimens at 4°C (ice water bath) during pipetting.

III. Procedure

A. Production of Antiserum

Note: Nonhuman calcitonins differ so markedly from hCT that antisera to them generally do not cross-react with hCT, and vice versa.

Emulsify 50–100 μg of CT in 1 ml of complete Freund's adjuvant. Inject intradermally dividing the 1 ml into 20–30 sites on both sides of the neck of a goat (this species seems to yield good antisera with unusual regularity). Inject 15 ml of pertussis vaccine intramuscularly at a distant site (i.e., the rump). Repeat antigen injections subcutaneously at 3-week intervals, with pertussis, for at least 6 months. Sores will develop at some injection sites. Screen the goat serum for specific binding before each immunization; do not combine the yields from each separate bleeding because of possible rapid and marked changes in antiserum characteristics between bleedings. The highest antibody titers may occur within the week after each immunization. The current antiserum we use was generated by this technique after five injections, five months from the first dose of CT.

When an antiserum of suitable titer is obtained, characterize it for minimum usable titer, sensitivity, and response to variation in assay conditions, especially pH and incubation intervals. Carefully assess sensitivity of the antiserum to nonspecific effects of human plasma proteins.[5,16,17]

B. Assay Setup

1. Plan the assay as outlined in Table 1. To permit calculation of controls for nonspecific binding, at many points in the assay use "no-antibody" tubes containing diluent plus hypocalcitoninemic plasma or sample. Each complete set or "board" of 80 tubes represents the maximum number held by the GLC-2B centrifuge. Place quality-control plasmas in each 80-tube set. Number and color-code each tube with indelible markers (i.e., Sharpie No. 3000; Sanford Corp., Belleville, Illinois). It is particularly useful to color-code the control tubes, to avoid inadvertent addition of antiserum.

2. Thaw plasma samples, standards, standard diluent, assay diluent, and internal reference standards for quality control; keep at 4°C on ice-water baths. Dilute an appropriate amount of antiserum in assay diluent, from the 100-fold diluted stock, and place in ice bath.

3. Place labeled tubes in a plastic rack, in an ice-water bath.

4. By pipetting, add reagents in the following order: diluent, hypocalcitoninemic plasma, standards, plasma samples, and antiserum. Repipets are useful for addition of diluent and antiserum. *Note:* It is appropriate first to measure all new specimens at a single dilution, because most will be within normal limits. Specimens seeming to have a high iCT value should be repeated at multiple dilutions. For such samples, "trace" tubes should contain a volume of diluted hypocalcitoninemic plasma equal to the sample addition volume.

5. Vortex-mix all tubes, either in large groups with the SMI Model 2600, or in pairs with the Vortex-Genie. Place in prechilled 80-tubed boards, attach sealing tops with rubber bands, and place in refrigerator or cold room (4°C). We agitate the tube continuously with an Eberbach reciprocating shaker during incubation, although some workers believe this is unnecessary. Preincubate in this fashion for 3–5 days without tracer.
   Note: Some antisera yield importantly greater sensitivity with very long preincubations,[22] so this possibility must be checked with each new antiserum.

6. Preparation of tracer CT. Note: This description presupposes that technical personnel are skilled in safe handling of radionuclides. Add 50 μl of 0.3 mol/l phosphate buffer to the 12 × 75-mm borosilicate glass reaction tube. Add 0.8–1.0 mCi of ^{125}I to the tube; shake gently. Add 5 μl (5 μg) of hormone to the tube; shake gently. Add 20 μl of Chloramine-T solution,

Table 1. Typical Calcitonin Assay Plan: Samples at a Single Dilution (200-μl Plasma Addition)

Tube Contents and No.	Contents (total volume, 300 μl)[a]	Comments
Standard controls (1–4)	Assay diluent, 200 μl BSA diluent, 100 μl	Control for nonspecific binding of tracer
Standard "trace" (5–8)	Assay diluent, 100 μl BSA diluent, 100 μl Antiserum, 100 μl	
Standards (9–32)	Standard, 100 μl Antiserum, 100 μl Assay diluent, 100 μl	12 pairs, with CT ranging from 1.25 to 200 pg/tube
Standard controls (33–36)	Same as 1–4	Average the values with those from tubes 1–4
Standard "trace" (37–40)	Same as 5–8	Average the values with those from tubes 5–8
Plasma controls (41–44)	Hypocalcitoninemic plasma, 200 μl Assay diluent, 100 μl	
Plasma "trace" (45–58)	Hypocalcitoninemic plasma, 200 μl Antiserum, 100 μl	
Control for sample 1 (49)	Plasma 1, 200 μl Assay diluent, 100 μl	Control for nonspecific binding by this individual sample
Sample 1 (50,51)	Plasma 1, 200 μl Antiserum, 100 μl	
Seven more samples (52–72)[b]	Same as for plasma 1 (tubes 49–51)	
Plasma controls (73–76)	Same as 41–44	Average with the values from tubes 41–44
Plasma "trace" (77–80)	Same as 45–58	Average with the values from tubes 45–48

[a]Delayed addition of tracer CT to all tubes not tabulated here.
[b]In subsequent sets, 80 tubes minus 16 for hypocalcitoninemic plasma control and trace tubes leaves 64 for 21 samples, one or two of which should be internal reference standards.
BSA, bovine serum albumin.

shake, then IMMEDIATELY ($<$10 s) add 50 μl sodium metabisulfite and shake gently. This last step is crucial: prolonged exposure to the Chloramine-T results in poor-quality tracer. Next add 1.5 ml of the egg albumin mixture, vortex–mix, and remove 50 μl for chromatoelectrophoresis and calculation of specific activity (usually 200–400 Ci/g).

Divide the remaining mixture into two 12 × 75-mm glass tubes, each containing 6 mg of Quso G-32; vortex–mix, then centrifuge in a small table top centrifuge for 3 min. Discard the supernates (transfer pipette), count the radioactivity of the Quso pellets, and record. Add 2 ml of mercaptoethanol wash to each of the two tubes containing the pellets, vortex–

mix, centrifuge for 3 min, then again discard supernates; count the radio-activity of the pellets and record. Add 1 ml of acetone–acetic acid wash to each tube and vortex–mix; then add 1 ml of mercaptoethanol wash to each tube, vortex, and centrifuge for 3 min. Aspirate supernates containing labeled CT, count their radioactivity, and transfer to the laboratory on ice. Note the residual counts of the Quso. Before use in the assay, pass about 0.5 ml of raw tracer through the 1×50 cm Bio-Gel P-10 column to elim-inate free iodine and aggregated tracer. Collect eluate fractions in tubes containing assay diluent, to reduce adsorptive losses. Determine where the peak elutes by gamma counting and save material from the central region at $4°C$.

Material should not be used more than 2–3 days after this column puri-fication step. We use ^{125}I-labeled CT from one iodination for 4 weeks, per-forming column purification once weekly. On the day of use, dilute appro-priate amounts of pooled, purified tracer with cold assay diluent to 2000 cpm/200 μl for addition to the assay. With our reagents, high sensitivity of the assay depends on addition of a small amount of tracer. With other antisera, one might be able to add more tracer and use shorter counting times.

7. To each assay tube, add 200 μl of tracer CT, prepared as described above, with a precalibrated Repipet. Vortex–mix, re-cover the tubes, and return to the shaker for a final incubation of 48–72 h.

8. Prepare dextran-coated charcoal. From 4 to 18 h before use, add 5 g of charcoal and 0.5 g of dextran to 200 ml of dextran-charcoal buffer per 80 assay tubes. Then stir with a magnetic stirring bar at $4°C$, covered with Parafilm. At the same time, make a separate solution of 0.2 g of human albumin in 200 ml of dextran-charcoal buffer per 80 tubes. Keep in the cold room ($4°C$) for 4–18 h. On the morning designated for phase sepa-ration, centrifuge the dextran-charcoal suspension at 430–450 g for 10 min (Sorvall RC-2B, HS-4 rotor, 1500 rpm). Carefully pour off supernate, which contains undesirable small charcoal particles ("fines"). Reconstitute the charcoal residue in 200 ml of the human albumin phase-separation buffer solution per 80 tubes. Stir with magnetic stirrer, on ice, at a speed less than that which induces frothing, for 15–30 min before use.

9. With a 10-ml graduated pipette premarked at 2-ml intervals, rapidly add 2 ml of the charcoal suspension to each assay tube, in groups of 80. Vor-tex–mix, then centrifuge each group of 80 tubes at 2000 rpm in the GLC-2B at 733 g for 15 min. Then as *rapidly and reproducibly as possible,* decant the supernate into one counting tube and place the precipitate in another. Note: Use of a refrigerated centrifuge might reduce stringency of time requirement for phase separation. Count the radioactivity of both the supernates and the precipitates for 4 min.

10. Calculate results of the assay. The first computations are for "damage" (no-antibody controls for nonspecific binding) and "trace" (maximum or no-cold binding, or B_0). "Damage" is

$$\frac{B}{B + F}$$

Where

B = bound radioactivity (supernate) and
F = free radioactivity (charcoal pellet).

"Trace" calculations correct for nonspecific binding:

$$\text{Trace} = \frac{B - (B + F) \text{ ("damage")}}{F}$$

Where

B = bound radioactivity from the "trace" tube
F = free radioactivity from the "trace" tube

and "damage" is the mean value for appropriate no-antibody controls, expressed as a fraction.

Binding in standard and sample tubes is then calculated with reference to appropriate "trace" tubes: bovine serum albumin-buffer-containing tubes for standards, and hypocalcitoninemic plasma-containing tubes for samples. Sample tube binding is calculated as described above, with correction of each sample for nonspecific binding, then expressed as a percent of maximum binding:

$$\frac{B/F \text{ patient}}{\text{"trace" } B/F} \times 100$$

The standard curve is likewise expressed as percent of maximum binding, in a log-logit transformation (Figure 1). The assay limit of detection is determined by statistical analysis of "trace" tubes. Calculate the mean of "trace" values for both the standard curve and sample tubes minus 2 SD, expressed as percent of maximal binding. Generally, both 95% confidence limits (for standards and samples) will agree within a few percent, and will almost always lie between 85 and 95%. Because we think conservatism is warranted in stating detection limits, we accept no less than a 10% decrease of tracer binding as evidence of detectable CT in a sample, and prefer at least 15% decrease. Similarly, we generally do not report a value if the specimen decreases tracer binding to less than 20% of maximum; such a sample is instead reassayed at multiple dilutions to obtain two or three points within the working region of the standard curve (20–90%).

IV. Quality Control and Standardization

Because both circulating CT and the antisera to it are heterogeneous, no absolute performance criteria are applicable to all CT assays. Therefore, workers performing this radioimmunoassay must determine their own performance criteria, and use accumulated experience and internal reference standards to

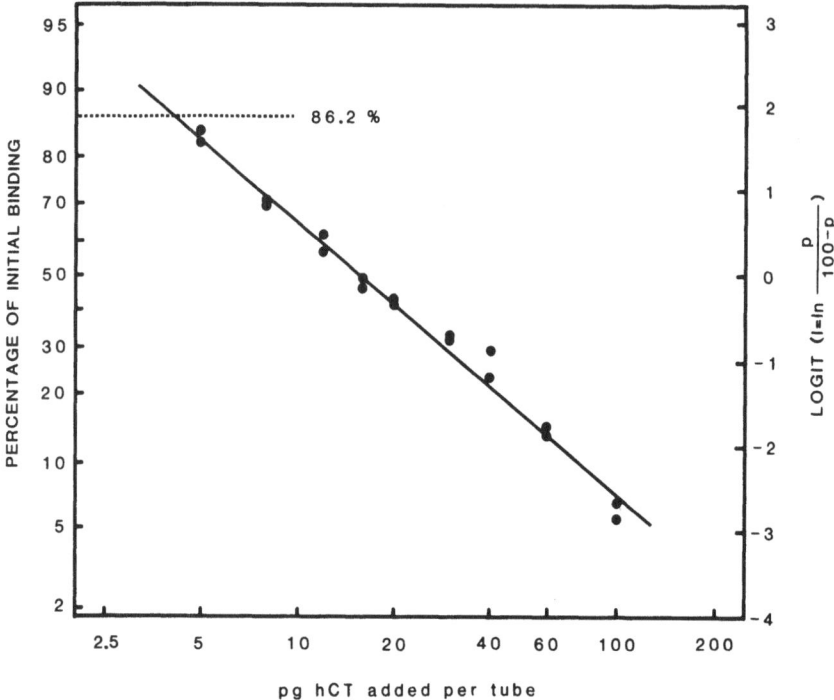

Figure 1. Typical standard curve for assay of human calcitonin (hCT) with 60,000-fold diluted antiserum G-1701, ^{125}I-labeled hCT tracer, and hCT monomer as standard. The ordinate represents binding as percent of maximum in the absence of standard. The *dashed line* depicts the assay limit of detection (see text). The logit (*l*) is calculated as ln (percent of initial binding/100 minus percent of initial binding). The *solid line* represents the least-squares regression line for the data (r = 0.99).

monitor quality. Comparison of paired-sample results in an established assay can be of some value, however.

Quality control is achieved primarily through frequent assessment of "damage" and "trace" values, standard curve performance, and reproducibility of internal reference standards.

In our laboratory, the "damage" or nonspecific binding of ^{125}I-labeled CT is generally less than 2% of total counts with standard diluent, and less than 2–4% for 200 μl of human plasma sample (plasma vol/incubation mixture vol, 2–5).[5] We regard "damage" of greater than 5% as evidence for unsatisfactory tracer or improper phase separation; such results are generally disregarded, and the assay repeated. Rarely, individual plasma specimens yield persistently high nonspecific binding; iCT cannot be quantitated in such samples. Repeat fasting samples from the same patients usually behave normally.

With antiserum G-1701 at a final dilution of 60,000-fold, "trace" binding with freshly prepared tracer occasionally is as high as 80%, decreasing to a low of 40% 4–6 weeks after iodination. However, "trace" in most assays is 45–55%;

this is, of course, strictly a function of the antiserum in use, quality of the tracer, etc.

The limit of detection, determined as stated above, is 20–25 pg equivalents of CT monomer/ml plasma, or 20–25 ng/l, when the volume of sample addition is 200 μl. A grossly low or high limit of detection may alert one to errors in calculations or to substandard assay performance. With this antiserum, 50% displacement of tracer occurs with standard additions of 10–20 pg/tube. Half-maximum displacement outside these limits usually denotes errors in standard dilution or degradation of the standards. As in any assay, each new standard set *must be cross-checked* in parallel with the "old" standard at least three times before switching to the new one. Failure to adhere to such rigorous criteria for standards will result in excessive variability and a spuriously wide normal range.

Internal reference standard plasmas offer a continuing check on assay performance, and a running calculation of within- and between-assay variability is essential. One must be cautious about this, however, becuase CT content of such quality plasmas may decrease with time. Scrupulous storage at $-80\,°C$ in small aliquots is essential. The within-assay coefficient of variation (CV) for samples even in the low-normal range must be less than 20%. With meticulous technique, within- and between-assay CV $\leq$ 5% is attainable for the internal reference standard or quality-control plasmas.[8]

The interpretation of radioimmunoassay results must be based on normative data developed with the assay in use. This is particularly a problem for CT, for which there is considerable variation with age and sex.[5,23,24] We recommend determination of CT in at least 20 persons of each sex in the age ranges 5–17, 18–30, 31–50, and over 50 years, to obtain a first approximation of normal values for fasting CT. If the assay is to be used to measure CT after stimulation of secretion by pentagastrin or calcium, then it is absolutely necessary to make the appropriate stimulation tests in normal volunteers.[5,10,12] *Failure to do so will lead to the reporting of totally uninterpretable clinical data.*

Interpretation of apparently high immunoreactive CT values must be circumspect;[25] at a minimum, one must show parallelism with the standard curve when the specimen is assayed at multiple dilutions. Serious divergence from linearity almost always indicates nonspecific interference in the tracer–antibody reaction (e.g., hemolysis).

V. Discussion

The calcitonin radioimmunoassay described here is capable of measuring CT immunoractivity (iCT) in plasma of normal people, and of quantitating increases and decreases of CT concentration in response to physiologic and pharmacologic manipulation.[5–8] Its use has eluciated several important points in human CT physiology,[26] and has proved highly effective in the presurgical diagnosis of CT-secreting medullary carcinoma of the thyroid gland.[25] The

Table 2. Normal Plasma iCT Concentrations (pg/ml) in Healthy Adults, Ages 21–46 yr[5,7]

	Male		Female	
Condition	Range	Mean	Range	Mean
Basal (fasting)	<25–87	48 (25)	<25–66	31 (30)
Calcium infusion[a]	<25–205	106 (20)	<25–135	55 (25)
Pentagastrin injections[b]	<25–182	111 (10)	<25–73	36 (14)

[a]Calcium gluconate, 15 mg of elemental Ca/kg body weight infused intravenously over 4 h. Values shown are peak concentrations (third or fourth hour).
[b]Pentagastrin (Peptavlon, Ayerst), 0.5 µg/kg infused intravenously over 5 s. Values shown represent peak concentrations, usually from the first sample at 1–2 min.
Note: These data apply only to the assay with the reagents described in this paper, and must not be used to judge results from assays incorporating other reagents. Number in parentheses is n for that study. For calculation, undetectable values were assigned the value of the usual assay limit of detection, 25 pg/ml. Median values were not significantly different.

assay's major usefulness has been in research, because clinical states in which iCT measurement is of *proved* value are rather uncommon.

There are several problems in the use of the CT radioimmunoassay: some biologic, some technical, some practical. The major biological problems are, first, that the actual concentration of CT in human plasma appears to be very low; and, secondly, that circulating iCT is not homogeneous. The plasma iCT concentration determined in most newer assays is well under 100 pg/ml (<30 pmol/l), and many older assays for CT have lower limits of detection higher than that. Even in assays that do detect iCT in normal plasma, such as the one described in this chapter, the hormone is undetectable in many samples.[5] For example, our assay detects iCT in plasma from 90% of normal males, but only 40% of normal females.[5] Therefore, it is not possible to define a *subnormal* concentration of plasma iCT. In both sexes, plasma iCT declines with age,[23,24] so the assay is even more limited in the elderly.

The normal ranges for basal and stimulated plasma iCT values are given in Table 2.[5] It must be emphasized that such normative data must be obtained for any CT radioimmunoassay, even one using precisely the methods described herein, because of variation among antisera.

One of the many possible causes of disagreement about iCT results among various groups appears to be immunochemical heterogeneity of circulating CT.[18,27–31] The hormone yields five or more discrete peaks by gel elution chromatography[27–31] and isoelectric focusing,[32] peaks variably recognized by different CT assays.[18,28, 29] The structures of these iCT species are still under study, but almost all appear to be larger than monomeric CT; they are not "fragments." Some may be biosynthetic precursors of monomeric CT.[33] In any case, this heterogeneity of circulating CT at least partly explains different iCT values in different assay systems.[18,28–30] It is also possible that differences in susceptibility of some antisera to nonspecific interference with tracer binding, and other technical factors, contribute importantly to discrepant results among various assays. For example, some assays have shown "normal" plasma iCT values as great as 1000 pg/ml, 10 times the values obtained in other systems. The practical consequence of these observations is that each CT radioimmunoassay must be carefully standardized, and normative data must be collected for that assay. "Literature normals" are totally invalid.

Another "biological" problem with use of CT assays is that an increased iCT value is not specific for medullary carcinoma of the thyroid gland (MTC).[25] Plasma or serum iCT is reportedly high in many patients with other cancers,[34-36] as well as those with pancreatitis, pernicious anemia, lung disease, and renal failure[37-39] Proper use of the assay in immunodiagnosis of MTC is nonetheless possible; guidelines for such use have been published.[25] Briefly, iCT is most useful in screening family members of persons already known to have MTC, and who may have multiple endocrine neoplasia, type 2a or 2b.[10,25] In this context, a palpable thyroid nodule, as well as elevated basal plasma iCT, is strong presumptive evidence for MTC. However, in screening studies, many affected persons will have normal basal iCT levels and no palpable thyroid abnormality; in this situation, provocative testing with the secretagogues calcium and/or pentagastrin is essential to early diagnosis of MTC.[5,9-15] Persistent elevation of plasma iCT after total thyroidectomy is strong evidence for residual MTC. In addition, in the long-term follow-up of patients who have undergone surgery, iCT measurement may indicate recurrences before there is clinical evidence.[15]

Some authors recommend routine use of iCT measurement in posttreatment follow-up of patients with certain cancers (e.g., of the lung).[34-36] In our current view, this recommendation is premature, primarily because the sensitivity, specificity, and utility of the test relative to conventional measures have not been adequately determined.

The technical quality of the CT assay is in most respects (specificity, accuracy, precision) as good as that of any peptide hormone assay. Recovery of synthetic CT added to plasma is excellent (Figure 2). The problem of inability

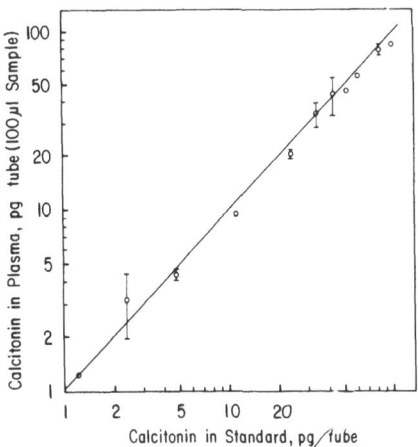

Figure 2. Recovery of synthetic hCT monomer added to hypocalcitoninemic (Hypo-CT) plasma. The abscissa represents actual CT concentrations per tube of standards prepared in standard diluent (containing bovine serum albumin; see text). The ordinate represents apparent concentrations of CT per 100 μl addition; CT was added to plasma from a thyroidectomized person to yield nominal concentrations identical to those of the standards. The *solid line* represents a line of identity. As shown, recovery was essentially 100%. *Bars* indicate mean $\pm$ SD (n = 4).

to measure at or below the lower limits of normal afflicts many, if not most, peptide hormone assays. The major technical problem with CT radioimmunoassay is that many antisera seem to be susceptible to nonspecific interference with tracer binding.[16,17] Assays incorporating such antisera may yield spuriously high iCT values in some patients. The importance of varying amino acid sequence-specificity of CT antisera[18,29] is not clear in the way that it is for parathyrin antisera.[40] Because CT standards decay even at $-70°$C, the best assay performance requires meticulous attention to standard curve coefficients and results of quality-control plasma measurements. Hemolysis is another factor that we and others have noted as causing variable nonspecific interference in the assay. We are aware of several patients who have received inappropriate treatment (e.g., total thyroidectomy) because of reported increased iCT concentrations in other assays, whereas the values were normal with this assay. Several approaches have been tried to reduce such nonspecific effects, including affinity chromatography,[17] immunoextraction,[41] and nonspecific adsorbent extraction.[42] Each of these methods has limitations, and for the moment they are primarily research techniques.

In summary, success with CT radioimmunoassay first of all requires good fortune in raising an antiserum. Only a few have the requisite sensitivity combined with freedom from susceptibility to nonspecific interference with tracer binding. Next, meticulous technique in radioiodination, standard preparation, and other elements of technique are essential. The assay must be characterized,[5] and age- and sex-specific normative data obtained.[5,24] Finally, the physicians using the test must choose carefully the two settings in which CT measurement is of established value: posttreatment evaluation and follow-up of any person who has had medullary carcinoma of the thyroid; and preoperative screening of persons suspected of having multiple endocrine neoplasia, type 2.[9-15]

Acknowledgments　The radioimmunoassay described here relies heavily on the work of Dr. C. D. Arnaud in parathyroid hormone assay. Dr. C. S. Anast and Ms. Sammy Langeluttig collaborated in production of the antiserum; Dr. W. Rittel (Ciba-Geigy, Basle) donated the CT; and several research technologists participated in this work, but we are particularly indebted to Mrs. Joan M. Larson. Financial support came from NIH (CA-11911, AM-00983, and RR-585) and the Mayo Foundation.

References

1. Copp DH, Davidson AGF, Cheney BA: Evidence for a new parathyroid hormone which lowers blood calcium. Proc Canad Fed Biol Sci 4:17, 1961.
2. Hirsch PF, Voelkel EF, Munson PL: Thyrocalcitonin: hypocalcemic hypophosphatemic principle of the thyroid gland. Science 146:412–413, 1964.

3. Kumar MA, Slack E, Edwards A, Soliman HA, Baghdianz A, Foster GV, MacIntyre I: A biological assay for calcitonin. J Endocrinol 33:469–475, 1965.
4. Queener SF, Bell NH: Calcitonin: a general survey. Metabolism 24:555–566, 1975.
5. Heath H III, Sizemore GW, Plasma calcitonin in normal man: differences between men and women. J Clin Invest 60:1135–1140, 1977.
6. Owyang C, Heath H III, Sizemore GW, Go VLW: Comparison of the effects of pentagastrin and meal-stimulated gastrin on plasma calcitonin in normal man. Am J Dig Dis 23:1084–1088, 1978.
7. Lambert PW, Heath H III, Sizemore GW, Pre- and post-operative studies of plasma calcitonin in primary hyperparathyroidism. J Clin Invest 63:602–608, 1979.
8. Austin LA, Heath H III, Go VLW: Regulation of calcitonin secretion in normal man by changes of serum calcium within the physiologic range. J Clin Invest 64:1721–1724, 1979.
9. Gagel RF, Melvin KEW, Tashjian AH Jr., Miller HH, Feldman ZT, Wolfe HJ, De Lellis RA, Cervi-Skinner S, Reichlin S; Natural history of the familial medullary thyroid carcinoma—pheochromocytoma syndrome and the identification of preneoplastic stages by screening studies: a five-year report. Trans Assoc Am Phys 88:177–191, 1975.
10. Sizemore GW, Heath H III, Carney JA: Multiple endocrine neoplasia, type 2. Clin Endocrinol Metab 9:299–315, 1980.
11. Hennessy JF, Wells SA Jr., Ontjes DA, Cooper CW: A comparison of pentagastrin injection and calcium infusion as provocative agents for the detection of medullary carcinoma of the thyroid. J Clin Endocrinol Metab 39:487–495, 1974.
12. Sizemore GW, Go VLW: Stimulation tests for diagnosis of medullary thyroid carcinoma. Mayo Clin Proc 50:53–56, 1975.
13. Verdy M, Cholette JP, Cantin J, Lacroix A, Sturtridge WC: Calcium infusion and pentagastrin injection in diagnosis of medullary thyroid carcinoma. Canad Med Assoc J 119:29–35, 1978.
14. Baylin SB, and Wells SA Jr., Management of medullary thyroid cancer. In Greenfield LD (ed.): Thyroid Cancer. C.R.C. Press, West Palm Beach, FL, pp 151–163, 1978.
15. Stepanas AV, Samaan NA, Hill CS Jr., Hickey RC: Medullary thyroid carcinoma. Importance of serial serum calcitonin measurement. Cancer 43:825–837, 1979.
16. Deftos LJ, Bury AE, Habener JF, Singer FR, Potts JT Jr.: Immunoassay for human calcitonin. II. Clinical studies. Metabolism 20:1129–1137, 1971.
17. Tashjian AH Jr., Voelkel EF, Human calcitonin: application of affinity chromatography. In Jaffe BJ, Behrman HR (eds.): Methods of Hormone Radioimmunoassay. Academic Press, New York pp 199–214, 1974.
18. Sizemore GW, Heath H III: Immunochemical heterogeneity of calcitonin in plasma of patients with medullary thyroid carcinoma. J Clin Invest 55:1111–1118, 1975.
19. Argemi B, Hours MC, Even F, Garguilo G, Hollard JM: Re-examination of human calcitonin radioimmunoassay. Acta Endocrinol 88:75–86, 1978.
20. Arnaud CD, Tsao HS, Littledike T: Radioimmunoassay of human parathyroid hormone in serum. J Clin Invest 50:21–34, 1971.

21. David L, Cohn DV, Anast CS: Identification of artifacts introduced to radioimmunoassays by the treatment of sera with charcoal in order to obtain "dehormonized" sera. Applications to the radioimmunoassays of parathormone and calcitonin. Pathol Biol (Paris) 23:833–834, 1975.

22. Parthemore JG, Deftos LJ: Calcitonin secretion in normal human subjects. J Clin Endocrinol Metab 47:184–188, 1978.

23. Samaan NA, Anderson GD, Adam-Mayne ME: Immunoreactive calcitonin in the mother, neonate, child and adult. Am J Obstet Gynecol 121:622–625, 1975.

24. Deftos LJ, Weisman MH, Williams GW, Karpf DB, Frumar AM, Davidson BJ, Parthemore JG, Judd HL: Influence of age and sex on plasma calcitonin in human beings. N Engl J Med 302:1351–1353, 1981.

25. Sizemore GW, Heath H, III: Calcitonin and medullary carcinoma of the thyroid gland. Curr Probl Cancer 6:1–68, 1981.

26. Austin LA, Heath H, III: Calcitonin: physiology and Pathophysiology. N Engl J Med 304:269–278, 1981.

27. Singer FR, Habener JF: Multiple immunoreactive forms of calcitonin in human plasma. Biochem Biophys Res Commun 61:710–716, 1974.

28. Deftos LJ, Roos BA, Bronzert D, Parthemore JG, Immunochemical heterogeneity of calcitonin in plasma. J Clin Endocrinol Metab 40:409–412, 1975.

29. Snider RH, Silva OL, Moore CF Becker KL, Immunochemical heterogeneity of calcitonin in man: effect on radioimmunoassay. Clin Chem Acta 76:1–14, 1977.

30. Goltzman D, Tischler AS: Characterization of the immunochemical forms of calcitonin released by a medullary thyroid carcinoma in tissue culture. J Clin Invest 61:449–458, 1978.

31. Heath H III, Sizemore GW, Immunochemical heterogeneity of calcitonin in tumor, tumor venous effluent and peripheral blood of patients with medullary thyroid carcinoma. J Lab Clin Med 93:390–401, 1979.

32. Dermody WC, Rosen MA, Ananthaswamy R, McCormick WM, Levy AG: Characterization of the major forms of human calcitonin in tissue and serum. J Clin Endocrinol Metab 52:1090–1098, 1981.

33. Goodman RH, Jacobs JW, Habener JF: Cell-free translation of messenger RNA coding for a precursor of human calcitonin. Biochem Biophys Res Commun 91:932–938, 1979.

34. Coombes RC, Hillyard C, Greenberg PB, MacIntyre I, Plasma-immunoreactive-calcitonin in patients with non-thyroid tumors. Lancet 1:1080–1083, 1974.

35. Silva OL, Broder LE, Doppman JL, Snider RH, Moore CF, Cohen MH, Becker K L, Calcitonin as a marker for bronchogenic cancer. A prospective study. Cancer 44:680–684, 1979.

36. Schwartz KE, Wolfsen AR, Forster B, Odell WD, Calcitonin in nonthyroidal cancer. J Clin Endocrinol Metab 49:438–444, 1979.

37. Sizemore GW, Go VLW, Kaplan EL, Sanzenbacher LJ, Holtermuller KH, Arnaud CD: Relations of calcitonin and gastrin in the Zollinger–Ellison syndrome and medullary carcinoma of the thyroid. N Engl J Med 288:641–644, 1973.

38. Heynen G, Franchimont P, Human calcitonin radioimmunoassay in normal and pathological conditions. Eur J Clin Invest 4:213–222, 1974.

39. Becker KL, Nash D, Silva OL, Snider RH, Moore CF: Increased serum and urine calcitonin in patients with pulmonary disease. Chest 79:211–216, 1981.

40. Arnaud CD, Goldsmith RS, Bordier PJ, Sizemore GW, Influence of immunoh-eterogeneity of circulating parathyroid hormone on results of radioimmunoassays of serum in man. Am J Med 56:785–793, 1974.
41. Parthemore JG, Deftos LJ: The regulation of calcitonin in normal human plasma as assessed by immunoprecipitation and immunoextraction. J Clin Invest 56:835–841, 1975.
42. Hillyard CJ, Cooke TJC, Coombes RC, Evans IMA, MacIntyre I: Normal plasma calcitonin: circadian variation and response to stimuli. Clin Endocrinol 6:291–298, 1977.

Chapter 14

Adenylate Cyclase Bioassay for Parathyroid Hormone

Robert A. Nissenson

I. Background and Principles

A. Need for a PTH Bioassay

The application of immunochemical techniques to the measurement of circulating peptide hormones has dramatically advanced our understanding of endocrine homeostasis in health and disease. However, two sources of difficulty in the interpretation of radioimmunoassay (RIA) measurements of peptide hormones have become increasingly evident. First, circulating peptide hormones are immunochemically heterogeneous. This contributes to the well-known discrepancies in peptide hormone RIA values obtained with antisera possessing different peptide-region specificities. Second, antigenic determinants of peptide molecules bear no necessary relationship to sites required for their biologic action. Thus, depending on the antiserum employed, RIAs may measure only biologically active hormone, only biologically inactive hormone, or a combination of these.

Such difficulties are particularly apparent in RIA measurements of PTH. The major circulating species of immunoreactive PTH in patients with hyperparathyroidism are hormonal fragments containing the mid- and COOH-regions of human PTH.[1-4] Because the NH_2-terminal 1–34 peptide reproduces the known physiologic actions of native PTH(1–84),[5] circulating mid-and COOH-region PTH fragments are presumed to be biologically inactive. Available data support this assumption.[6] Thus, mid- and COOH-region-specific RIAs, while often providing clinically useful information, do not measure levels of biologically active hormone. Because the half-life of circulating COOH-region PTH fragments is greater than that of biologically active hormone, such RIA measurements may fail to reflect accurately acute parathyroid secretory status. A new technique[7] capable of quantitating circulating, biologically active PTH in selected hyperparathyroid patients is described below.

B. Basis of the PTH Bioassay

Ever since the discovery of a PTH-stimulated adenylate cyclase in renal cortical plasma membranes,[8,9] investigators have used this enzyme system for the bioassay of PTH in vitro. There is considerable justification for the assumption that this activity reflects a physiologic relevent action of PTH in vivo. Thus, exogenous administration of cyclic AMP (or its dibutyryl derivative) can reproduce a number of PTH-dependent physiological effects on the kidney.[10-14] Furthermore, under some circumstances, the quantity of urinary cyclic AMP contributed by the renal tubular cells (nephrogenous cyclic AMP) is a valid index of parathyroid status in humans.[15] However, renal adenylate cyclase, as generally assayed, requires nanogram to microgram quantities of PTH for activation, whereas subnanogram levels of the hormone appear to circulate in vivo.[16] Thus the bioassay utility of renal cyclase has been limited to the measurement of PTH activity in parathyroid gland extracts or in synthetic preparations of PTH analogs. Fortunately, recent advances in our knowledge of hormone receptor-adenylate cyclase coupling (see below) have suggested new ways to augment hormonal sensitivity of the enzyme.

There is substantial evidence that most, if not all, hormone-responsive adenylate cyclase systems consist of at least three intrinsic plasma-membrane proteins: catalytic adenylate cyclase, hormone receptor, and a coupling protein (N) which mediates the catalytic activation produced by guanosine triphosphate (GTP), fluoride (F^-), cholera toxin, and hormones.[17] Studies of the β-adrenergic receptor–adenylate cyclase system in genetic variants of S49 lymphoma cells strongly suggest that these three components represent distinct gene products.[18,19] Wild-type S49 cells possess adenylate cyclase which is activated by β-adrenergic agents, F^-, cholera toxin, GTP, and Mg^{2+}. Clonal genetic variants of these cells (termed "cyc$^-$") lack adenylate cyclase responsiveness while retaining β-adrenergic receptors as evidenced by the binding of labeled β-adrenergic antagonists.[18] These variants do not lack the catalytic moiety of adenylate cyclase per se, because Mn^{2+}, which acts directly on the catalytic unit, elicts enzyme activity.[20] However, cyc$^-$ cells lack the N protein which mediates enzyme activation produced by regulatory ligands.[19]

Considerable progress has been made towards understanding how hormone receptors, N protein, and catalytic unit interact to produce hormonal stimulation of adenylate cyclase. Data obtained largely in the β-adrenergic system have led to the following proposed mechanism[21,22]:

$$
\begin{array}{ccc}
GDP & GTP \\
\uparrow & \downarrow \\
\end{array}
$$

$$H + R \rightarrow HR + N_{GDP} \rightarrow HRN \rightarrow HR + N_{GTP} + C_i \rightarrow$$

$$\downarrow$$

$$H + R$$

$$Pi$$
$$\uparrow$$
$$N_{GTP}C_a \rightarrow N_{GDP}C_i \rightarrow N_{GDP} + C_i$$

where H = hormone, R = receptor, N = guanyl nucleotide-dependent regulatory protein, C_i = catalytic adenylate cyclase in inactive conformation, and C_a = catalytic adenylate cyclase in active conformation.

In this formulation, binding of the hormonal agonist leads to a conformational change in the receptor protein resulting in a direct association of the receptor with the N protein. During formation of the ternary HRN complex, GDP dissociates from the N protein, permitting GTP to bind to the regulatory site. Concomitantly, the N_{GTP} complex dissociates from HR and H dissociates from R. N_{GTP} then interacts with a free (inactive) catalytic subunit, resulting in formation of the holocatalytic (active) state $N_{GTP}C_a$. Enzyme activation is terminated by the hydrolysis of GTP to GDP, presumably due to GTPase activity intrinsic to the N protein. The major function of the hormone is thus to promote the interaction of R with N, leading to displacement of inactive GDP by active GTP at the regulatory site.

One piece of evidence consistent with this model is that analogs of GTP that are resistant to terminal phosphate hydrolysis (e.g., 5′-guanylimidodiphosphate [Gpp(NH)p]) greatly potentiate hormonal stimulation of adenylate cyclase in several systems. The bioassay described below utilizes this effect of Gpp(NH)p to potentiate PTH-activation of renal adenylate cyclase.[23,24]

II. Practical Aspects

The three major procedures required for the guanyl nucleotide-amplified renal adenylate cyclase bioassay for PTH are (1) isolation of canine renal cortical plasma membranes having PTH-responsive adenylate cyclase, (2) incubation of membranes with substrate (ATP) in the presence of samples or standards containing PTH, and (3) quantitation of the enzymatic product (cyclic AMP).

A. Isolation of Canine Renal Cortical Plasma Membranes

Required Instrumentation

1. Standard surgical and dissection instruments
2. Polytron homogenizer (Brinkmann Instruments)
3. Potter-Elvehjem homogenizer with loose-fitting Teflon pestle
4. High-speed refrigerated centrifuge
5. Ultracold ($-80\,°C$) freezer

Required Solutions

1. 0.9% NaCl/1.0 mM EDTA/5 mM Tris-HCl, pH 7.5 (solution A)
2. 0.25 M sucrose/1.0 mM EDTA/ 5 mM Tris–HCl, pH 7.5 (solution B)
3. 2.0 M sucrose/1.0 mM EDTA/5 mM Tris–HCl, pH 7.5 (solution C)
4. 1.0 mM EDTA/5 mM Tris–HCl, pH 7.5 (solution D)

Protocol

Canine renal cortical plasma membranes are isolated by a modification of the procedure of Fitzpatrick et al.[25]:

1. Adult mongrel dogs (20–30 kg) are killed by intravenous pentobarbital, and the kidneys rapidly removed and placed into ice-cold solution A. All further steps are performed at 3–4°C.
2. Whole kidneys are sliced longitudinally, and cortical tissue dissected away from the medulla and corticomedullary regions.
3. Cortical sections are stripped of capsular tissue and fat, and dispersed in 3 volumes (relative to wet weight) of solution B with three 5-s bursts of Polytron homogenizer (setting 4).
4. Complete homogenization is effected with 10 strokes of a motor-driven loose Teflon pestle.
5. The homogenate is centrifuged at 1475 g (10 min), and the pellets resuspended in solution C with three strokes of a motor-driven loose Teflon pestle and recentrifuged at 13,300 g (10 min).
6. The 13,300 g supernatant is diluted eightfold with solution D and centrifuged at 20,000 g (15 min). This sediments plasma membranes (top 'fluffy" layer) and mitochondria (bottom "compact" layer).
7. The plasma membranes are resuspended by swirling gently in solution B, followed by recentrifugation at 20,000 g (15 min). This cycle of resuspension–centrifugation is repeated at least three times.
8. Final resuspension of membranes is in solution B at about 3 mg protein/ml. Plasma membranes are frozen in dry ice-acetone in multiple aliquots, and stored at $-80°C$. In our experience, PTH-sensitive adenylate cyclase activity is stable for at least 6 months under these storage conditions. The yield of purified plasma membranes is variable, averaging about 40 mg protein per kidney. It is recommended that each preparation be tested for enrichment (relative to homogenate) in the basolateral membrane marker enzyme Na^+–K^+ ATPase.[26] The specific activity of this enzyme should be at least 6–8-fold greater in the membrane preparation than in the homogenate.

B. Measurement of Adenylate Cyclase: Incubation Procedures

Adenylate cyclase is a magnesium-dependent enzyme which catalyzes the irreversible conversion of ATP to the second messenger cyclic AMP plus pyrophosphate:

$$\text{ATP} \xrightarrow[\text{AC}]{\text{Mg}^{2+}} \text{cyclic AMP} + \text{PP}_i$$

In the presence of the appropriate acute hormonal stimulus, enzyme activity is enhanced due to an increase in V_{max} without change in the K_m for ATP. Adenylate cyclase can be assayed in either of two ways. In our laboratory, ATP labeled with ^{32}P in the alpha-phosphate position is used as substrate, and formation of [^{32}P]cyclic AMP is measured. Alternatively, unlabeled ATP can be used as substrate, and the unlabeled cyclic AMP formed can be quantitated by either RIA[27] or competitive protein-binding assay.[28] In our experience, the ^{32}P method offers somewhat greater precision than the other methods which require an additional assay step. This enhanced precision is important for detecting small increases in adenylate cyclase produced by threshold concentrations of PTH, and is worth the precautions necessary for handling moderate quantities (≈ 0.1 mCi/100-tube assay) of $[\alpha-^{32}P]$ATP.

Two requirements for the accurate assessment of adenylate cyclase activity are maintenance of a constant $[\alpha-^{32}P]$ATP level despite the presence of membrane ATPase and prevention of the metabolism of $[^{32}P]$ cyclic AMP to $[^{32}P]5'$-AMP by membrane phosphodiesterase. The former problem is solved by including in the assay an ATP-regenerating system consisting of creatine phosphate and creatine phosphokinase:

ADP + creatine phosphate $\rightarrow$ ATP + creatine

In the presence of a high concentration of creatine phosphate, (e.g., 10 mM), the reaction is essentially irreversible, and will quantitatively convert ADP to ATP. Under the adenylate cyclase assay conditions described below, $[^{32}P]$ATP is maintained at $>90\%$ of initial levels for at least 2 h of incubation. The phosphodiesterase problem is easily solved either by including enzyme inhibitors such as 1-methyl,3-isobutylxanthine (1.0 mM) or by adding sufficient unlabeled cyclic AMP (1.0 mM) to saturate the enzyme. In our laboratory we use 1.0 mM unlabeled cyclic AMP.

Required Solutions (All Stored at $-20°$C)

1. 125 mM Tris–Hepes, pH 7.5, containing 0.25% BSA (Miles Laboratories, fatty acid free), 5 mM MgCl$_2$, and 2.5 mM unlabeled cyclic AMP (solution E).
2. 0.1 M creatine phosphate and 3.0 mg/ml creatine phosphokinase (both from Sigma Chemical Co.) in deionized water (solution F).
3. 1.0 mM Gpp(NH)p; (Boehringer Mannheim Biochemicals) in deionized water (solution G).
4. $[\alpha-^{32}P]$ATP (Amersham Corp.; specific activity $>$ 400 Ci/mmol) diluted with 5–10 ml of 1.0 mM ATP (Sigma Chemical Co. catalog No. A 6144) to a concentration of 100 μCi/ml (solution H).
5. 50 mM Tris–HCl, pH 7.5, containing 2% (v/v) sodium dodecyl sulfate, 10 mM ATP, and 1.0 mM [^{3}H]cyclic AMP, 0.5 μCi/ml ("Stop" solution).
6. Synthetic bovine PTH(1–34) (Beckman Bioproducts, 6000 U/mg) 1 mg/ml in 10 mM acetic acid. Assay standards are freshly prepared by serial dilution with 10 mM acetic acid/0.1% BSA.

Protocol

Unless otherwise noted, all steps are carried out at 2–4°C.

1. In preparation for assay, solutions E–G are combined in the proportions 4:1:1 (E/F/G).
2. To each assay tube (12 $\times$ 75-mm borosilicate glass) are added 60 μl of the combined solution.

3. Solution H (10 μl) is added to each tube.
4. Ten microliters of freshly prepared dilutions of purified bovine PTH(1–34) ranging from 10 pg/ml to 100 ng/ml are added as standards where appropriate.
5. Ten microliters of the bovine PTH(1–34) vehicle (10 mM acetic acid containing 0.1% BSA) are added to tubes not receiving bovine PTH(1–34) standards.
6. Ten microliters of the serum samples (or appropriate dilutions) are added. Tubes containing PTH standards receive 10 μl of diluent serum, either hypoparathyroid (preferred) or euparathyroid human serum.
7. Assays are initiated with the addition of 10 μl of a 0.5–1.25 mg/ml suspension of canine renal membranes in 0.25 M sucrose/1 mM EDTA/5 mM Tris, pH 7.5.
8. Incubation is carried out for 30 min at 30°C, and is terminated by the addition of 100 μl of "Stop" solution, followed by heating in a boiling water bath for 3 min. Samples may then be either stored at −20°C or immediately processed as described below.

C. Quantitation of [^{32}P]Cyclic AMP Formed

A relatively small fraction (<1%) of the substrate [α−^{32}P]ATP is converted to [^{32}P]cyclic AMP during the incubation. Assessment of adenylate cyclase thus requires efficient separation of the cyclic AMP from even trace quantities of labeled ATP or other nucleotides generated by the assay procedure. The procedure of Salomon et al.[29] involving sequential chromotography on AG 50 cation exchange columns followed by aluminum oxide columns is highly effective in this regard, allowing greater than 50% recovery of cyclic AMP with assay blanks less than 0.005% of the added [α−^{32}P]ATP.

Preparation of Columns

Slurries of AG 50W-X4 (200–400 mesh) and neutral aluminum oxide in deionized H$_2$O are added to 0.8 cm (inner diameter, ID) polypropylene columns. Resins and columns are obtained from Bio-Rad. AG 50 columns are poured to a height of 3.0 cm, and are put through a regeneration cycle prior to use, as described below. Aluminum oxide columns are poured to a height of 2.0 cm, and are ready for use.

Protocol

1. To each 0.2-ml sample is added 0.8 ml of deionized water.
2. Samples are carefully poured onto coluns of AG 50W-X4.
3. The effluent from this and three successive 1.0-ml washes with deionized water are carefully discarded into a carboy suitable for liquid radioactive

waste disposal. This effluent contains the unreacted $[\alpha-^{32}P]ATP$ which binds minimally to the cation exchange column.

4. Cyclic AMP is eluted from these columns with an additional 3.0 ml of water. This effluent is allowed to drip directly onto neutral aluminum oxide columns which have been pretreated with 8.0 ml of 0.1 M imidazole-HCl, pH 7.5.

5. The effluent from this and two successive washes of the aluminum oxide columns with 1.0 ml of imidazole-HCl, pH 7.5 are discarded.

6. Cyclic AMP is eluted from these columns with an additional 5.0 ml of imidazole-HCl, pH 7.5, and collected directly into 20-ml glass scintillation vials.

7. Samples are counted in 12 ml of Dynagel scintillant (J T Baker Chemical Co.), using a program suitable for dual $(^{32}P/^3H)$ isotope counting.

8. For each sample, the formation of $[^{32}P]$cyclic AMP is corrected for the recovery of $[^3H]$cyclic AMP (average recovery 60%). Results are expressed as picomoles cyclic AMP formed per 30 min/mg membrane protein.

Both the AG 50 and aluminum oxide columns can be reused at least 10 times. AG 50 columns should be regenerated after each use as follows: wash the columns successively with 10 ml H_2O, 5 ml 2.0 N NaOH, 10 ml H_2O, 5 ml 2.0 N HCl, 10 ml $H_2O \times 3$. Aluminum oxide columns need only be washed with 10 ml H_2O after each use. Columns are stored partially immersed in H_2O between assays to prevent them from drying out.

D. Assay Properties

Figure 1 demonstrates typical adenylate cyclase activation curves for synthetic bovine PTH(1–34), purified human PTH(1–84), and parathyroid venous effluent sera obtained from patients with primary hyperparathyroidism. The PTH dose-response curve is generally linear over a 10-fold concentration range, corresponding to a three- to fourfold increase in adenylate cyclase activity. Of considerable importance is the fact that the activation curve produced by synthetic bovine PTH(1–34) is superimposable on that of purified human PTH(1–84) (when expressed on a molar basis), permitting the routine use of the readily available synthetic peptide as the bioassay standard. The absolute sensitivity of the assay is 100–300-pg-equivalents of bovine PTH(1–34)/ml serum. PTH-specificity of the assay can be demonstrated in two ways: (1) other potential activators of renal adenylate cyclase (glucagon, epinephrine, calcitonin, PGE[1], and AVP) produce little or no cyclase stimulation, and (2) bioactivity produced by sera or test samples is inhibited by [[8]Nle, [18]Nle, [34]Tyr]bPTH(3–34)amide, a potent specific antagonist of PTH-stimulated renal adenylate cyclase.[30]

III. Applications

The concentration of biologically active PTH circulating in normal man has been estimated to be 10 pg/ml.[16] This is approximately one order of magnitude

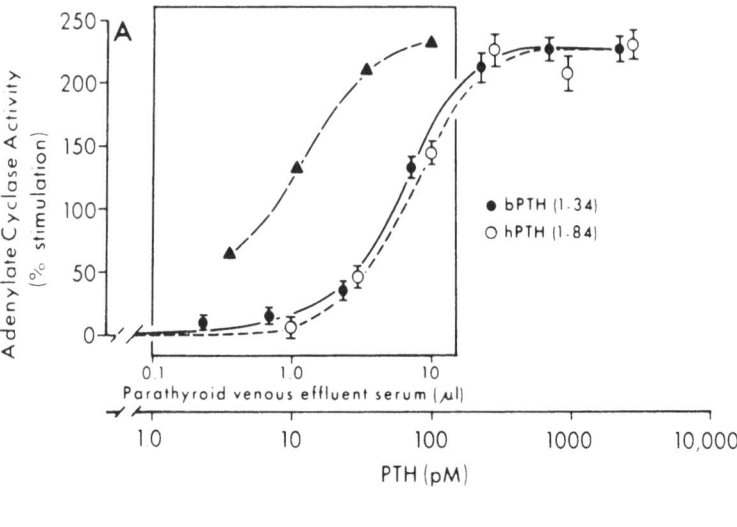

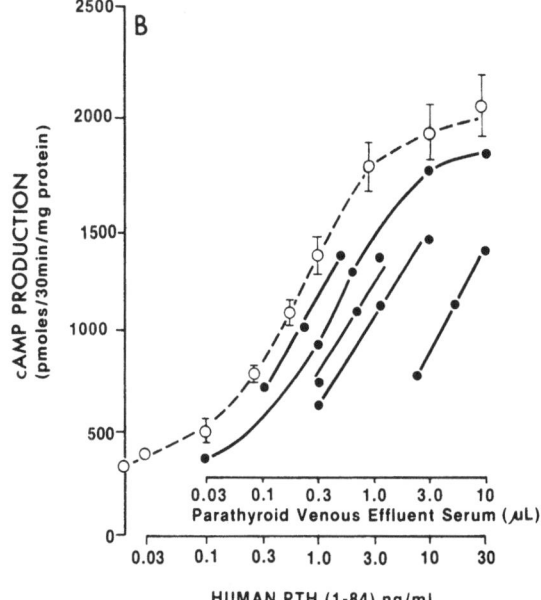

Figure 1. Activation of guanyl nucleotide-amplified canine renal cortical adenylate cyclase by parathyroid venous effluent sera from patients with primary hyperparathyroidism. **A:** Comparison of adenylate cyclase activation produced by bPTH(1–34) (●—●), hPTH(1–84) (○--○), and dilutions of a parathyroid venous effluent serum (▲--▲). **B:** Activation curves produced by five parathyroid venous effluent sera (●—●) compared to that produced by the hPTH(1–84) standard (○--○).

Table 1. Biologically Active and Immunoreactive PTH in
Parathyroid Venous Effluent Sera from Patients with Primary
Hyperparathyroidism

Patient No.	PTH [ngEq hPTH(1-84)/ml]		
	Bioassay	Intact RIA (CH-12)	C-Region RIA (GP-1M)
1	220	200	370
2	120	59	160
3	64	100	160
4	26	11	23
5	15	11	24
6	13	9.1	15
7	7.5	3.9	11
8	3.0	1.7	1.9

Spearman rank correlation coefficient: r_S = 0.97 and $P < .01$ for bioassay versus each RIA.

lower than the limit of detection of the adenylate cyclase bioassay. Sensitivity is thus the factor which limits the clinical utility of the bioassay. Nevertheless, in the short time since its development, the assay has been used in addressing several important problems.

As mentioned previously, circulating PTH is immunochemically heterogeneous. This is in part due to the fact that PTH fragments as well as intact hormone are primary secretory products of the parathyroid gland.[31] Little is known about the biologic activity of these PTH fragments. We have examined the biologic activity of PTH in venous effluents draining hyperfunctioning parathyroid glands from patients with primary hyperparathyroidism[7] (Table 1). A close correlation was observed between bioassay values and RIA values using antisera specific for intact PTH(1-84) (CH-12) and the COOH-region of the molecule (GP-1M). Gel filtration chromotography (Bio-Gel P-150) of four freshly obtained parathyroid venous effluent sera demonstrated a single peak of bioactivity corresponding to intact PTH(1-84). We thus find no evidence for secretion of biologically active PTH fragments by adenomatous parathyroid glands.

The ability to quantitate biologically active PTH in parathyroid venous effluent serum prompted us to evaluate the utility of the assay in localizing parathyroid tumors. We performed PTH bioassays on serum samples obtained by selective venous catheterization of the thyroid veins as well as several levels of the internal jugular, innominate, and subclavian veins in five patients with hypercalcemia and elevated serum immunoreactive PTH who had previously undergone unsuccessful neck explorations. In four cases, significant "step-ups" in concentrations of bioactive PTH (relative to peripheral values) were found. Peak bioactive PTH levels were, in these cases, predictive of tumor localization.

We have also used the PTH bioassay to measure activity in the peripheral

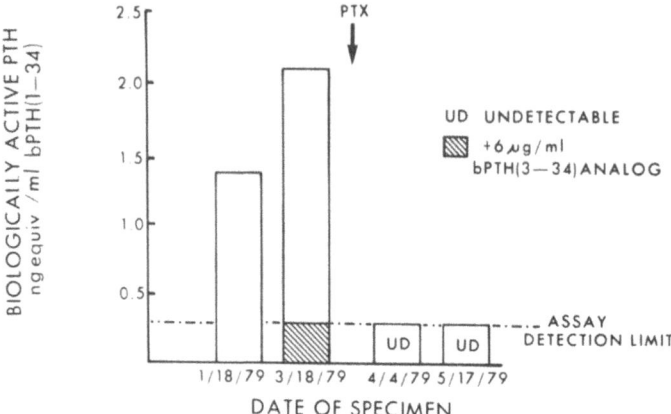

Figure 2. Biologically active PTH in the peripheral serum of a patient with severe renal osteodystrophy and hyperparathyroidism secondary to chronic renal failure. Corresponding immunoreactive PTH concentrations were: preoperatively 3400–4000 μlEq/ml (antiserum GP1M), 1.4–5.3 ngEq/hPTH(1–84) (antiserum CH-12) and, following subtotal parathyroidectomy (PTX), 140–160 μlEq/ml (GP-1M), 0.13–0.14 ngEq/ml hPTH(1–84) (CH-12).

serum of patients with hyperparathyroidism secondary to chronic renal failure. In one patient, the bioassay values were validated by the loss of measurable PTH bioactivity after the addition of the PTH antagonist [8Nle, 18Nle, 34Tyr]bPTH(3–34)amide and after parathyroidectomy (Figure 2). Of serum samples from 8 patients with chronic renal failure and markedly elevated COOH-region immunoreactive PTH, 5 contained biologically active PTH at concentrations ranging from 0.3 to 4.8 ng Eq human (PTH(1–84)/ml (Table 2). Biologically active PTH was undetectable in three other sera. When measured against the same hPTH(1–84) standard, COOH-region immunoreactive PTH was present at a much higher concentration (10–100 fold, in most cases) than biologically active PTH. This presumably reflects the accumulation of circulating biologically inactive COOH-region PTH fragments which are normally cleared by a process requiring glomerular filtration.[32] In contrast, values obtained with the RIA specific for intact PTH(1–84) agreed much more closely with bioassay values. We believe that whereas COOH-region PTH RIA values may accurately reflect chronic PTH hypersecretion in patients with chronic renal failure, values obtained with PTH bioassay (or with intact PTH RIAs) may better reflect acute parathyroid secretory status.

Finally, an exciting potential application of the PTH bioassay is in the study of malignancy-associated hypercalcemia (MAH). A recent study demonstrated that as many as 80% of unselected patients with MAH had elevated nephrogenous cyclic AMP excretion, despite normal or low PTH levels by RIA.[33] Furthermore, as a group, these patients had a significant reduction in

Table 2. Serum Concentrations of Biologically Active and Immunoreactive PTH in Eight Patients with Chronic Renal Failure

| Patient No. | Serum PTH [ngEq hPTH(1–84)/ml] | | | Serum PTH (μlEq/ml) by C-Region RIA (GP-1M)[c] |
	Bioassay	Intact RIA (CH-12)[a]	C-Region RIA (GP-FM)[b]	
1	4.8	5.3 (1.1)[d]	12 (2.5)[e]	3,400
2	0.8	2.3 (2.9)	56 (70)	12,000
3	0.8	1.0 (1.3)	26 (33)	3,100
4	0.7	0.80 (1.1)	8.0 (11)	5,600
5	0.3	0.80 (2.7)	48 (160)	7,300
6	<0.3	0.81 (>2.7)	26 (>87)	1,400
7	<0.3	0.55 (>1.8)	11 (>37)	1,700
8	<0.3	1.3 (>4.3)	78 (>260)	8,400

[a]Normal value, <0.15 ngEq hPTH(1–84)ml.
[b]Normal value, <0.4 ngEq hPTH(1–84)/ml.
[c]Normal value, <40 μlEq/ml.
[d]The ratio of intact RIA (CH-12) to bioassay results is in parentheses.
[e]The ratio of C-region RIA (GP-FM) to bioassay results is in parentheses.

renal phosphate threshold. The data suggest the presence in these patients of a circulating factor which has PTH-like biological effects on the kidney, but has reduced PTH immunoreactivity. We[34] and others[35] have recently described secretion by human cancers of a protein(s) which stimulates guanyl nucleotide-amplified renal adenylate cyclase via the PTH receptor but is distinct from PTH both immunochemically and by its greater molecular size. Thus, it is evident that the bioassay will prove useful in the isolation and characterization of tumor-derived PTH-like factors potentially important in producing hypercalcemia in patients with MAH.

Acknowledgement   I wish to express my gratitude to Ms. Linda Zitzner and Mr. Kong Pua for expert technical support, and to Ms. Sally Danekas for skillful preparation of the manuscript. This work was supported by Grant AM 27755 from the NIH.

References

1. Berson SA, Yalow RS: Immunochemical heterogeneity of parathyroid hormone in plasma. J Clin Endocrinol Metab 28:1037–1047, 1968.
2. Canterbury JM, Reiss E: Multiple immunoreactive molecular forms of parathyroid hormone in human serum. Proc Soc Exp Biol Med 140:1393–1398, 1972.
3. Segre GV, Habener JF, Powell D, Tregear GW, Potts Jr JT: Parathyroid hormone in human plasma. Immunochemical characterization and biological implications. J Clin Invest 51:3163–3172, 1972.
4. Arnaud CD, Goldsmith RS, Bordier PJ, Sizemore GW: Influence of immuno-

heterogeneity of circulating parathyroid hormone on the results of radioimmunoassay of serum in man. Am J Med 56:785–793, 1974.

5. Potts JT, Jr, Tregear GW, Keutmann HT, Niall HD, Saver R, Deftos LJ, Dawson BF, Hogan ML, Aurbach GD: Synthesis of a biologically active N-terminal tetratriacontapeptide of parathyroid hormone. Proc Natl Acad Sci USA 68:63–67, 1971.

6. Canterbury JM, Levey GS and Reiss E Activation of renal cortical adenylate cyclase by circulating immunoreactive parathyroid hormone fragments. J Clin Invest 52:524–527, 1973.

7. Nissenson RA, Abbott SR, Teitelbaum AP, Clark OH, Arnaud CD: Endogenous biologically active human parathyroid hormone: measurement by a guanyl nucleotide-amplified renal adenylate cyclase assay. J Clin Endocrinol Metab 52:840–846, 1981.

8. Chase LR, Aurbach GD: Renal adenyl cyclase: anatomically separate sites for parathyroid hormone and vasopressin. Science 159:545–547, 1968.

9. Dousa T, Rychlik I: The effect of parathyroid hormone on adenyl cyclase in rat kidney. Biochim Biophys Acta 158:484–486, 1968.

10. Rasmussen H, Pechet M, Fast D: Effect of dibutyryl cyclic adenosine 3', 5'-monophosphate, theophylline, and other nucleotides upon calcium and phosphate metabolism. J Clin Invest 47:1843–1850, 1968.

11. Augus ZS, Puschett JB, Senesky D, Goldberg M: Mode of action of parathyroid hormone and cyclic adenosine 3', 5'-monophosphate on renal tubular phosphate reabsorption in the dog. J Clin Invest 50:617–626, 1971.

12. Kuntziger H, Amiel C, Roinel N, Morel F: Effects of parathyroidectomy and cyclic AMP on renal transport of phosphate, calcium, and magnesium. Am J Physiol 227:905–911, 1974.

13. Burnatowska MA, Harris CA, Sutton RAL, Dirks JH: Effects of PTH and cAMP on renal handling of calcium, magnesium, and phosphate in the hamster. Am J Physiol 223:F514–F518, 1977.

14. Horiuchi N, Suda T, Takahashi H, Shimazawa E, Ogata E: In vivo evidence for the intermediary role of 3', 5'-cylic AMP in parathyroid hormone-induced stimulation of 1α,25-hydroxyvitamin D synthesis in rats. Endocrinology 101:969–974, 1977.

15. Broadus AE, Mahaffey JE, Bartter FC, Neer RM: Nephrogenous cyclic adenosine monophosphate as a parathyroid function test. J Clin Invest 60:771–783, 1977.

16. Goltzman D, Henderson B, Loveridge N: Cytochemical bioassay of parathyroid hormone. J Clin Invest 65:1309–1317, 1980.

17. Ross EM, Gilman AG: Biochemical properties of hormone-sensitive adenylate cyclase. Annu Rev Biochem 49:533–564, 1980.

18. Insel PA, Maguire ME, Gilman AG, Bourne HR, Coffino P, Melmon KL: Beta adrenergic receptors and adenylate cyclase: products of separate genes? Mol Pharmacol 12:1062–1069, 1976.

19. Ross EM, Gilman AG: Reconstitution of catecholamine-sensitive adenylate cyclase activity: interaction of solubilized components with receptor-replete membranes. Proc Natl Acad Sci USA 74:3715–3719, 1977.

20. Ross EM, Howlett AC, Ferguson KM, Gilman AG: Reconstitution of hormone-sensitive adenylate cyclase activity with resolved components of the enzyme. J Biol Chem 253:6401–6412, 1978.

21. Cassel D, Levkovitz H, Selinger Z: Kinetic evaluation of the regulatory GTPase cycle in the catecholamine-stimulated adenylate cyclase of turkey erythrocyte membrane. J Cyclic Nucleotide Res 3:393–406, 1977.

22. Levinson SL, Blume AJ: Altered guanine nucleotide hydrolysis as basis for increased adenylate cyclase activity after cholera toxin treatment. J Biol Chem 252:3766–3774, 1977.

23. Michelangeli VP, Hunt NH, Martin TJ: States of activation of chick kidney adenylate cyclase induced by parathyroid hormone and guanyl nucleotides. J Endocrinol 72:69–79, 1977.

24. Goltzman D, Callahan EN, Tregear GW, Potts Jr JT: Influence of guanyl nucleotides on parathyroid hormone-stimulated adenylyl cyclase activity in renal cortical membranes. Endocrinology 103:1352–1360, 1978.

25. Fitzpatrick DF, Davenport GR, Forte L, Landon EJ: Characterization of plasma membrane proteins in mammalian kidney I. Preparation of a membrane fraction and separation of the protein. J Biol Chem 244:3561–3569, 1969.

26. Post RL, Sen AK: Sodium and potassium-stimulated ATPase. In Estabrook RW, Pullman ME (Eds.): Methods in Enzymology, vol X. Academic Press, New York, pp 762–768, 1967.

27. Steiner AL, Kipnis DM, Utiger R, Parker C: Radioimmunoassay for the measurement of adenosine 3′, 5′-cyclic monophosphate. Proc Natl Acad Sci USA 64:367–373, 1969.

28. Gilman AG: A protein binding assay for adenosine 3′:5′-cyclic monophosphate. Proc Natl Sci USA 67:305–312, 1970.

29. Salomon Y, Londos C, Rodbell M: A highly sensitive adenylate cyclase assay. Anal Biochem 58:541–548, 1974.

30. Rosenblatt M, Callahan EN, Mahaffey JE, Pont A, Potts Jr JT: Parathyroid hormone inhibitors: design, synthesis and biologic evaluation of hormone analogues. J Biol Chem 252:5847–5851, 1977.

31. Flueck JA, DiBella FP, Edis AJ, Kehrwald JM, Arnaud CD: Immunoheterogeneity of parathyroid hormone in venous effluent serum from hyperfunctioning parathyroid glands. J Clin Invest 60:1367–1375, 1977.

32. Martin KJ, Hruska KA, Lewis J, Anderson C, Slatopolsky E: The renal handling of parathyroid hormone. Role of peritubular uptake and glomerular filtration. J Clin Invest 60:808–814, 1977.

33. Stewart AF, Horst R, Deftos L, Cadman EC, Lang R, Broadus A: Biochemical evaluation of patients with cancer-associated hypercalcemia. N Engl J Med 303:1377–1383, 1980.

34. Stewler GJ, Williams RD, Nissenson RA: Human renal carcinoma cells produce hypercalcemia in the nude mouse and a novel protein recognized by parathyroid hormone receptors. J Clin Invest 71:769–774, 1983.

35. Stewart AF, Insogna KL, Goltzman D, Broadus AE: Identification of adenylate cyclase-stimulating activity and cytochemical glucose-6-phosphate dehydrogenase-stimulating activity in extracts of tumors from patients with humoral hypercalcemia of malignancy. Proc Natl Acad Sci USA 80:1454–1458, 1983.

Chapter 15

The Cytochemical Bioassay for Parathyroid Hormone

David Goltzman

I. Introduction

A. Limitations of Radioimmunoassay for Parathyroid Hormone Measurement

Although radioimmunoassays represent the standard for measurement of parathyroid hormone (PTH), and their high through-put will continue to ensure their place as the major mode of routine measurement of PTH, increasing knowledge of PTH metabolism and increasing sophistication with radioimmunoassay technology have illuminated the limitations of this mode of assay. Thus, most radioimmunoassays are relatively insensitive and frequently cannot detect circulating levels in normal individuals. Furthermore, none can detect all normal levels and, therefore, no current radioimmunoassay can distinguish normal from hypoparathyroid levels. In part, this relates to the use of heterologous assay systems, frequently employing bovine PTH as a tracer and antiserum to bovine PTH or its fragments in attempting to measure human PTH, although recently successful homologous assay systems have been introduced. Another major factor limiting biological specificity of the radioimmunoassay is the presence in the circulation of chemical forms of the hormone of varying degrees of biological activity and of varying half-lives. Thus, the complex biosynthesis of PTH is known to proceed via precursor forms. Although none has as yet been proven to circulate, such a possibility does exist, and, at least ProPTH is known to have reduced bioactivity. Of more proven importance is the presence in the circulation of middle and carboxyl hormonal fragments which are believed to have no bioactivity and yet have prolonged half-lives relative to the major glandular form of the hormone, PTH(1–84). These fragments seem to contribute a high noise-to-signal ratio in attempting to determine biologically significant hormone concentrations.

B. Factors Limiting Bioassay Sensitivity, and Advantages of Cytochemical Bioassays

At least three major factors limit the sensitivity of most bioassays. In in vivo assays, between-animal variation is a particular problem which can alter hormonal sensitivity. In in vitro assays, alteration of target cells consequent to isolation may occur. Thus, in isolated cell systems used for bioassay, alteration in cell surfaces may occur consequent to the procedures (generally mechanical and/or enzymatic separation) used for isolation. Additionally, the potentially important role of supporting structures for the target cell response is eliminated. In the case of monolayer cultures, these problems may be complicated by alteration of the characteristic of the target cell or by overgrowth with non-target cells. Finally, in those bioassays utilizing whole tissue homogenates or cell fractions, target cell components are generally diluted with nontarget cell components (as in renal adenylate cyclase assays), frequently raising the background level of the response measured and, thereby, limiting sensitivity.

The cytochemical bioassay circumvents these particular problems in that only a single animal is used per assay, organ culture is employed after a short-term preincubation to maintain tissue architecture and to minimize the trauma of excision, and responses are measured directly within specific target cells of the tissue.

II. Background

A. Histochemistry Versus Quantitative Cytochemistry

Histochemistry is an extension of histology and involves morphology and localization only, for example, as seen with immunohistochemistry. Quantitative cytochemistry, on the other hand, involves measurement of biochemical reactions in morphologic sites.

B. Advances Permitting Measurement

One of the basic advances permitting quantitative cytochemistry was the development of the microspectrophotometer, which is a spectrophotometer working through a microscope. However, quantitative cytochemistry requires the measurement of chromophores precipitated in nonhomogeneous regions rather than in homogeneous solutions as can be measured by conventional spectrophotometers. Consequently, scanning and integrating microdensitometry (or microspectrophotometry) was developed which, through a scanning spot, breaks up the area to be measured in each cell into many small areas, measures the extinction of each separately, and integrates all the measured extinctions to give the total extinction of the chosen area.

Since tissues cannot be "fixed" and enzymes still measured, a method of "stabilizing" tissues so as to retain soluble enzymes within cells and also retain enzyme responsiveness had to be developed. Colloid stabilizers were thus developed for this purpose.

III. General Procedures for Cytochemical Bioassays[1,2] (Segment Assays)

A. Remove Appropriate Target Organ and Cut into Segments

Most cytochemical assays, including those for PTH,[3,4] ACTH,[5] gastrin,[6] and thyroid-stimulating hormone (TSH)[7] employ guinea pig target tissues, although the age and weight of the animal may vary from assay to assay. Thus for TSH assays, younger animals are preferred since cystic changes are more frequent in thyroids of older guinea pigs; whereas, older animals are preferred for ACTH assays in view of the larger size of the adult guinea pig adrenals. For the LH assay,[8] the superovulated rat ovary is employed.

Animals are sacrificed by asphyxiation in nitrogen, appropriate target tissues are quickly removed, and divided into approximately six to eight segments.

B. Maintain Each Segment Separately in Vitro

Each segment is then placed separately on a defatted lens tissue lying on a metal grid table standing in a culture dish. Culture medium is added to the level of the lens tissue. The dish sits in a culture chamber which is gassed with a mixture of 95% oxygen, 5% carbon dioxide. The gas inlet and outlet of the culture chamber are then closed off and the tissues are incubated for 5 h at 37°C. The culture medium generally employed for cytochemical bioassays is Trowell's T8 medium. When gassed with 95% O_2, 5% CO_2, the medium, containing neutral red, achieves a pH of 7.6. For different assays, this standard medium may be modified, for example, by the addition of ascorbate in the ACTH assay or acid in the gastrin assay.

Each segment is then maintained for 5 h in the Trowell's nonproliferative culture medium to reduce the endogenous endocrine influence on the tissue and to allow the cells to replenish essential metabolites and restabilize membranes.

C. Treat with Standard Hormone or Unknown

The culture medium of each segment is then removed with a pasteur pipette and replaced with fresh medium containing a given concentration of standard hormone or unknown. If a four-point standard curve is desired, only two to four segments remain for incubation with unknowns.

The incubation with standard or unknown is allowed to proceed for only a short time (minutes) dependent on the experimental protocol and the individual hormone, and the incubation is then stopped.

D. Chill and Section

Incubations are abruptly stopped by immersing segments individually in *n*-hexane in a beaker, kept at −70°C by an outer bath containing chunks of dry ice in ethanol. Each segment is removed within 30–60 s, placed in a dry glass tube, and can be kept then at −70°C for up to 1 week.

Sectioning is generally performed within 48 h. The chilled tissue is mounted on a microtome chuck, preferably within a cryostat cabinet kept at −25°C or −30°C. Sectioning is then done by means of a cryostat microtome with a knife cooled to −70°C. Cutting should proceed at a constant speed to ensure sections of uniform thickness. Sections are removed from the knife onto a glass slide kept at room temperature.

E. React for Cytochemical Response

Sections from each segment are reacted on the glass slides for the specific cytochemical reaction to be assessed. The reaction is dependent upon the biochemical effect induced by the hormone, for example, ascorbate depletion in the
adrenal is assessed in the ACTH assay, carbonic anhydrase stimulation in the
gastric fundus is assessed in the gastrin assay, and increased lysosomal membrane permeability of thyroidal acinar cells is assessed in the TSH reaction.
The capacity to stimulate glucose-6-phosphate dehydrogenase (G6PD) in renal
tubules is the response measured in the PTH assay.

F. Quantitate by Microdensitometry

The color reaction in individual target cells is quantitated in individual sections
for each standard dose of hormone and for each concentration of unknown by
scanning and integrating microdensitometry. When using a suitable instrument, such as the Vickers M85 microdensitometer, the histology of the specimen is first assessed by normal microscopy, and an optical "mask" is then
placed over the cell or region to be measured.

All the light absorptions within the mask area are integrated and recorded
on a digital meter. The integrated absorption recorded is the summed relative
absorption from all points inspected by the "flying spot" of light emitted by the
spectrophotometer and scanning within the mask area.

IV. Specific Aspects of the Cytochemical Bioassay for PTH

A. Target Organ

The cytochemical bioassay for PTH is a renal assay employing kidney segments from guinea pigs.[3,4] Hartley strain, albino guinea pigs weighing 450–
550 g are generally employed. Animals are treated for approximately 1 week
prior to assay with 0.6 ml daily of a multivitamin solution containing 600 IU/
ml vitamin D_2, 100 IU/ml vitamin A, and 50 mg/ml vitamin C.

After removal of kidneys, renal capsules are stripped off, kidneys are divided
sagittally, and then each half is cut into three to four equal segments (approximately 5 × 5 mm) along a corticomedullary plane.

B. Maintenance

Segments are placed on defatted lens tissues on metal grid tables in 35 × 10-
mm Petri dishes. Sufficient medium (generally 8–10 ml) is added to reach the
level of the lens tissues. After 5-h incubation in Trowell's medium, culture
medium containing metabolites released from the tissue is removed from each
segment and replaced with fresh medium for 8 min prior to incubating with
standard hormone or unknowns. Others have preferred Eagle's Minimum
Essential Medium with Earle's salts and Hepes buffer[9] rather than Trowell's.

C. Treatment with Standard or Unknown

Each segment is treated with either a single graded dose of PTH standard or with unknowns. Generally, standard PTH concentrations of 5 fg/ml, 50 fg/ml, 5 pg/ml, and 500 pg/ml are employed. Plasma samples are usually assayed at 1:100 and 1:1000 dilution. All hormone and plasma samples are bioassayed in culture medium containing a final concentration of 1% (v/v) plasma. Where necessary PTH-free plasma is used to achieve the required plasma concentration. The optimal time of incubation for PTH standards or unknowns was previously found by us to be 6 min.[10]

D. Sectioning

Chilled segments are cut within 3 days (and generally within 24 h) into 16-μm sections. A Bright cryostat (Bright Instrument Co. Ltd., Huntingdon, England) with the knife cooled to −70° was found to be suitable for this purpose. Sections of 16-μm are cut in a corticomedullary plane. At least four sections are obtained per segment. This is technically the most difficult part of the assay and requires the most skill in achieving sections of uniform thickness.

E. Cytochemical Reaction

Unfixed sections are then reacted for glucose-6-phosphate dehydrogenase (G6PD) activity,[11] the index of response of the cytochemical bioassay for PTH.

The reaction medium, contains 5 mM glucose-6-phosphate, 3 mM NADP$^+$, 0.67 mM phenazine methosulfate (an intermediate hydrogen acceptor), 5 mM neotetrazolium chloride, 10 mM KCN, 12% polyvinyl alcohol (a colloid stabilizer) in 0.05 M glycyl glycine buffer, pH 8.0. A stock solution of 12% polyvinyl alcohol containing 0.3% neotetrazolium chloride in glycyl glycine buffer, pH 8.0, is stored at 4°C. On the day of the assay 50 mg NADP$^+$, 30 mg glucose-6-phosphate, and 13 mg KCN are added to 20 ml of stock solution, which is then gassed with nitrogen and warmed to 37°C. Immediately before reacting, 4mg of phenazine methosulfate in 100 μl of distilled water is added, mixed, and used immediately. The solution is added to the section on the glass slide within the boundaries of 1-cm diameter plastic ring put over the section, and is permitted to react with the section on the glass slide for approximately 5 min. The colorless neotetrazolium, reduced by the generation of NADPH from the action of tissue G6PD, is converted to an intensely colored red-purple formazan which precipitates in the target cells at the site of the reaction. At the end of the reaction period the slide is immersed in water to wash off the medium.

F. Quantitation

The reduced neotetrazolium resulting from the G6PD reaction is measured in individual cells of intensely staining renal tubules with a Vickers (Vickers

Instruments, Inc., Woburn, Massachusetts) M85 scanning and integrating microdensitometer at 585 nM (the isobestic wave-length for the red intermediate half-reduced formazan and the purple fully reduced diformazan), with a $\times 40$ objective and a mask size of 17-μm diameter. Generally, the reaction is quantitated in one cell of each of 10 tubules per section and in 2 sections per dose of standard hormone or per dilution of unknown. Consequently, 20 readings are obtained per dose of standard hormone or per dilution of unknown.

V. Validation of the Assay

The assay has now been validated in a number of ways independently by several laboratories.

A. Precision

This is an index of the reproducibility of the measurement, involves the slope of the dose-response curve and the deviation of points above this regression line. It is calculated as λ (index of precision) = s (standard deviation about the regression line) $\div$ b (slope of the line).[12] For a bioassay, a λ of less than 0.3 is generally considered adequate. The published values of λ for the cytochemical bioassay of PTH are 0.09, 0.19, and 0.18.

B. Parallelism and Sensitivity

Cytochemical bioassays employ a 4 + 2 (4 doses of standard + 2 dilutions of unknowns) form of assay. Plasma is usually diluted 1:100 and 1:1000 and log–dose responses are parallel with those of the standard. Generally, bovine PTH(1–84) has been used as a standard in the assay. One of the major features of the assay is sensitivity. Lower limits of hormone detection may range from 0.1 to 50 fg/ml of bovine PTH(1–84) but most commonly are 5 fg/ml. Dose-response curves are generally linear for 3 to 4 orders of magnitude.

C. Specificity

In addition to the nature of the response measured, the specificity of the assay for PTH appears to lie also in the time of measurement of the response and in the site of measurement. Whatever the factors involved, a variety of unrelated hormones have been shown either to not stimulate in the assay at all, or to produce nonparallel responses, or to stimulate only at excessively high concentrations.[3,9,10] Additionally, pre-incubation with PTH antiserum[3,9,10] has been shown to eliminate the hormonally induced response. On the other hand, synthetic analogs of the hormone have been shown to react in the assay.

D. Physiologic Validation

Several physiologic studies have demonstrated that the activity measured in plasma seems to correlate with the physiologic or pathophysiologic status of the individual such as the calcium status and the state of the parathyroid glands.[10]

VI. Summary

The cytochemical bioassay for PTH represents the most sensitive method of measuring PTH available to date. Furthermore, as a bioassay, it detects only that hormone capable of initiating a physiologic response and not inert fragments. Consequently, it represents a highly useful adjunct to our armamentarium for assessment of PTH physiology and pathophysiology. However, it is a fastidious assay, requiring careful attention to detail and excellent technical ability. Furthermore, a very major drawback is the limited through-put of the assay, providing a serious temporal restriction on the ability to process numerous samples. Perhaps these problems will be overcome by the development of "section" assays, in which sections are cut from segments prior to incubation with hormone or unknown, substantially increasing assay capability. Nevertheless, at the moment, the assay remains a highly useful research tool, but a research tool only.

References

1. Chayen J, Daly JR, Loveridge N, Bitensky L: The cytochemical bioassay of hormones. Recent Prog Horm Res 32:33–39, 1976.
2. Chayen J: The Cytochemical Bioassay of Polypeptide Hormones. Springer-Verlag, Berlin, 1980.
3. Chambers DJ, Dunham J, Zanelli JM, Parsons JA, Bitensky L, Chayen J: A sensitive bioassay of parathyroid hormone in plasma. Clin Endocrinol (Oxf) 9:375–379, 1978.
4. Goltzman D, Stewart AF, Broadus AE: Hypercalcemia associated with malignancy: Evaluation with a cytochemical bioassay for parathyroid hormone. J Clin Endocrinol Metab, 53:899, 1981.
5. Chayen J, Loveridge N, Daly JR: A sensitive bioassay for adrenocorticotrophic hormone in human plasma. Clin Endocrinol (Oxf). 1:219–233.
6. Loveridge N, Bloom SR, Welbourn RB, Chayen J: Quantitative cytochemical estimation of the effect of pentagastrin (0.005–5pg/ml) and of plasma gastrin on the guinea-pig fundus in vitro. Clin Endocrinol (Oxf) 3:389–396, 1974.
7. Bitensky L, Alaghband-Zadeh J, Chayen J: Studies on thyroid stimulating hormone and the long-acting thyroid stimulating hormone. Clin Endocrinol (Oxf) 3:363–374.
8. Rees LH, Holdaway IM, Kramer RM, McNeilly AS, Chard T: A new bioassay for luteinizing hormone. Nature 244:232–234, 1973.

9. Fenton S, Somers S, Heath DA: Preliminary studies with the sensitive cytochemical assay for parathyroid hormone. Clin Endocrinol (Oxf) 9:381–384, 1978.

10. Goltzman D, Henderson B, Loveridge N: Cytochemical bioassay of parathyroid hormone: characteristics of the assay and analysis of circulating hormonal forms. J Clin Invest. 65:1309–1317, 1980.

11. Chayen J, Bitensky L, Butcher RG: Practical Histochemistry, Wiley, London, 1973.

12. Gaddum JH: Bioassays and mathematics. Pharmacol Rev 5:87–134, 1953.

Index